Catalysis for the Production of Sustainable Fuels and Chemicals

Catalysis for the Production of Sustainable Fuels and Chemicals

Special Issue Editors

Flora T. T. Ng
Ajay K. Dalai

MDPI • Basel • Beijing • Wuhan • Barcelona • Belgrade • Manchester • Tokyo • Cluj • Tianjin

Special Issue Editors

Flora T. T. Ng
200 University Avenue West
Canada

Ajay K. Dalai
University of Saskatchewan
Canada

Editorial Office
MDPI
St. Alban-Anlage 66
4052 Basel, Switzerland

This is a reprint of articles from the Special Issue published online in the open access journal *Catalysts* (ISSN 2073-4344) (available at: https://www.mdpi.com/journal/catalysts/special_issues/Sustainable_Fuels_and_Chemicals).

For citation purposes, cite each article independently as indicated on the article page online and as indicated below:

LastName, A.A.; LastName, B.B.; LastName, C.C. Article Title. *Journal Name* **Year**, *Article Number*, Page Range.

ISBN 978-3-03936-040-6 (Hbk)
ISBN 978-3-03936-041-3 (PDF)

Cover image courtesy of of Arunima Soma Dalai.

Contents

About the Special Issue Editors . . . . **vii**

Flora T. T. Ng and Ajay K. Dalai
Catalysis for the Production of Sustainable Fuels and Chemicals
Reprinted from: *Catalysts* **2020**, *10*, 388, doi:10.3390/catal10040388 . . . . **1**

**Muhammad Abdus Salam, Derek Creaser, Prakhar Arora, Stefanie Tamm,
Eva Lind Grennfelt and Louise Olsson**
Influence of Bio-Oil Phospholipid on the Hydrodeoxygenation Activity of
NiMoS/Al$_2$O$_3$ Catalyst
Reprinted from: *Catalysts* **2018**, *8*, 418, doi:10.3390/catal8100418 . . . . **5**

Sandeep Badoga, Prachee Misra, Girish Kamath, Ying Zheng and Ajay K. Dalai
Hydrotreatment Followed by Oxidative Desulfurization and Denitrogenation to Attain Low
Sulphur and Nitrogen Bitumen Derived Gas Oils
Reprinted from: *Catalysts* **2018**, *8*, 645, doi:10.3390/catal8120645 . . . . **23**

**Petr Zeman, Vladimír Hönig, Martin Kotek, Jan Táborský, Michal Obergruber, Jakub Mařík,
Veronika Hartová and Martin Pechout**

Reprinted from: *Catalysts* **2019**, *9*, 337, doi:10.3390/catal9040337 . . . . **43**

Yuanqing Liu, Xiaoming Guo, Garry L. Rempel and Flora T. T. Ng
The Promoting Effect of Ni on Glycerol Hydrogenolysis to 1,2-Propanediol with In Situ
Hydrogen from Methanol Steam Reforming Using a Cu/ZnO/Al$_2$O$_3$ Catalyst
Reprinted from: *Catalysts* **2019**, *9*, 412, doi:10.3390/catal9050412 . . . . **59**

**Serguei Alejandro-Martín, Adán Montecinos Acaricia, Cristian Cerda-Barrera and
Hatier Díaz Pérez**
Influence of Chemical Surface Characteristics of Ammonium-Modified Chilean Zeolite on Oak
Catalytic Pyrolysis
Reprinted from: *Catalysts* **2019**, *9*, 465, doi:10.3390/catal9050465 . . . . **81**

**Zhipeng Tian, Chenguang Wang, Zhan Si, Chengyan Wen, Ying Xu, Wei Lv, Lungang Chen,
Xinghua Zhang and Longlong Ma**
Enhancement of Light Olefins Selectivity Over N-Doped Fischer-Tropsch Synthesis Catalyst
Supported on Activated Carbon Pretreated with KMnO$_4$
Reprinted from: *Catalysts* **2019**, *9*, 505, doi:10.3390/catal9060505 . . . . **95**

Seyyedmajid Sharifvaghefi, Babak Shirani, Mladen Eic and Ying Zheng
Application of Microwave in Hydrogen Production from Methane Dry Reforming: Comparison
Between the Conventional and Microwave-Assisted Catalytic Reforming on Improving the
Energy Efficiency
Reprinted from: *Catalysts* **2019**, *9*, 618, doi:10.3390/catal9070618 . . . . **105**

**Hui Wang, Kyle Rogers, Haiping Zhang, Guoliang Li, Jianglong Pu, Haoxuan Zheng,
Hongfei Lin, Ying Zheng and Siauw Ng**
The Effects of Catalyst Support and Temperature on the Hydrotreating of Waste Cooking Oil
(WCO) over CoMo Sulfided Catalysts
Reprinted from: *Catalysts* **2019**, *9*, 689, doi:10.3390/catal9080689 . . . . **119**

Bamidele Victor Ayodele, Siti Indati Mustapa, May Ali Alsaffar and Chin Kui Cheng
Artificial Intelligence Modelling Approach for the Prediction of CO-Rich Hydrogen Production
Rate from Methane Dry Reforming
Reprinted from: *Catalysts* **2019**, *9*, 738, doi:10.3390/catal9090738 . . . . **133**

**Xianming Zhang, Shuang Chen, Fengjiao Wang, Lidan Deng, Jianmin Ren, Zhaojie Jiao and
Guilin Zhou**
Effect of Surface Composition and Structure of the Mesoporous Ni/KIT-6 Catalyst on Catalytic
Hydrodeoxygenation Performance
Reprinted from: *Catalysts* **2019**, *9*, 889, doi:10.3390/catal9110889 . . . . **153**

Jinglin Gao, Philip Effah Boahene, Yongfeng Hu, Ajay Dalai and Hui Wang
Atomic Layer Deposition ZnO Over-Coated Cu/SiO$_2$ Catalysts for Methanol Synthesis from
CO$_2$ Hydrogenation
Reprinted from: *Catalysts* **2019**, *9*, 922, doi:10.3390/catal9110922 . . . . **171**

Afees A. Ayandiran, Philip E. Boahene, Ajay K. Dalai and Yongfeng Hu
Hydroprocessing of Oleic Acid for Production of Jet-Fuel Range Hydrocarbons over Cu and
FeCu Catalysts
Reprinted from: *Catalysts* **2019**, *9*, 1051, doi:10.3390/catal9121051 . . . . **189**

**Wanpeng Hu, Hui Wang, Hongfei Lin, Ying Zheng, Siauw Ng, Manlin Shi, Ying Zhao and
Ruoqian Xu**
Catalytic Decomposition of Oleic Acid to Fuels and Chemicals: Roles of Catalyst Acidity and
Basicity on Product Distribution and Reaction Pathways
Reprinted from: *Catalysts* **2019**, *9*, 1063, doi:10.3390/catal9121063 . . . . **211**

**Ivan V. Shamanaev, Irina V. Deliy, Evgeny Yu. Gerasimov, Vera P. Pakharukova and
Galina A. Bukhtiyarova**
Enhancement of HDO Activity of MoP/SiO$_2$ Catalyst in Physical Mixture with Alumina
or Zeolites
Reprinted from: *Catalysts* **2020**, *10*, 45, doi:10.3390/catal10010045 . . . . **225**

Tetiana Kulik, Borys Palianytsia and Mats Larsson
Catalytic Pyrolysis of Aliphatic Carboxylic Acids into Symmetric Ketones over Ceria-Based
Catalysts: Kinetics, Isotope Effect and Mechanism
Reprinted from: *Catalysts* **2020**, *10*, 179, doi:10.3390/catal10020179 . . . . **239**

About the Special Issue Editors

Flora T.T. Ng obtained her B.S. at the University of Hong Kong, M.S. and Ph.D. from the University of British Columbia, Vancouver, BC, Canada. She joined the University of Waterloo, Canada in 1987, appointed University Research Chair in 2006 and has held the prestigious title of University Professor since 2008. She is a Fellow of the Royal Society of Canada. Professor Ng has expertise in catalysis, bitumen oilsands upgrading, green chemistry, and engineering, and is a world leading expert on catalytic distillation, a green reactor technology. In 2008, she received the Catalysis Award from the Chemical Institute of Canada (CIC). She was Chair of the Catalysis Division, CIC, represented Canada to the International Association of Catalysis Societies and served as President of the Canadian Catalysis Foundation. She received the 2013 Green Chemistry and Engineering Award (Individual) from CIC. She has been involved in successful collaborative research with various international oil and chemical companies. She was recently awarded a Visiting Professorship for Senior International Scientist at the Institute of Process Engineering Chinese Academy of Sciences, Beijing, China. Her current research projects include the development of new green processes for the production of biodiesel from waste oils using solid catalysts, the conversion of glycerol to valuable chemicals, and the production of bio-jet fuel and high octane gasoline using catalytic distillation.

Ajay K. Dalai is Distinguished Professor of University of Saskatchewan (UofS) and Canada Research Chair in Bioenergy (Tier 1) in the Department of Chemical and Biological Engineering at the UofS. In 2012, Professor Dalai worked as Special Adviser to VP Research Signature Areas of Energy and Mineral Resources. He has served as Associate Dean, Research and Partnership in Engineering (2009–2014) and Head of Chemical Engineering (2007–2009) at the UofS. His research focus is the novel catalyst development for gas to liquid technologies, biodiesel production, hydrogen/syngas production, hydroprocessing of gas oils and vegetable oils, value-added products from biomass, and nanomaterials development for various industrial processes. Dr. Dalai has published over 490 research papers mostly in heterogeneous catalysis and catalytic processes in international journals and conference proceedings. He has submitted several patent applications. His ground-breaking research has led to over 25,000 citations of his work and H-index of 74. He serves on the Editorial Board of several prestigious journals. Professor Dalai has received several national and international awards including JSPS fellowship with Okayama University (1999), Kentucky Colonel Award (2004), DAAD Fellowship with Karlsruhe Institute of Technology (2010), Canadian Catalysis Lectureship Award (2011), Glory of India award (2012), Fulbright Fellowship with UC Davis (2012), and Bantrel Award in Design and Industrial Practice (2014). He is a life member of many professional societies and is an active member of the American Institute of Chemical Engineers (AIChE), American Chemical Society (ACS), Indian Institute of Chemical Engineers (IIChE), the Chemical Institute of Canada (CIC), and Royal Society of Canada (RSC). He has received many Distinguished Speaker awards from IIChE. Professor Dalai is a fellow of Engineering Institute of Canada (2011), The Chemical Institute of Canada (2012), The Canadian Academy of Engineering (2012), American Institute of Chemical Engineers (2013), Indian Institute of Chemical Engineers (2013), Royal Society of Canada (2014), and Royal Society of Chemistry (2016).

 *catalysts*

Editorial

Catalysis for the Production of Sustainable Fuels and Chemicals

Flora T. T. Ng [1,*] and Ajay K. Dalai [2,*]

[1] Department of Chemical Engineering, University of Waterloo, 200 University Avenue West, Waterloo, ON N2L 3G1, Canada

[2] Department of Chemical and Biological Engineering, University of Saskatchewan, Saskatoon, SK S7N 5A9, Canada

* Correspondence: fttng@uwaterloo.ca (F.T.T.N.); ajay.dalai@usask.ca (A.K.D.)

Received: 24 March 2020; Accepted: 1 April 2020; Published: 2 April 2020

Keywords: sustainable fuels and chemicals; hydrogenolysis; deoxygenation; desulfurization and denitrogenation; CO_2 utilization; hydroprocessing; pyrolysis and cracking; syngas and hydrogen; light olefins; biomass and bio-oil; catalysis

---

The emission of green-house gases and environmental concerns have led to recent research in the use of renewable feedstocks derived from biomass, waste oils and fats as a source for fuels and chemicals. However, these feedstocks contain a large amount of oxygen functional groups. Processing these feedstocks generally requires esterification and transesterification, deoxygenation, hydrogenation, hydrogenolysis, aldol condensation and cracking reactions. The bio-oils produced through pyrolysis and hydrothermal treatment are treated through catalytic processes for producing clean fuels. In addition, syngas produced through the gasification of biomass is converted to liquid fuels. Petroleum crude oils contain large amounts of heteroatoms and contaminants. Due to the global mandate of cleaner fuels, these materials require severe processing with a large energy input, and hence require novel and advanced processes for creating environmentally friendly fuels for societal needs. The main focus of this issue was to solicit recent advances in the catalytic processing of renewable and non-renewable feedstocks for sustainable fuels and chemicals.

In this Special Issue of *Catalysts*, 15 high-quality papers, which are externally and thoroughly reviewed, are published, and the catalytic conversion of biomass and biomass-derived liquids and hydrocarbon-based fuels is focused on. The novelty and contributions of these papers are briefly highlighted below for the benefit of the readers. The performance of sulfided NiMo/γ-alumina for the hydrodeoxygenation (HDO) of a fatty acid to alkanes was performed in a batch reactor by Abdus-Salem et al. (2018), and the catalyst's deactivation was due to the blockage of active sites and pore blockage and largely contributed to product coke and the irreversible production of aluminum phosphate. The oxidative stability of hydrotreated vegetable oils as compared to fatty acid methyl ester and their suitability as fuels in a mixture with different transportation fuels is studied in detail by Zeman et al. (2019). The properties of fuel blend, such as viscosity, density, and cold filter plugging point, confirmed that the hydrotreated vegetable oil blend with fuel can meet the fuel standard for engines without modification. The hydrotreating of waste cooking oil to produce green diesel using unsupported and supported CoMoS on a mixed oxide was investigated by Wang et al. (2019). The support favored the metal dispersion and surface area of CoMoS, a decrease in reaction temperature for HDO, enhancement of hydrogenation capabilities and reduction in polymerization activities. A series of 25 wt.% Ni/KIT6 reduced at different temperatures (400–550 °C) was evaluated by Zhang et al. (2019) for the HDO of ethyl acetate to ethane and methane. The catalyst reduced at 450 °C was most suitable for higher product selectivity due to its optimum metal support interaction, Ni dispersion, crystallite size, and ability to adsorb and activate hydrogen. Ayandiran et al. (2019) investigated several Fe promoted

Cu/SiO$_2$-Al$_2$O$_3$ catalysts for jet fuel production from oleic acid in an autoclave. The impact of support acidity by using ZSM-5 and HZSM-5 on the jet-fuel range hydrocarbons' yield and selectivity was investigated. These catalysts were extensively characterized and the higher yield and selectivity for Fe(3)-Cu(15)/SiO$_2$-Al$_2$O$_3$ was attributed to their excellent textural properties, high oxophilic iron content, high metal dispersion, and mild Bronsted acidity. The catalytic decomposition of oleic acid to organic liquid product (OLP) containing mostly alkenes and cyclo-alkenes was studied by Hu et al. (2019) using various acidic and basic catalysts. Lewis acidic catalysts were beneficial for deoxygenation and secondary cracking, whereas CaO, among all basic catalysts, favored dehydrogenation activity. The enhancement of the HDO activity of MoP on SiO$_2$ with a physical mixture with alumina or zeolites for methyl palmitate (MP) to palmitic acid was investigated by Shamanaev et al. (2020). Additional acid sites to MoP/SiO$_2$ enhanced HDO reaction, and isomerization and the cracking of alkanes were favored by the addition of zeolites. The catalytic pyrolysis of Chilean Oak to value-added chemicals and fuels was studied by Alejandro-Martin et al. (2019) using ammonium-modified Chilean Zeolite. Bronsted acidity was favorable for ketones, aldehydes and hydrocarbons, and for a decrease in esters, ethers and acids, and was responsible for upgraded bio-oil with lower oxygen content and higher value-added chemicals. The kinetics, isotope effect and mechanism of catalytic pyrolysis of aliphatic carboxylic acids into ketones over Ce-based catalysts were investigated by Kulik et al. (2020). The study showed that it is a sterically controlled reaction and demonstrated a method for studying the catalytic pyrolysis mechanism.

Liu et al. (2019) investigated the promotional effect of Ni on glycerol hydrogenolysis to 1,2- propanediol (1,2-PD) with in-situ hydrogen from methanol present in crude glycerol using a Cu/ZnO/Al$_2$O$_3$ catalyst. Ni contributed to the generation of in situ hydrogen with higher methanol conversion, reduced the conversion of glycerol, and improved the selectivity of 1,2-PD. Ni assisted in the slower dehydration of glycerol to acetol and higher hydrogenation activity of acetol to 1,2-PD. Hydroprocessing (HT) followed by oxidative desulfurization (ODS) and denitrogenation (ODN) of gas oils was investigated by Badoga et al. (2019). HT was performed using NiMo/γ-Al$_2$O$_3$, and alumina, alumina–titania supported Mo, P, Mn and W catalysts were tested for sustainable ODS and ODN of gas oils using tert–butyl hydroperoxide as an oxidant, followed by the adsorption of oxidized compounds on activated carbon. MnPMo/γ-Al$_2$O$_3$ was the most suitable for the HT process. This research showed that the combination of HT, ODS and ODN resulted in the most efficient removal of S and N compounds from gas oils.

A novel artificial intelligence modelling for the prediction of a CO-rich hydrogen production rate from methane dry reforming was successfully conducted by Ayodele et al. (2019). Light olefin selectivity with a higher O/P ratio via the Fischer Tropsch Synthesis (FTS) process was investigated by Tian et al. (2019) for a FeN catalyst supported on 10MnK-AC. Light olefin selectivity was improved due to the electronic donor effect of nitrogen and the suppression effect on the second hydrogenation over 10MnK-AC support. Microwave and conventional heating systems were used by Sharifvaghefi et al. (2019) for hydrogen production from the dry reforming of methane using Ni and Ni-MnO supported on activated carbon. Microwave heating improved the reactants' conversion and hydrogen selectivity, due to the enhancement of energy efficiency of the reaction as compared to conventional heating. The mechanism of methanol synthesis from CO$_2$ hydrogenation using Cu-ZnO-based catalysts was investigated by Gao et al. (2019) using a combination of strong electronic adsorption (SEA) and atomic layer deposition (ALD) techniques. These materials were extensively characterized. The catalyst with 5 wt.% Cu on one atomic layer of ZnO produced smaller and more Cu sites, and ZnO basic sites, and a stronger interaction between them, which possibly led to CO$_2$ activation and hydrogenation to form methanol.

In summary, many of these 15 publications deal with the production of sustainable fuels and chemicals and have a higher level of novelty related to an improved catalytic process and chemical product efficiency, catalytic reaction mechanism and process sustainability. The Guest Editors sincerely thank all the authors, reviewers and the publication team from *Catalysts*, including Ms. Vivian Niu and

Catalysts **2020**, *10*, 388

Ms. Cora Huang and others, who made this publication possible. Also, the Editors thank Ms. Arunima Soma Dalai for preparing the cover page for this special issue.

Conflicts of Interest: The author declares no conflict of interest.

Article

Influence of Bio-Oil Phospholipid on the Hydrodeoxygenation Activity of NiMoS/Al₂O₃ Catalyst

Muhammad Abdus Salam [1,2], Derek Creaser [1,*], Prakhar Arora [1], Stefanie Tamm [1], Eva Lind Grennfelt [3] and Louise Olsson [1]

[1] Competence Centre for Catalysis, Chemical Engineering, Chalmers University of Technology, SE-41296 Gothenburg, Sweden; mabdus@chalemrs.se (M.A.S.); prakhar@chalmers.se (P.A.); stamm@chalmers.se (S.T.); louise.olsson@chalmers.se (L.O.)

[2] Department of Chemical Engineering and Polymer Science, Shahjalal University of Science and Technology, Sylhet 3114, Bangladesh; salam-cep@sust.edu

[3] Preem, SE-41296 Gothenburg, Sweden; eva.lind-grennfelt@preem.se

[*] Correspondence: derek.creaser@chalmers.se; Tel.: +46-317-723-023

Received: 15 August 2018; Accepted: 20 September 2018; Published: 25 September 2018

Abstract: Hydrodeoxygenation (HDO) activity of a typical hydrotreating catalyst, sulfided NiMo/γ-Al₂O₃ for deoxygenation of a fatty acid has been explored in a batch reactor at 54 bar and 320 °C in the presence of contaminants, like phospholipids, which are known to be present in renewable feeds. Oleic acid was used for the investigation. Freshly sulfided catalyst showed a high degree of deoxygenation activity; products were predominantly composed of alkanes (C17 and C18). Experiments with a major phospholipid showed that activity for C17 was greatly reduced while activity to C18 was not altered significantly in the studied conditions. Characterization of the spent catalyst revealed the formation of aluminum phosphate (AlPO₄), which affects the active phase dispersion, blocks the active sites, and causes pore blockage. In addition, choline, formed from the decomposition of phospholipid, partially contributes to the observed deactivation. Furthermore, a direct correlation was observed in the accumulation of coke on the catalyst and the amount of phospholipid introduced in the feed. We therefore propose that the reason for the increased deactivation is due to the dual effects of an irreversible change in phase to aluminum phosphate and the formation of choline.

Keywords: HDO; sulfide catalyst; NiMo/Al₂O₃; phospholipid; fatty acid; choline

---

1. Introduction

Fossil fuel depletion, caused by increasing demands and finite resources, as well as concern about greenhouse gas (GHG) emissions, has led to a search for alternative renewable sources of energy [1,2]. Biomass is considered a possible alternative future fuel source to minimize CO₂ emissions. The type of biomass employed can be of different kinds, but mainly: carbohydrates, lignocellulosic material, and waste animal fats or waste/non-edible oils have been considered [3]. All of these renewable resources are suitable for upgrading to liquid biofuel in many different processes, for example, to bio-ethanol by fermentation of sugar or starch, to bio-diesel by transesterification of fat/oil, and to renewable diesel through catalytic hydrodeoxygenation of triglycerides and fatty acids [4,5]. Chemically, animal fats and non-edible oils are composed of triglycerides, with high amounts of oxygen. They also can contain minor, but varying, amounts of free fatty acids, phospholipids, sterols, pigments, and waxes. Hence, the challenge for bio-oil upgrading depends on its viability, properties, composition, competition with the food grade biomass, and presence of impurities.

In the last few years, extensive studies have been performed concerning the catalytic upgrading of triglyceride feedstocks to diesel range fuel with high cetane numbers [4,6–8]. Several catalysts have already been reported to be able to perform effectively, but the most common are Ni or Co promoted MoS_2 on alumina [5,9]. Despite the type of catalyst employed, upgrading involves a cascade of catalytic reactions, e.g., hydrogenation, hydrodeoxygenation, decarboxylation/decarbonylation (DCO), isomerization, dehydration, cracking, etc., which predominantly produces a mixture of liquid hydrocarbons. Thereby, the process reduces the oxygen content and increases the stability and quality of the fuel to be used as transportation fuel. However, the process involves high pressure, temperature, and high H_2 consumption [10]. On the other hand, the ultimate advantages of these processes are the reduction of GHG emissions, more flexibility of feedstocks, and higher carbon yields of the liquid hydrocarbon fraction. The physical and chemical properties of the products resemble the quality of conventional diesel; thus, they are named as green or renewable diesel. Although being commercially employed [11–13], such a process still deserves continued attention since the selectivity and deactivation of the catalyst may be highly sensitive to variations in biomass feedstocks during the actual processing conditions [14]. The amount and types of impurities present in a biomass feedstock depends on the prior production process for, e.g., liquefaction/pyrolysis/supercritical oxidation, its conditions, and treatment of the feedstocks [15]. Bio-oil impurities include sulfur, nitrogen, phosphorus bearing compounds, alkali, and alkaline earth materials, such as Na, Ca, Fe, Mg, etc. [12,16,17].

Despite the feedstock pretreatment (e.g. degumming, refining etc.), varying amounts of phosphorus (P) can be present in the bio-oil/biomass feedstocks (e.g. vegetable/algal oils) in the form of phospholipids and other cell forming constituents [18–20]. Phospholipids are amphipathic molecules having both a polar phosphate group and non-polar long fatty acid tails. Major phospholipids identified in naturally occurring bio/algal-oil are phosphatidylcholine, phosphatidic acid, phosphatidylethanolamines, phosphatidyl-glycerols, etc. [20,21]. Additionally, some of the most common algae species, with the potential of biofuel production, contain as much as 40% of the cell weight as phosphatidylcholine [22]. It should be noted that removal of phospholipid is an energy intensive step and consumes a large amount of water in current industrial practice. In crude form, the bio/algal oil phospholipid content can be up to 5% or higher, whereas after modern degumming/refining processes, it is expected to be much lower (<0.05%) in the refined oil to be hydrotreated [23]. Although some of the phospholipids (non-hydratable) are difficult to remove from oil and may not be removed completely, it is critical to investigate to what extent they affect the catalytic hydrotreatment process. There are only a few studies available in the literature regarding this topic [17,21].

Nevertheless, phosphorus has mostly been studied as a second promoter in the bimetallic CoMo or NiMo/γ-Al_2O_3 catalyst system for hydrodesulphurization (HDS) and hydrodenitrogenation (HDN) [24–28]. Interestingly, a promoting effect of P was reported for low loading (typically in the range of 1 to 3 wt.%) and it is considered to enhance the dispersion of the active phases (NiMoS/MoS_2) by weakening the metal/support interaction and changing the geometry/dispersion of the active components from tetrahedral Ni/Mo to more active octahedral Ni/Mo [27–30].

As known, heterogeneous catalysts can be deactivated by several reasons, but mainly by poisoning, fouling or physical blockage, thermal degradation, vapor formation and/or leaching, attrition or crushing, and vapor-solid and/or solid-solid interactions [14,17,31]. There are excellent reviews/articles on the deactivation of hydroprocessing catalysts [32,33]. Several deactivation studies of HDO catalysts have been carried out considering possible bio-oil impurities, like trap grease phospholipid, chlorine (Cl), potassium (K), and iron (Fe) [12,16,17]. Mortensen et al. [16] concluded that irreversible K deactivation of $NiMoS_2$/ZrO_2 catalyst was due to the occupancy of vacant sites near the MoS_2 slab edges. Recently, Arora et al. [12] showed that Fe preferentially blocks the Ni sites to shift the activity and selectivity of NiMoS/Al_2O_3. Kubička and Horáček [17] exemplified that the deactivation of a sulfided commercial CoMo/γ-Al_2O_3 by rapeseed oils and trap grease was primarily due to synergic effects between phospholipids and alkali metals causing coking and pore plugging. Murzin and co-workers claimed that phosphorus blocks the active Pd sites during decarboxylation of behenic acids [34]. The effect of hydrotreating products, such as H_2O, NH_3, and H_2S, on such catalyst systems has also been the subject of many studies [6,8,35–37]. Laurent and Delmon [6] reported that ammonia inhibits more strongly than water, although the presence of H_2S promoted deoxygenation of esters. Such effects by H_2S and different sulfiding agents have also been studied in detail for various oxygenate upgrading catalysts [38–40].

However, there are to our knowledge no studies available where a detailed investigation of the bio-oil phospholipids on the HDO activity of $NiMoS/Al_2O_3$ has been explored. Therefore, the focus of this study is to systematically examine the effect of a major bio-oil phospholipid, phosphatidylcholine (PC), toward the HDO activity of a presulfided NiMo/γ-alumina catalyst while upgrading a major fatty acid (oleic acid), which, for example, is present in canola oil (61%), Jatropha oil (42%), microalgal oil (23%), tall oil (15%), etc. [12]. We considered oleic acid a good model compound because it includes both a double bond on the carbon chain and a carboxylic group as other unsaturated fatty acids, like linoleic acid. Activity and selectivity changes observed during the reaction led us to investigate the spent catalyst to examine the cause of deactivation. In this regard, transmission electron microscopy (TEM), X-ray photoelectron spectroscopy (XPS), Inductively Coupled Plasma (ICP)-Sector Field Mass Spectroscopy (SFMS), and Temperature Programmed Oxidation (TPO) and Elemental Analysis (EM) were performed with the spent catalysts.

2. Results and Discussions

2.1. HDO Activity of NiMo/Al₂O₃

2.1.1. Activity Measurements without Phospholipid

Figure 1 presents the reactant and product profiles throughout the course of the reaction. The overall oxygenate conversion (sum of oleic acid and stearic acid) was >99% after 240 min of HDO. As seen from Figure 1, heptadecane (C17) and octadecane (C18) formation increases over time and encompasses 70% of the final yield of products from oleic acid.

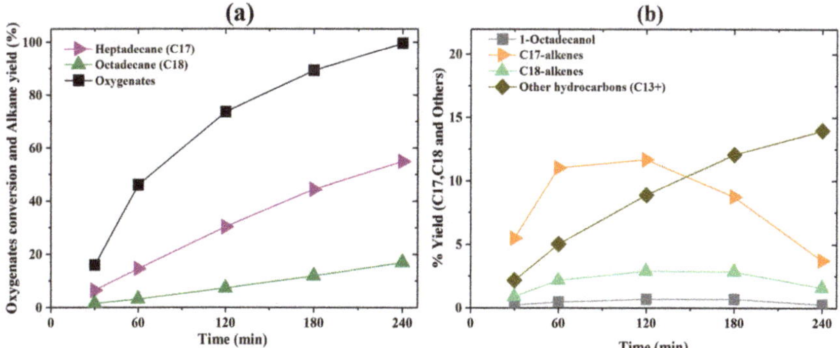

Figure 1. Products and reactant profile during the reaction at 54 bar, 320 °C, and 0.3 mL dimethyl disulfide (DMDS), (**a**) oxygenates conversion, C17 and C18 yield, (**b**) other products and intermediates.

With the reaction conditions employed, oleic acid (OA) is first converted to stearic acid (SA) through hydrogenation of its double bond. Further hydrogenation of it to fatty alcohol is reported to occur very slowly [39,41]. Instead, it can be reduced readily at the catalyst surface via hydrogen to form an aldehyde, 1-octadecanal, which undergoes keto-enol tautomerism to form 1-octadecanol [42]. Interestingly, deoxygenation of both 1-octadecanal and 1-octadecanol yields C17 and C18 alkenes, which are further hydrogenated to form their corresponding alkanes. Clearly, C17 formation implies loss of a carbon atom either as CO or CO_2, whereas C18 formation occurs without carbon loss, but oxygen removal as H_2O. The former route is so-called decarboxylation/decarbonylation (DCO_x) and the latter direct hydrodeoxygenation (HDO). These are the typical overall reactions over sulfided NiMo catalysts in the presence of a sulfiding agent, $DMDS/H_2S$ [12,39,43]. We have observed very small amounts of 1-octadecanol (Figure 1b) as the reaction intermediate. This indicates its rapid conversion to hydrocarbons via protonation of the hydroxyl group.

The trend shows that DCO_x dominated during the entire period whereas the direct-HDO route had a lower rate of reaction. It also indicates that the process is less selective for the enol formation route in this case. It should be mentioned that the DMDS used decomposes to form H_2S to maintain the sulfidity of the catalyst and prevent possible metal-oxide formation. These results are in line with previous deoxygenation studies of oleic acid with sulfided NiMo/γ-alumina [4,12,39,42,44].

Figure 1b summarizes the other products and intermediates formed during the reaction. Alkenes (C17 and C18) formed via DCO_x and direct-HDO are converted to their corresponding alkanes upon hydrogenation. Formation of shorter hydrocarbons, like tri-, tetra-, penta-, and hexadecane (C13–C16), and longer alkanes, like nonadecane (C19) and eicosane (C20), were due mainly to the presence of other acids in the feedstock. We grouped them here as 'other hydrocarbons (C13+)'. Although C17 and C18-alkenes are congregated together, several isomers of them were identified by GC-MS analysis. Furthermore, small amounts of an esterification product (stearyl stearate) was observed in very low quantities (less than 0.5%) due to the reaction between stearic acid and 1-octadecanol. Trace amounts of S-bearing compounds identified by the GC-MS in the liquid phase were predominantly thiols, like tert-hexadecanethiol ($C_{16}H_{34}S$) and 2,4,6-tri-t-butylbenzenethiol ($C_{18}H_{30}S$). This indicates that minor sulfur contamination of the liquid phase occurred.

The overall mole balance for each experiment was found to be in the range of $100 \pm 10\%$. For the sake of brevity, C17/C18 alkanes and alkenes are grouped together as C17+ and C18+ in the following sections unless otherwise stated.

2.1.2. Activity Measurements with Phospholipid

In the presence of phosphatidylcholine (PC), a notable impact in the HDO activity and selectivity was observed (Figure 2). It is evident from Figure 2a that oxygenate conversion decreases with the increasing PC concentration in the feed. Overall, hydrocarbon yield (Figure 2b,c) decreases with an increasing amount of PC, leading to higher amounts of oxygenates remaining after 240 min of HDO.

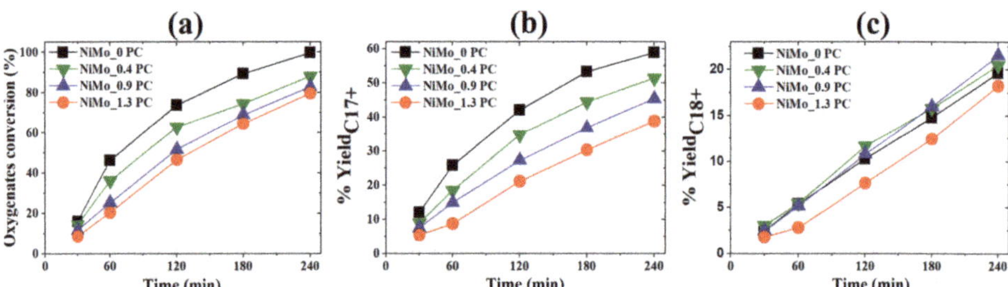

Figure 2. Effect of phospholipid (PC) on the HDO of oleic acid at 54 bar and 320 °C using 0.3 mL DMDS; (**a**) oxygenates conversion, (**b**) yield of C17+, and (**c**) yield of C18+.

A lower amount of PC in the feed (0.2%, not shown in figure) in the absence of DMDS appears to have very little effect on the HDO of OA. However, higher concentrations of PC in the feed or catalyst that are run two times in the presence of PC (Figures 2 and 3) exhibit significant deactivation, even for saturation of the double bonds (hydrogenation), which is a rather quick reaction in these conditions. Consequently, the conversion of oxygenates to any other products is delayed. The initial rate of oxygenate conversion drops noticeably form 59.1 mmol/g catalyst/h (without PC) to 27.6 mmol/g catalyst/h for the highest amount of PC tested in this study (Table 1). PC addition also caused a higher amount of SA compared to OA to be observed after 240 min of HDO. This implies that the influence on OA hydrogenation is less severe than on the intermediate SA conversion by HDO.

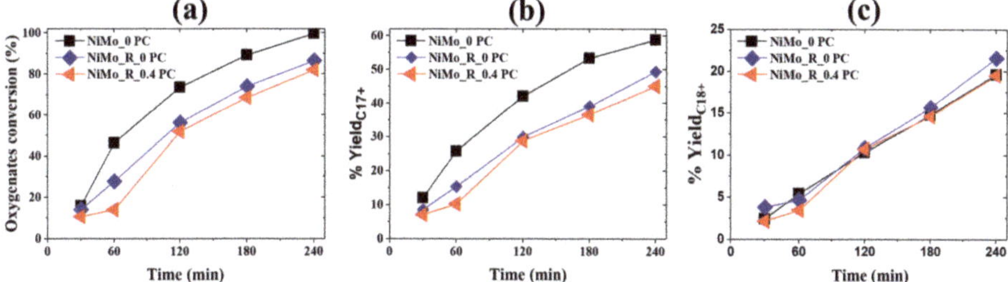

Figure 3. Effect of phospholipid (PC) on the HDO of oleic acid at 54 bar and 320 °C using 0.3 mL DMDS; (**a**) oxygenates conversion, (**b**) yield of C17+, (**c**) yield of C18+.

Table 1. Initial rates of oxygenates and alkane formation measured at 1 h.

Catalyst	$r_{oxygenates}$ (mmol/g/h)	r_{C17+} (mmol/g/h)	r_{C18+} (mmol/g/h)
NiMo_0 PC	59.1	65.8	13.8
NiMo_0.4 PC	46.3	47.3	13.9
NiMo_0.9 PC	32.2	38.2	13.2
NiMo_1.3 PC	27.6	23.8	7.6
NiMo_Choline	45.9	39.5	14.9
NiMo_R_0 PC	35.7	39.4	11.9
NiMo_R_0.4 PC	18.0	26.3	8.8

Likewise, the formation rate and yield of C17+ decreased with the gradual increase of PC feed (Table 1 and Figure 2b). Interestingly, the loss in the C17+ yield is directly proportional to the amount of PC introduced (Figure 4). Conversely, the yield of C18+ is not altered significantly except for the experiment with the highest phospholipid (PC) addition. The overall selectivity for heptadecane and its isomers dropped from 0.8 to 0.4 while selectivity for octadecane and its isomers remained constant at ca. 0.2. Moreover, the amount of C17 and C18 alkenes and their isomers were relatively higher in experiments with PC (Figure 4). This also indicates the inhibition of the hydrogenation step by PC. These results suggest that the DCO_x route is affected most severely in the presence of phospholipid (PC).

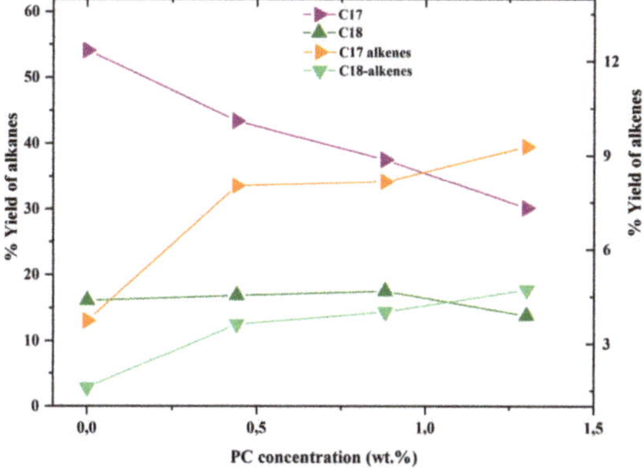

Figure 4. Effect of phospholipid (PC) on the formation of C17 and C18 alkanes/alkenes at the end of a 4 h reaction.

The underlying cause for these observations can be attributed to the products formed by the decomposition of the PC at the reaction conditions. As mentioned earlier, PC is comprised of a non-polar hydrocarbon tail, a hydrophilic phosphate group, and an organic choline group containing nitrogen. It can undergo decomposition via acid hydrolysis to form mainly fatty acids, glycerol, phosphoric acid, and choline with an intermediate glycerophosphate/ phosphorylcholine [45,46]. It is stated that the presence of oleic acid and phosphoric acid can even enhance the PC decomposition [45]. Hence, we consider that PC has been decomposed fully with the reaction conditions and time employed with no glycerophosphate/phosphorylcholine intermediates remaining. The fatty acid contribution from the PC can be as high as 6.7 wt.% of OA in the feed and has been accounted for in the above experiments. Glycerol formed can be converted to propane [47]. It has been reported that phosphoric acid can act as a polymerization/ oligomerization catalyst and lead to the formation of carbonaceous compounds, and thereby hinder access to the active sites [17]. However, the effect of choline on such a catalyst system has not been explored, and rather has been ignored. Laurent and Delmon [6] reported that increasing amounts of di-aminopropane affects severely the decarboxylation selectivity during deoxygenation of di-ethyldecanedioate. Like amine compounds, choline has a strong Lewis base character and can affect the acidity of the catalyst and/or neutralize the effect of H_2S. Also, choline, upon heating, can be converted to trimethylamine, which can strongly bind to the catalyst surface or can react with the fatty acids/intermediates. To understand the extent of the effect of choline, experiments with choline hydroxide addition (48 wt.% in water) have been performed. To explicate, the experiment was performed with an equivalent amount of choline present as 1.3% PC in the feed and the results are shown in Figure 5.

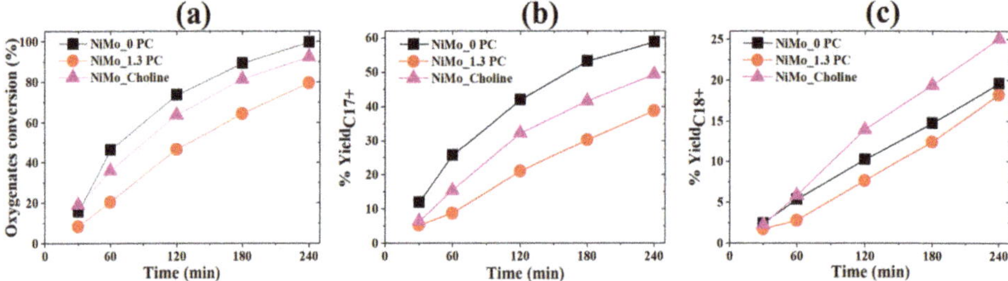

Figure 5. Effect of choline on the deoxygenation of oleic acid 54 bar and 320 °C using 0.3 mL DMDS; (**a**) oxygenates conversion, (**b**) yield of C17+, (**c**) yield of C18+.

It is apparent from Figure 5 that the presence of choline decreases the C17+ yield by 16% relative to the baseline experiment (NiMo_0 PC) whereas for the phospholipid (1.3% PC) it is about 34%. This indicates again higher deactivation in the presence of phospholipid (PC) rather than choline only. Interestingly, about a 26% increase in C18+ yield has been noticed. In addition, liquid phase analysis by GC-MS revealed the formation of small amounts of octadecanamides and octadecanitrile, which would be due to the reaction between fatty acids/intermediate alcohols with the amines formed from choline. Such products can bind to the catalytic surfaces as well. Gas phase analysis from the experiment with PC and choline showed no ammonia formation. Henceforth, we discard the possibility that choline decomposed into ammonia at these reaction conditions.

Since choline also contains water, it is possible that the water also influenced the reaction. To exclude this possibility an additional experiment was performed with an equivalent amount of water and the results are shown in Figure 6. The results clearly show that these water levels do not influence the results.

Based on the above results, we conclude that the observed deactivation can be related to the combined effects of phosphoric acid and choline formed upon decomposition of the phospholipid (PC). Moreover, the results also demonstrate that poisoning caused by the phosphorus part is more severe than that from the choline moiety. To further understand the cause of these changes, a detailed characterization of the recovered catalysts has been performed, which will be discussed in the next section.

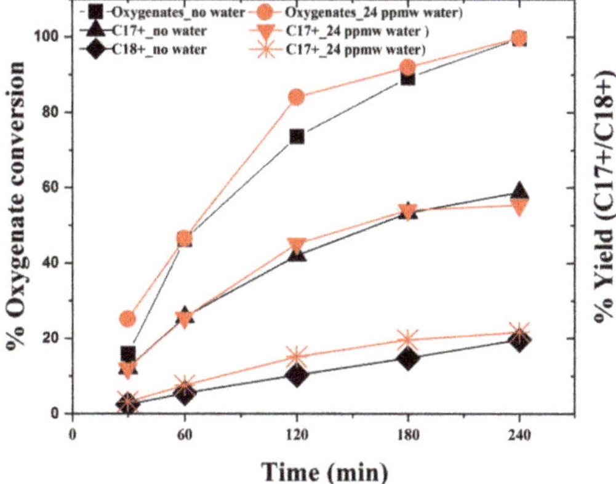

Figure 6. Effect of an equivalent amount of water in choline hydroxide on the HDO of oleic acid at 54 bar, 320 °C, and 1000 rpm of stirring.

2.2. Catalyst Characterization

2.2.1. Textural Properties of the Catalyst

Nitrogen physisorption results shows that the catalyst surface area, pore volume, and pore size drop pointedly even in the presence of small amounts of PC in the feed (Table 2). It is well known that carbon deposition occurs mainly due to chemisorbed species formed by undesirable side reactions. For example, formation of alkenes can enhance the coke formation by oligomerization in the presence of phosphoric acid [48]. It has been observed that about 11% of the BET surface area has been blocked by coke formed in the experiment without phospholipid whereas this increases to 27% in presence of the lowest amount of PC (NiMo_0.4 PC) tested. This infers that severe pore blocking, and subsequent pore volume and pore size reduction, occurs in the presence of PC.

Table 2. Surface area (BET), pore volume, and pore size of the catalyst.

Catalyst	BET Surface Area (m^2/g)	Pore Volume (cm^3/g)	Average Pore Size (Å)
Synthesized	143	0.33	62.3
NiMo_0 PC	126	0.27	62.1
NiMo_0.4 PC	105	0.22	55.2

Elemental analysis shows the carbon, hydrogen, sulfur, and nitrogen content of the studied catalysts in Table 3. Sulfur loss relative to the sulfided and unused catalyst can be ascribed to oxygen incorporation into the active sulfide phase during the reaction or the loss of loosely bound sulfur species on the support to the reaction mixture. Since all the recovered catalyst samples have undergone a similar amount of sulfur loss and all values are above the minimum stoichiometric amount required for complete sulfidation of the Ni promoted MoS_2 phase (5.1%), deactivation due to sulfur loss or sulfur loss promoted by PC addition can be ruled out. However, the carbon content of the catalysts treated with PC mostly increased with an increasing addition of PC. Additionally, a comparable amount of nitrogen can be seen in all experiments with phospholipid (PC). This validates the presence of nitrogen as amines/amides on the catalyst surface, which is likely due to choline formation.

Table 3. Elemental contents of carbon, hydrogen, sulfur, and nitrogen on the freshly sulfided and recovered catalyst samples.

Catalyst	C, wt.%	H, wt.%	S, wt.%	N, wt.%
Sulfided	0.4	1.3	9.22	<0.05
NiMo_0 PC	2.96	1.57	7.30	<0.05
NiMo_0.4 PC	5.34	1.65	7.39	0.36
NiMo_0.9 PC	5.90	1.3	7.40	0.30
NiMo_1.3 PC	7.74	1.73	7.47	0.37
NiMo_R_0 PC	7.26	1.87	7.12	<0.05
NiMo_R_0.4 PC	6.31	1.42	7.05	0.26

2.2.2. ICP-SFMS on the Recovered Catalyst and Liquid Samples

To verify the phosphorus content, the spent catalyst and liquid phase samples after the reaction were further analyzed by ICP-SFMS. The results are reported in Table 4. The atomic ratio of $Ni/(Ni + Mo)$ for the spent catalysts remained close to those for the as-synthesized catalyst and thus we consider that no leaching of metals occurred. Analysis shows that lower amounts of P were deposited on the spent catalyst compared to the maximum value estimated considering complete decomposition of the phosphatidylcholine and assuming all P added to the liquid phase becomes deposited on the catalyst. Moreover, the amount of P remaining in the liquid phase after reaction is very small.

Table 4. ICP-SFMS data for spent catalyst and liquid phase.

Catalyst	P Added in Liquid, ppm	Theoretical Maximum P in Catalyst, wt.%	Actual P in Catalyst (ICP), wt.%	Atomic Ratio of Ni/(Ni +Mo)	P Remaining in Liquid Phase (ICP) ppm
As Synthesized	-	-	-	0.37	-
NiMo_0.9 PC	532	12.16	1.27	0.41	11.1
NiMo_0.4 PC	179	4.09	1.98	0.39	8.37
NiMo_R_0.4 PC *	179	8.18	3.18	0.39	10.6

* Refers to recovered catalyst from the first experiment.

Surprisingly, the amount of P on the catalyst from ICP analysis is not proportional to the total P added initially to the liquid phase. This might be related to different amounts of loosely bound phosphorus species were washed away by ethanol during treatment of the recovered catalysts. However, the phosphorus concentration for the spent catalyst that had been run two times is significantly higher compared to that for the NiMo_0.4 PC sample, which is reasonable. The elemental analysis indicates that catalysts exposed to PC are enriched in P more than N, which shows that more phosphorus is remaining on the surface.

2.2.3. XPS and TEM Analysis

XPS analysis of the catalyst samples recovered after being exposed to the PC are presented in Figure 7. P2p signals from the samples exposed to PC indicates the presence of phosphates. Such a peak at 133.8 ± 0.2 eV implies the formation of metal-phosphate complexes, which can affect the activity of the catalyst in several ways. However, broadening of the phosphate peak area validates an increasing P deposition on the catalyst with the increasing phospholipid (PC) in the feed. Phosphate species can affect the active surface area, pore plugging, active phase dispersion, or morphology of the catalyst, which, consequently, can reduce the active site density and imposes steric hindrance for the incoming molecule to adsorb on them [49]. Indeed, pore plugging and surface area reduction is confirmed by the BET results (Table 2). Furthermore, no additional peak below the binding energy of 130 eV ruled out the formation of metal phosphides at these reaction conditions.

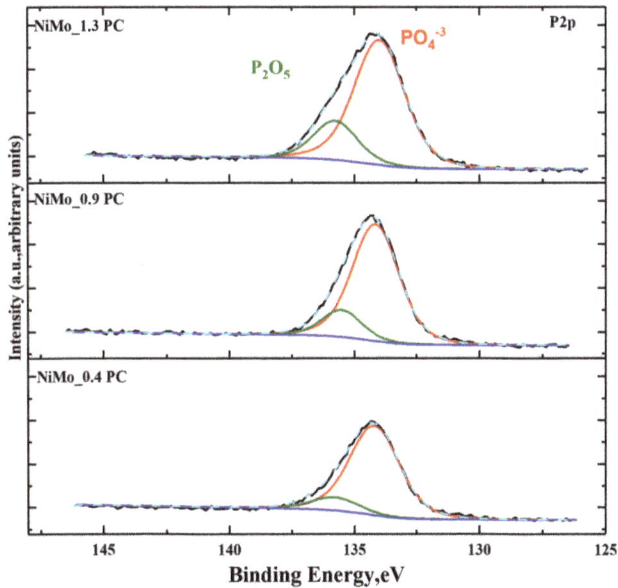

Figure 7. P2p core level spectra for the recovered catalysts with PC.

However, to understand the active phase dispersion and morphology, TEM analysis was performed on the catalyst (NiMo_1.3 PC) collected from the experiment with the highest amount of PC (Figure 8a). Bright Field (BF) micrographs (Figure 8a) showed the characteristic lamellar pattern of the Ni promoted active MoS_2 phase [16]. Comparing this pattern to the catalyst (NiMo_0 PC) without PC (Figure S1) indicates an increase in the stacking degree of MoS_2 to 4.4 $\pm$ 1.7 with PC exposure compared to 2.3 $\pm$ 1 without. Similarly, an average slab length of 5.3 $\pm$ 3.2 nm compared to 3.8 $\pm$ 1.7 nm in NiMo_0 PC has been measured based on 150 slabs. Since higher stacking leads to more inactive basal planes of MoS_2 to be accessed by edge atoms (Ni), the changes observed indicate a lower dispersion of the active MoS_2 phases on the support caused by PC [50]. Such an effect gives lower coordination between Ni and MoS_2 to maintain the active NiMoS phase relative to the baseline catalyst. This can be one possible reason for the activity and selectivity changes observed with these PC addition studies.

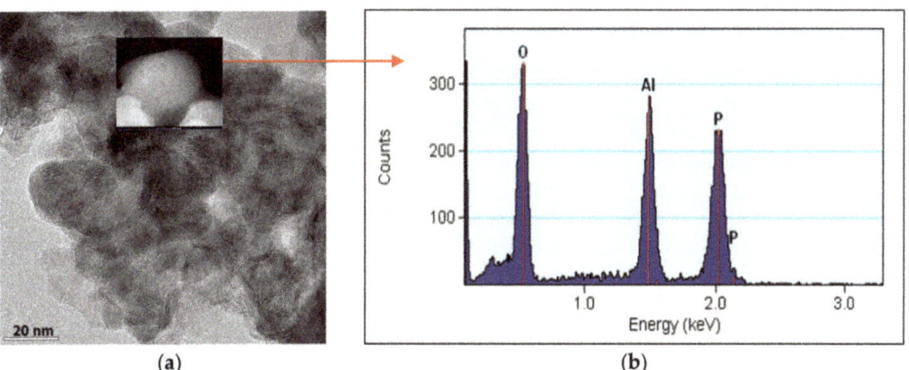

Figure 8. TEM analysis on the spent catalyst representing (**a**) BF-TEM micrograph, inset in (**a**) shows a characteristic area of the support and (**b**) energy dispersive X-ray (EDX) analysis on the characteristic area in (**a**).

Furthermore, STEM-EDX analysis on the magnified image in Figure 8a showed a particle with only Al_2O_3, where P was present. In other TEM images, P has been observed with all other elements in the catalyst. NiMoS, NiS phases were observed without interference from phosphorus while some areas showed only MoS_2 (Figure S2). It is reported that γ-alumina can easily interact with phosphoric acid through surface hydroxyl groups to form phosphates [51]. Eijsbouts et al. [28] mention that alumina has a strong affinity for polar phosphate groups. Therefore, from the decomposition of PC, phosphoric acid thus formed can interact with the support to form aluminum phosphates ($AlPO_4$), which is consistent with the XPS studies and TEM-EDX mapping (Figure 8b). Moreover, elemental maps over a particle shown in HAADF-STEM imaging and EDX mapping is shown in Figure 9. It is clear from the mapping that phosphorous is well distributed over the Al_2O_3 particle. This result suggests that phosphorus is strongly bound to the support alumina.

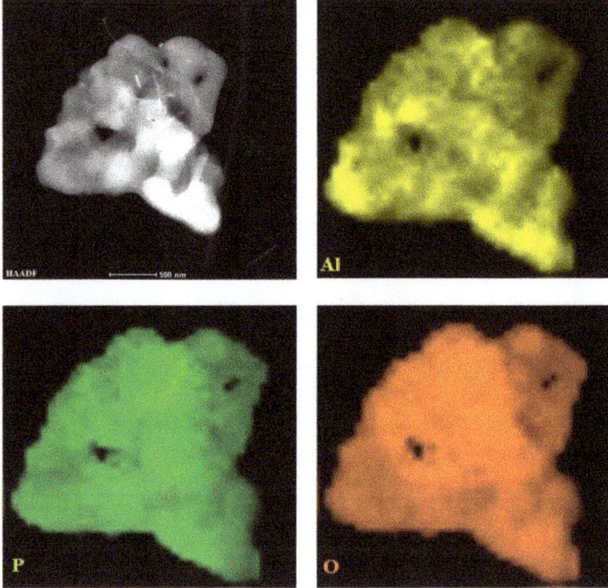

Figure 9. HAADF-STEM imaging and EDX mapping on a single particle showing Al, O, and P elemental distribution.

2.2.4. Temperature Programmed Oxidation (TPO) of Recovered Catalysts

Coke formation on the catalyst surfaces can block the active site and reduce pore volumes to cause deactivation. Diez et al. [33] concluded that coke species can block the edges and corners of the active MoS_2 crystallites to hinder its activity. Figure 10a shows the profile of combined carbon oxides ($CO + CO_2$) and water (H_2O) formation during the TPO of recovered catalyst samples. The H_2O profile for the spent catalyst, NiMo_1.3 PC, is shown in Figure 10a for comparison. A similar trend for H_2O has been observed for all samples. Two distinct regions of coke formation are apparent. In the first region, A (200–300 °C), water formation was found to be much higher than in the second region, B (300–600 °C). This fraction of coke in region A, which will be referred to here as soft coke, is mainly composed of residual adsorbed reactants and intermediate species deposited on the catalyst surface with higher hydrogen content that is combusted readily at low temperatures. Whereas the coke in region B is more of a refractory/advanced nature. Refractory coke is also more likely to be strongly bound to the catalyst/support surface (chemisorbed) with a lower hydrogen content and hence requires higher temperature for combustion [52]. Formation of such coke is also known to be influenced by the support acidity [53]. On the other hand, soft coke is at least partially reversibly adsorbed to the catalyst surface and burns off easily at lower temperatures. Hence, the initial coke formation on all spent catalyst samples can be attributed to the formation of both soft and refractory coke.

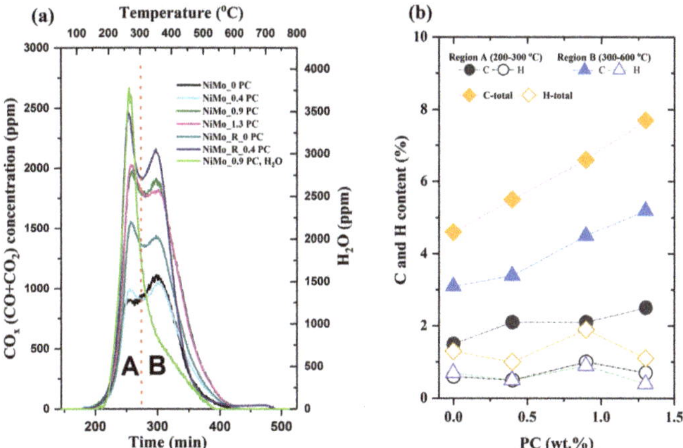

Figure 10. (a) CO, CO_2, and H_2O profile during TPO of the recovered catalyst; (b) measurement of carbon (C) and hydrogen (H) content (wt.%) of the recovered catalysts from TPO.

Table S1 and Figure 10b summarizes the amount of coke formed on each recovered catalyst sample with the distribution of carbon (C) and hydrogen (H) contents in the two regions above. The C content is higher in the higher temperature region than the low temperature one. Likewise, carbonaceous deposits are rich in H in the low temperature region while leaner in H in the high temperature region. Both TPO and elemental C and H analysis (Table 3 and Table S1) are in very good agreement. By comparison with the spent catalysts, it is apparent that the carbon deposition increases with the quantity of PC added to the reactants (Figure 10b). It can be presumed that the higher amount of coke deposition is due to the adsorption of more reactants and intermediates on the catalyst surface from lower oxygenate conversion. However, in Figure 8 and Table S1, carbon, even in the high temperature region with low H content, increases with PC exposure. Hence, the higher carbon deposition stems directly from PC introduction, which contributes more to the formation of refractory coke.

On the other hand, coke deposition (Table S1) was found to be lower for the recovered catalyst after a second experiment with PC (NiMo_R_0.4 PC) compared to the same catalyst without PC (NiMo_R_0 PC). Interestingly, the oxygenate conversion was significantly lower on the PC poisoned catalyst compared to the experiment without PC (Figure 3). It is possible that the large amount of phosphorus on the spent catalyst (3.18%, Table 4) blocked more active sites and thereby lowered the coke deposition due to lower activity.

3. Materials and Methods

3.1. Catalyst Preparation

The catalyst (NiMo/γ-Al$_2$O$_3$) was prepared by a sequential wet impregnation process [44] using pre-calcined γ-alumina as the support material and deionized water as the solvent. Ammonium molybdate tetrahydrate (81–83% MoO_3 basis, Sigma-Aldrich, St. Louis, MO, USA) and nickel (II) nitrate hexahydrate (98%, Sigma-Aldrich, St. Louis, MO, USA) were used for the impregnation of Mo and Ni, respectively. Initially γ-alumina was calcined in air at 550 °C for 2 h. Molybdenum was then impregnated in a pH stabilized aqueous solution of γ-alumina by adding an aqueous solution of ammonium molybdate tetrahydrate using a pH of 4.0. The resulting solution was then freeze dried under vacuum to remove the moisture. The dried sample was then calcined in air at 450 °C for 2 h.

Likewise, Ni was impregnated using the same procedure mentioned above, but with a pH of 9.0. The solution pH was maintained using 10 wt.% HNO_3 (Merck, Darmstadt, Germany) and 2 M NH_4OH (Merck, Darmstadt, Germany), respectively. The final solution was freeze dried, followed by calcining the catalyst powder in air for 2 h at 450 °C.

3.2. Catalytic Activity Measurements

3.2.1. Sulfidation

The synthesized catalyst was sulfided before each activity test using DMDS ($\geq$99.5%, Sigma-Aldrich, Poole, Dorset, UK) and hydrogen (AGA, 99.9%, Air Liquide Gas AB, Malmö, Sweden) in the batch reactor (300 mL, Parr Inc., Moline, IL, USA). About 500 mg of $NiMo/Al_2O_3$ catalyst was sulfided using 0.5 mL of DMDS and hydrogen following a procedure described in the bibliography [44]. The reactor was then cooled, depressurized, and kept under a nitrogen atmosphere.

3.2.2. Hydrodeoxygenation of Oleic Acid

Activity experiments were examined in three parts. In the first segment, activity of the freshly sulfided catalyst was explored. In the second step, the effects of PC were investigated. Finally, recovered catalysts from the first two parts were further tested to gain more insight into the phospholipid effect. All the experiments were performed in the autoclave reactor mentioned above.

With the freshly sulfided catalyst (ca. 500 mg), the reactor was filled with 150 mL of liquid consisting of 10 wt.% oleic acid (tested according to Pharmacopoeia Europaea, Ph. Eur., Fluka) in dodecane ($\geq$99%, Sigma Aldrich, St. Louis, MO, USA). GC-MS analysis confirmed the purity of the oleic acid feed was 85 wt.%. Remaining fraction constituents included palmitoleic, myristic, palmitic, eicosanoic, and nonadecanoic acids. The reaction conditions were maintained at 320 °C, with 54 bar of H_2 at a stirring rate of 1000 rpm, which are typical conditions for hydrotreating reactions in the literature [12,44,54]. Reaction samples were collected during the reaction (30, 60, 120, 180, 240 min) to analyze the liquid composition. The sampling line was purged with N_2 after each sample withdrawal. Sample withdrawal caused a temporary pressure-drop of ca. 2 bar in the reactor. The reactor was repressurized to 54 bar with H_2 immediately following each sample extraction.

After the reaction, the catalyst was recovered and washed with warm (50 °C) absolute ethanol (150 mL) to remove the adhering reactants, residual products, and intermediates. The collected catalyst was subsequently dried and further characterized.

For the studies with phospholipid, L-α-Phosphatidylcholine ($\geq$99%, Sigma-Aldrich, St. Louis, MO, USA) was added to the feed. It contains a glycerol backbone, having a hydrophobic long hydrocarbon chain with fatty acid residues linked via an ester group to a polar phosphate group and an organic group, called choline containing nitrogen. Additional experiments were carried out with choline hydroxide (48% in water, Fluorochem, Glossop, UK) to elucidate the possible effects of choline produced from the decomposition of the PC. The approximate fatty acid composition in the PC is reported to be 33% 16:0 (palmitic), 13% 18:0 (stearic), 31% 18:1 (oleic), and 15% 18:2 (linoleic). A typical structure of it is shown below.

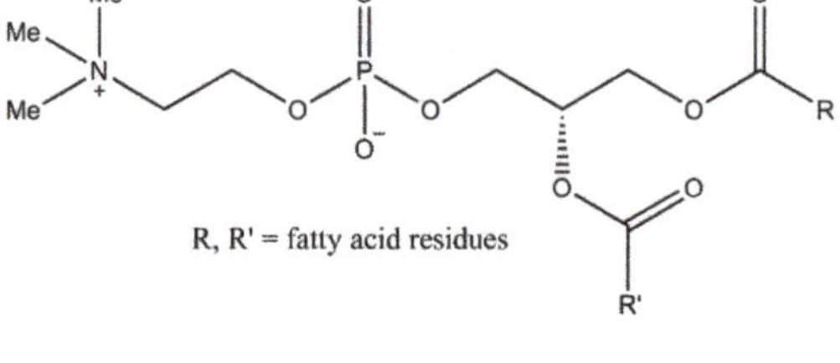

All other parts of the experiment were identical to the earlier described experiments. The amount of phosphatidylcholine added is presented here as a weight fraction based on the total liquid feed. The concentrations studied were ca. 0.4, 0.9, and 1.3 wt.% of PC, respectively. Table 5 shows these data, including the acronyms used in this study. The amount of fatty acid contribution from the PC was in the range of 2.2 to 6.7% of the oleic acid in the feed. Likewise, recovered catalysts from the first run of 0.4 wt.% PC were further tested with 0.4 wt.% PC to understand a more prolonged effect on the catalyst. Reproducibility of the experiments were also examined by repeated experiments with freshly sulfided and recovered catalysts, which gave a relative standard deviation of <5% for HDO activity based on the yields of the different alkane and olefin final products.

Table 5. Concentration of phospholipid studied and acronyms used in this article.

PC in Feed (wt.%)	Catalyst (NiMo/Al$_2$O$_3$)
0	NiMo_0 PC
0.44	NiMo_0.4 PC
0.88	NiMo_0.9 PC
1.3	NiMo_1.3 PC
0	NiMo_R_0 PC *
0.44	NiMo_R_0.4 PC *

* Refers to recovered catalyst from the first experiment.

3.3. Product Analysis

Liquid phase samples were analyzed by the method described by our group [44] using GC-MS (Agilent 7890–5977A, Agilent, Wilmington, DE, USA). The collected samples were first heated in a water bath (60 °C) to dissolve the solidified contents of the reaction mixture. Then, 1 mL of the mixture was taken out and 0.2 mL of pyridine (Sigma, 99.9%) was added to obtain a single liquid phase. The mixture was then centrifuged (WIFUG Lab centrifuges, 500 E, WIFUG, Bradford, UK) at 1500 rpm for 2 min to separate the catalyst particles from the liquid. 0.2 mL of the clear solution was further treated with 30 µL BSTFA (*N,O*-Bis(trimethylsilyl)-trifluoroacetamide, ($\geq$99.5%, Sigma-Aldrich, St. Louis, MO, USA) in a reaction vial to obtain a better separation of the products in the GC column. The mixture was then kept overnight (17 h) to complete the silylation reaction. Finally, the mixture was microfiltered to remove the fine catalyst particles and 1 µL of the filtrate was analyzed by GC-MS.

The oven ramp settings for the GC were as follows: 100 °C for 1 min, ramp at 10 °C/min to 190 °C, and continue to ramp at 30 °C/min to 300 °C. The GC was equipped with a HP-5 column, and a Flame Ionization Detector (set point was 335 °C). The injector temperature was kept at 325 °C. The retention times for the major products heptadecane and octadecane were 10.6 ± 0.2 and 11.4 ± 0.2 min, respectively.

Conversion and yield of the reactant and products were calculated based on the following equations:

$$\text{Oxygenate conversion (\%)} = \left(1 - \frac{\text{Moles of oxygenate left at any time}}{\text{Moles of oxygenate in feed}}\right) \times 100$$

$$\text{Yield (\%)} = \frac{\text{Moles of product produced}}{\text{Moles of reactant in feed}} \times 100$$

$$\text{Selectivity} = \frac{\text{Moles of product (C17 or C18) produced at 240 min}}{\text{Total moles of C17 and C18 produced at 240 min}}$$

$$\text{Initial reaction rate} = \frac{\text{Moles of oxygenate consumed (mmol)}}{\text{Amount of catalyst (g)} \times \text{Elasped time (h)}}$$

3.4. Catalyst Characterization

Freshly synthesized and spent catalyst was characterized by N$_2$-physisorption and Inductively Coupled Plasma (ICP)-Sector Field Mass Spectroscopy (SFMS, ALS Scandinavia AB, Luleå, Sweden).

Freshly sulfided catalyst was characterized by elemental analysis. Spent catalysts collected from each experiment were further characterized by XPS and TPO for their chemical states and to estimate the quantity of coke deposition, respectively. In addition, HAADF-STEM imaging was performed on the spent catalyst recovered from the experiment with 1.3 wt.% of PC.

The specific surface area, pore size, and pore volume of the synthesized catalyst were determined by nitrogen adsorption-desorption isotherms using a TriStar 3000 gas adsorption analyzer (Norcross, GA, USA). At first, the sample (~300 mg) was dried (degassing) at 225 °C under vacuum for 2 h in a flow of dry air. The sample was then placed in the apparatus for analysis. The specific surface area was calculated based on the Brunauer-Emmett-Teller equation (BET). The pore size was estimated based on the desorption isotherm using the Barret-Joyner-Halenda equation (BJH).

Elemental C, H, and N analysis was carried out on a CE Instruments elemental analyzer EA-1110 (CE Instruments, Milan, Italy), whereas a Fisons instrument elemental analyzer NA2000 (Fisons Instruments, Milan, Italy) was configured for S analysis. The analysis uses high temperature combustion followed by GC separation, detection by thermal conductivity, and quantification based on the calibration.

X-ray Photoelectron Spectroscopy (XPS) studies on sulfided and different spent catalysts were performed using a Perkin Elmer PHI 5000C ESCA system (Waltham, MA, USA). The monochromatic Al-Kα source with a binding energy of 1486.6 eV was used to record spectra for all samples. Samples were initially placed on a double-sided carbon tape on the sample holder. They were then placed on the manipulator fork of the pretreatment chamber under nitrogen atmosphere. The sample holder was then carefully transferred to the sample stage of the ultra-high vacuum chamber when the desired pressure had been reached (usually less than 2×10^{-8} torr). In the vacuum chamber, the sample was irradiated with the radiation source and emitted photoelectrons were detected by a spherical energy analyzer. The angle between the source and detector was 90°. In addition to the full scan data, high resolution spectra with a step of 0.125 eV were recorded for Ni, Mo, O, S, C, and P core level spectra. Data were analyzed by MultiPak software provided with the instrument and with CasaXPS. The C1s binding energy of 284.6 eV was taken as the reference for all spectra.

Samples for transmission electron microscopy (TEM) were prepared by applying a small amount of the dry powder onto a carbon coated copper grid. TEM analysis was performed using an FEI Titan 80-300 TEM (FEI, Inc., Hillsboro, OR, USA) operating at an accelerating voltage of 300 kV. Images were acquired in the scanning TEM (STEM) mode using a high angle annular dark field (HAADF) detector. Energy dispersive X-ray (EDX) analysis for chemical identification was also performed in the STEM mode using an Oxford X-sight detector. Spectrum acquisition and data analysis were performed using the TEM Imaging & Analysis (TIA). MoS_2 slab length and stacking degree were measured by ImageJ software.

Temperature Programmed Oxidation (TPO) experiments were performed in a setup consisting of a manifold of mass flow controllers (MFC) for feed gas mixing, a Setaram Sensys differential scanning calorimeter with a quartz tube containing the catalyst sample, and a mass spectrometer (Hiden HPR-20 QUI, Hiden Analytical, Warrington, UK) for measuring the quantities of oxidation products in the outlet carrier gas. Ar was used as the carrier gas for all experiments. The total gas flow rate was maintained at 20 NmL/min through quartz tubes. Approximately 25 mg of the spent catalyst was placed on the sintered bed of the quartz tube. The sample was then heated to 250 °C at 25 °C min^{-1}, followed by 2 h of isothermal operation, and cooling to 50 °C while the sample was continuously exposed to Ar only. Finally, the sample was heated to 800 °C at 2 °C min^{-1} with 21% O_2 in Ar. The coke deposition on each of the spent catalyst samples was calculated based on a calibration for CO, CO_2, and H_2O in the carrier gas. Water calibration was done by completely oxidizing a set feed concentration of hydrogen and excess oxygen over a Pt/Al_2O_3 catalyst at 300 °C.

4. Conclusions

In this study, NiMo/γ-Al$_2$O$_3$ catalyst has been examined for HDO of oleic acid in the presence of a major bio/algal oil phospholipid, phosphatidylcholine. Activity tests with the freshly sulfided catalyst revealed high deoxygenation activity mainly by DCO$_x$ and direct HDO. Studies with different amounts of PC in the liquid feed shows that overall oxygenate conversion, fatty acid/olefin hydrogenation (saturation of the double bond), and subsequent C=O hydrogenolysis were affected significantly. Moreover, the decarboxylation/decarbonylation route was more influenced by PC compared to the direct HDO route and a proportional decrease in C17+ yield was observed as a function of the PC concentration in the feed. We propose that the observed deactivation is due to formation of both phosphoric acid and choline (a strong base) from the decomposition of the phosphatidylcholine.

The elemental analysis revealed that the sulfur amount was similar in all recovered catalyst samples, thus, the deactivation is not originating from a loss of sulfur. However, comparable amounts of nitrogen were observed on each catalyst after experiments with phospholipids, demonstrating the interaction of choline/amines with the catalyst. ICP-SFMS data on the recovered catalyst and liquid phase clearly show that P selectively deposited on the catalyst. Moreover, TEM-EDX showed that P was evenly distributed over the support alumina. It did not appear that P associated preferentially with the active NiS or NiMoS phase, but was instead distributed over the support, which lowered the dispersion of the MoS$_2$ slabs and affected the Ni promotion. Hence, there was no direct evidence indicating P interaction with Ni and Mo species. This observation in combination with XPS led us to the conclusion that aluminum phosphate (AlPO$_4$) is formed, which blocks the actives sites and pores, and thereby results in the observed deactivation. Nitrogen physisorption data on the recovered catalyst confirms severe pore blocking even in the presence of small amounts of phospholipid in the feed.

TPO of spent catalysts was used to quantify and characterize the coke formed. Two clear peaks for the CO$_x$ (CO+CO$_2$) were evident, and, in addition, large water peaks were observed that coincided with the low temperature CO$_x$ peak. Analysis of these data clearly showed that the C/H ratio was much lower for the low temperature peak, and we therefore conclude that this peak mostly results from oxidation of residual adsorbed reactants and intermediate species, i.e. soft coke. The high temperature peak had a much lower hydrogen content, and this peak therefore originates from mainly advanced coke formation. However, an increasing trend of more advanced coke formation has been observed with the increasing PC concentration in the feed, likely resulting from the formation of phosphoric acid from the PC decomposition.

Supplementary Materials: The following are available online at http://www.mdpi.com/2073-4344/8/10/418/s1, Figure S1: BF-TEM micrograph of the catalyst collected from the baseline experiment (NiMo_0 PC) without PC, Figure S2: TEM-EDX data: The brighter areas show signal from Mo but no Ni, Table S1: Measurement of carbon (C) and hydrogen (H) content (wt.%) of the recovered catalyst by TPO.

Author Contributions: M.A.S., S.T. and L.O. planned the experiments; M.A.S. performed the experiments; M.A.S., D.C., P.A., S.T. and L.O. analyzed the data; M.A.S. wrote the paper; P.A., S.T., D.C., E.L.G. and L.O. reviewed and edited the paper; D.C. and L.O. supervised M.A.S.

Funding: This research was funded by [Preem] grant number [239-2012-1584] and [Formas] grant number [239-2014-164]

Acknowledgments: This work is performed at the Competence Centre for Catalysis (KCK), division of chemical reaction engineering at Chalmers University of Technology. We would like to acknowledge Preem, Formas (Contract: 239-2012-1584 and 239-2014-164) and Swedish Energy Agency for financial aid. We also like to acknowledge Stefan Gustafsson for his help with TEM analysis.

Conflicts of Interest: The authors declare no conflict of interest.

References

1. Dauenhauer, P.J. *Handbook of Plant-Based Biofuels*; Ashok, P., Ed.; CRC Press: Boca Raton, FL, USA, 2010; Volume 3, pp. 386–387. ISBN 978-1-56022-175-3.
2. Kubička, D.; Tukač, V. Hydrotreating of triglyceride-based feedstocks in refineries. *Adv. Chem. Eng.* **2012**, *42*, 141–194.
3. Srifa, A.; Viriya-empikul, N.; Assabumrungrat, S.; Faungnawakij, K. Catalytic behaviors of Ni/γ-Al$_2$O$_3$ and Co/γ-Al$_2$O$_3$ during the hydrodeoxygenation of palm oil. *Catal. Sci. Technol.* **2015**, *5*, 3693–3705. [CrossRef]
4. Kubička, D.; Kaluža, L. Deoxygenation of vegetable oils over sulfided Ni, Mo and NiMo catalysts. *Appl. Catal. A Gen.* **2010**, *372*, 199–208. [CrossRef]
5. Mortensen, P.M.; Grunwaldt, J.-D.; Jensen, P.A.; Knudsen, K.G.; Jensen, A.D. A review of catalytic upgrading of bio-oil to engine fuels. *Appl. Catal. A Gen.* **2011**, *407*, 1–19. [CrossRef]
6. Laurent, E.; Delmon, B. Study of the hydrodeoxygenation of carbonyl, carboylic and guaiacyl groups over sulfided CoMo/γ-Al$_2$O$_3$ and NiMo/γ-Al$_2$O$_3$ catalyst. *Appl. Catal. A Gen.* **1994**, *109*, 97–115. [CrossRef]
7. Furimsky, E. Catalytic hydrodeoxygenation. *Appl. Catal. A Gen.* **2000**, *199*, 147–190. [CrossRef]
8. Enol, O.İ.; Viljava, T.-R.; Krause, A.O.I. Hydrodeoxygenation of aliphatic esters on sulphided NiMo/γ-Al$_2$O$_3$ and CoMo/γ-Al$_2$O$_3$ catalyst: The effect of water. *Catal. Today* **2005**, *106*, 186–189.
9. He, Z.; Wang, X. Hydrodeoxygenation of model compounds and catalytic systems for pyrolysis bio-oils upgrading. *Catal. Sustain. Energy* **2012**, *1*, 28–52. [CrossRef]
10. Marafi, M.; Stanislaus, A.; Furimsky, E. *Handbook of Spent Hydroprocessing Catalysts: Regeneration, Rejuvenation, Reclamation, Environment and Safety*; Elsevier BV: Oxford, UK, 2010; ISBN 978-0-444-53556-6.
11. Kalnes, T.N.; Koers, K.P.; Marker, T.; Shonnard, D.R. A technoeconomic and environmental life cycle comparison of green diesel to biodiesel and syndiesel. *Environ. Prog. Sustain. Energy* **2009**, *28*, 111–120. [CrossRef]
12. Arora, P.; Ojagh, H.; Woo, J.; Grennfelt, E.L.; Olsson, L.; Creaser, D. Investigating the effect of Fe as a poison for catalytic HDO over sulfided NiMo alumina catalysts. *Appl. Catal. B Environ.* **2018**, *227*, 240–251. [CrossRef]
13. Laakkonen, M.; Myllyoja, J.; Toukoniitty, B.; Hujanen, M.; Saastamoinen, A.; Toivo, A. Process for Manufacture of Liquid Fuel Componenst from Renewable Sources. U.S. Patent 13,646,250, 4 November 2013.
14. Bartholomew, C.H. Mechanisms of catalyst deactivation. *Appl. Catal. A Gen.* **2001**, *212*, 17–60. [CrossRef]
15. Beckman, D.; Elliott, D.C. Comparisons of the yields and properties of the oil products from direct thermochemical biomass liquefaction processes. *Can. J. Chem. Eng.* **1985**, *63*, 99–104. [CrossRef]
16. Mortensen, P.M.; Gardini, D.; Damsgaard, C.D.; Grunwaldt, J.-D.; Jensen, P.A.; Wagner, J.B.; Jensen, A.D. Deactivation of Ni-MoS$_2$ by bio-oil impurities during hydrodeoxygenation of phenol and octanol. *Appl. Catal. A Gen.* **2016**, *523*, 159–170. [CrossRef]
17. Kubička, D.; Horáček, J. Deactivation of HDS catalysts in deoxygenation of vegetable oils. *Appl. Catal. A Gen.* **2011**, *394*, 9–17. [CrossRef]
18. Popov, S.; Kumar, S. Renewable fuels via catalytic hydrodeoxygenation of lipid-based feedstocks. *Biofuels* **2013**, *4*, 219–239. [CrossRef]
19. Dijkstra, A.J. About water degumming and the hydration of non-hydratable phosphatides. *Eur. J. Lipid Sci. Technol.* **2017**, *119*, 1600496. [CrossRef]
20. Paisan, S.; Chetpattananondh, P.; Chongkhong, S. Assessment of water degumming and acid degumming of mixed algal oil. *J. Environ. Chem. Eng.* **2017**, *5*, 5115–5123. [CrossRef]
21. Kiatkittipong, W.; Phimsen, S.; Kiatkittipong, K.; Wongsakulphasatch, S.; Laosiripojana, N.; Assabumrungrat, S. Diesel-like hydrocarbon production from hydroprocessing of relevant refining palm oil. *Fuel Process. Technol.* **2013**, *116*, 16–26. [CrossRef]
22. Bailey, D.S.; Northcote, D.H. Phospholipid composition of the plasma membrane of the green alga, hydrodictyon africanum. *Biochem. J.* **1976**, *156*, 295–300. [CrossRef] [PubMed]
23. Meng, X.; Pan, Q.; Ding, Y.; Jiang, L. Rapid determination of phospholipid content of vegetable oils by FTIR spectroscopy combined with partial least-square regression. *Food Chem.* **2014**, *147*, 272–278. [CrossRef] [PubMed]
24. Fitz, C.W.; Rase, H.F. Effects of phosphorus on nickel-molybdenum hydrodesulfurization/hydrodenitrogenation catalysts of varying metals content. *Ind. Eng. Chem. Prod. Res. Dev.* **1983**, *22*, 40–44. [CrossRef]

25. Chadwick, D.; Aitchison, D.W.; Badilla-Ohlbaum, R.; Josefsson, L. Influence of phosphorus on the HDS activity of Ni-Mo/γ-Al$_2$O$_3$ catalysts. *Stud. Surf. Sci. Catal.* **1983**, 323–332.

26. Bouwens, S.M.A.M.; Vissers, J.P.R.; de Beer, V.H.J.; Prins, R. Phosphorus poisoning of molybdenum sulfide hydrodesulfurization catalysts supported on carbon and alumina. *J. Catal.* **1988**, *112*, 401–410. [CrossRef]

27. Rayo, P.; Ramírez, J.; Torres-Mancera, P.; Marroquín, G.; Maity, S.K.; Ancheyta, J. Hydrodesulfurization and hydrocracking of maya crude with P-modified NiMo/Al$_2$O$_3$ catalysts. *Fuel* **2012**, *100*, 34–42. [CrossRef]

28. Eijsbouts, S.; Van Gestel, J.N.M.; Van Veen, J.N.M.; De Beer, V.H.J.; Prins, R. The effect of phosphate on the hydrodenitrogenation activity and selectivity of alumina-supported sulfided Mo, Ni, and Ni-Mo catalysts. *J. Catal.* **1991**, *131*, 412–432. [CrossRef]

29. Mangnus, P.J.; van Langeveld, A.D.; de Beer, V.H.J.; Moulijn, J.A. Influence of phosphate on the structure of sulfided alumina supported cobalt-molybdenum catalysts. *Appl. Catal.* **1991**, *68*, 161–177. [CrossRef]

30. Zhou, T.; Yin, H.; Liu, Y.; Han, S.; Chai, Y.; Liu, C. Effect of phosphorus content on the active phase structure of NiMoP/Al$_2$O$_3$ catalyst. *J. Fuel Chem. Technol.* **2010**, *38*, 69–74. [CrossRef]

31. Argyle, M.; Bartholomew, C. Heterogeneous catalyst deactivation and regeneration: A review. *Catalysts* **2015**, *5*, 145–269. [CrossRef]

32. Furimsky, E.; Massoth, F.E. Deactivation of hydroprocessing catalysts. *Catal. Today* **1999**, *52*, 381–495. [CrossRef]

33. Diez, F.; Gates, B.C.; Miller, J.T.; Sajkowski, D.J.; Kukes, S.G. Deactivation of a nickel-molybdenum/.gamma.-alumina catalyst: Influence of coke on the hydroprocessing activity. *Ind. Eng. Chem. Res.* **1990**, *29*, 1999–2004. [CrossRef]

34. Simakova, I.; Simakova, O.; Mäki-Arvela, P.; Murzin, D.Y. Decarboxylation of fatty acids over Pd supported on mesoporous carbon. *Catal. Today* **2010**, *150*, 28–31. [CrossRef]

35. Laurent, E.; Delmon, B. Influence of water in the deactivation of a sulfided NiMoγ-Al$_2$O$_3$ catalyst during hydrodeoxygenation. *J. Catal.* **1994**, *146*, 281–291. [CrossRef]

36. Froment, G.; Delmon, B.F. *Catalyst Deactivation*, 1st ed.; Elsevier: Amsterdam, The Netherlands, 1994; ISBN 9780080887388.

37. Badawi, M.; Paul, J.F.; Cristol, S.; Payen, E.; Romero, Y.; Richard, F.; Brunet, S.; Lambert, D.; Portier, X.; Popov, A.; et al. Effect of water on the stability of Mo and CoMo hydrodeoxygenation catalysts: A combined experimental and dft study. *J. Catal.* **2011**, *282*, 155–164. [CrossRef]

38. Şenol, O.İ.; Viljava, T.-R.; Krause, A.O.I. Effect of sulphiding agents on the hydrodeoxygenation of aliphatic esters on sulphided catalysts. *Appl. Catal. A Gen.* **2007**, *326*, 236–244. [CrossRef]

39. Coumans, A.E.; Hensen, E.J.M. A model compound (methyl oleate, oleic acid, triolein) study of triglycerides hydrodeoxygenation over alumina-supported nimo sulfide. *Appl. Catal. B Environ.* **2017**, *201*, 290–301. [CrossRef]

40. Ryymin, E.-M.; Honkela, M.L.; Viljava, T.-R.; Krause, A.O.I. Insight to sulfur species in the hydrodeoxygenation of aliphatic esters over sulfided NiMo/γ-Al$_2$O$_3$ catalyst. *Appl. Catal. A Gen.* **2009**, *358*, 42–48. [CrossRef]

41. Dupont, C.; Lemeur, R.; Daudin, A.; Raybaud, P. Hydrodeoxygenation pathways catalyzed by MoS$_2$ and NiMoS active phases: A DFT study. *J. Catal.* **2011**, *279*, 276–286. [CrossRef]

42. Donnis, B.; Egeberg, R.G.; Blom, P.; Knudsen, K.G. Hydroprocessing of bio-oils and oxygenates to hydrocarbons. Understanding the Reaction Routes. *Top. Catal.* **2009**, *52*, 229–240. [CrossRef]

43. Kubička, D. Future refining catalysis-introduction of biomass feedstocks. *Collect. Czechoslov. Chem. Commun.* **2008**, *73*, 1015–1044. [CrossRef]

44. Ojagh, H.; Creaser, D.; Tamm, S.; Arora, P.; Nyström, S.; Grennfelt, E.L.; Olsson, L. Effect of dimethyl disulfide on activity of NiMo based catalysts used in hydrodeoxygenation of oleic acid. *Ind. Eng. Chem. Res.* **2017**, *56*, 5547–5557. [CrossRef]

45. Changi, S.; Matzger, A.J.; Savage, P.E. Kinetics and pathways for an algal phospholipid (1,2-dioleoyl-*sn*-glycero-3-phosphocholine) in high-temperature (175–350 °C) water. *Green Chem.* **2012**, *14*, 2856–2867. [CrossRef]

46. de Koning, A.J.; McMullan, K.B. Hydrolysis of phospholipids with hydrochloric acid. *Biochim. Biophys. Acta, Lipids Lipid Metab.* **1965**, *106*, 519–526. [CrossRef]

47. Qu, L.; Prins, R. Hydrogenation of cyclohexene over in situ fluorinated NiMoS catalysts supported on alumina and silica-alumina. *J. Catal.* **2002**, *207*, 286–295. [CrossRef]

48. De Klerk, A. Oligomerization of 1-hexene and 1-octene over solid acid catalysts. *Ind. Eng. Chem. Res.* **2005**, *44*, 3887–3993. [CrossRef]

49. Iwamoto, R.; Grimblot, J. Influence of phosphorus on the properties of alumina-based hydrotreating catalysts. *Adv. Catal.* **1999**, *44*, 417–503.

50. Liu, G.; Robertson, A.W.; Li, M.M.-J.; Kuo, W.C.H.; Darby, M.T.; Muhieddine, M.H.; Lin, Y.-C.; Suenaga, K.; Stamatakis, M.; Warner, J.H.; et al. MoS_2 monolayer catalyst doped with isolated Co atoms for the hydrodeoxygenation reaction. *Nat. Chem.* **2017**, *9*, 810–816. [CrossRef] [PubMed]

51. Sigurdson, S.; Sundaramurthy, V.; Dalai, A.K.; Adjaye, J. Phosphorus promoted trimetallic NiMoW/γ-Al_2O_3 sulfide catalysts in gas oil hydrotreating. *J. Mol. Catal. A Chem.* **2008**, *291*, 30–37. [CrossRef]

52. Marafi, M.; Stanislaus, A. Effect of initial coking on hydrotreating catalyst functionalities and properties. *Appl. Catal. A Gen.* **1997**, *159*, 259–267. [CrossRef]

53. Marafi, M.; Stanislaus, A. Influence of catalyst acidity and feedstock quality on hydrotreating catalyst deactivation by coke deposition. *Pet. Sci. Technol.* **2001**, *19*, 697–710. [CrossRef]

54. Gong, S.; Shinozaki, A.; Shi, M.; Qian, E.W. Hydrotreating of jatropha oil over alumina based catalysts. *Energy Fuels* **2012**, *26*, 2394–2399. [CrossRef]

 *catalysts*

Article

Hydrotreatment Followed by Oxidative Desulfurization and Denitrogenation to Attain Low Sulphur and Nitrogen Bitumen Derived Gas Oils

Sandeep Badoga [1], Prachee Misra [1], Girish Kamath [1], Ying Zheng [2] and Ajay K. Dalai [1,*]

[1] Catalysis and Chemical Reaction Engineering Laboratories, Department of Chemical Engineering, University of Saskatchewan, Saskatoon, SK S7N 5A9, Canada; sab809@mail.usask.ca (S.B.); prm173@mail.usask.ca (P.M.); gik089@mail.usask.ca (G.K.)
[2] School of Engineering, University of Edinburgh, Mayfield Road, Edinburgh EH9 3DW, UK; yzheng@unb.ca
* Correspondence: ajay.dalai@usask.ca

Received: 28 September 2018; Accepted: 3 December 2018; Published: 10 December 2018

Abstract: To lower the sulphur content below 500 ppm and to increase the quality of bitumen derived heavy oil, a combination of hydrotreating followed by oxidative desulfurization (ODS) and oxidative denitrogenation (ODN) is proposed in this work. NiMo/γ-Al$_2$O$_3$ catalyst was synthesized and used to hydrotreat heavy gas oil (HGO) and light gas oil (LGO) at typical operating conditions of 370–390 °C, 9 MPa, 1–1.5 h^{-1} space velocity and 600:1 H$_2$ to oil ratio. γ-Alumina and alumina-titania supported Mo, P, Mn and W catalysts were synthesized and characterized using X-ray diffractions, N$_2$ adsorption-desorption using Brunauer–Emmett–Teller (BET) method, X-ray photoelectron spectroscopy (XPS) and Fourier transform infrared spectroscopy (FT-IR). All catalysts were tested for the oxidation of sulphur and nitrogen aromatic compounds present in LGO and HGO using tert-butyl hydroperoxide (TBHP) as oxidant. The oxidized sulphur and nitrogen compounds were extracted using adsorption on activated carbon and liquid-liquid extraction using methanol. The determination of oxidation states of each metal using XPS confirmed the structure of metal oxides in the catalyst. Thus, the catalytic activity determined in terms of sulphur and nitrogen removal is related to their physico-chemical properties. In agreement with literature, a simplistic mechanism for the oxidative desulfurization is also presented. Mo was found to be more active in comparison to W. Presence of Ti in the support has shown 8–12% increase in ODS and ODN. The MnPMo/γ-Al$_2$O$_3$-TiO$_2$ catalyst showed the best activity for sulphur and nitrogen removal. The role of Mn and P as promoters to molybdenum was also discussed. Further three-stage ODS and ODN was performed to achieve less than 500 ppm in HGO and LGO. The combination of hydrotreatment, ODS and ODN has resulted in removal of 98.8 wt.% sulphur and 94.7 wt.% nitrogen from HGO and removal of 98.5 wt.% sulphur and 97.8 wt.% nitrogen from LGO.

Keywords: oxidative desulfurization; oxidative denitrogenation; hydrotreating; XPS; activated carbon; tert-butyl hydroperoxide

---

1. Introduction

Worldwide increase in industrialization has led to an increase in consumption of petroleum oil and coal [1]. It has raised serious environmental concerns due to the rise in sulphur oxide (SOx) levels in air. Therefore, environmental protection agencies (EPA) such as US EPA limits the sulphur levels to 15 ppm in diesel fuel, which is expected to be further lowered to 10 ppm. On the other hand, the gas oil extracted from unconventional sources such as, oil sands and shale oil contain high amount of sulphur and nitrogen impurities. For instance the oil sands bitumen derived heavy gas oil (HGO) contains ~40,000 ppm sulphur and ~4000 ppm nitrogen [2]. The bitumen derived gas oils (HGO, light gas oil (LGO) and naphtha) are upgraded onsite via hydrotreating to lower the sulphur and nitrogen content before sending them to further processing in existing refineries. Hydrotreating is a catalytic process operating in the presence of hydrogen at high pressure (8–12 MPa) and temperatures (350–400 °C) to remove sulphur and nitrogen via processes known as hydrodesulfurization (HDS) and hydrodenitrogenation (HDN), respectively. Typically, for HGO the hydrotreating can lower the sulphur and nitrogen content to ~2200 ppm and 1700 ppm, respectively. To further lower the sulphur content from ~2200 ppm during upgrading requires severe hydrotreating operating conditions such as higher pressures, temperatures and hydrogen flowrates to remove sulphur from refractory molecules such as alkyl substituted dibenzothiophenes (DBT). Increasing the temperature leads to cracking of oil and higher pressures lead to an increase in saturation of aromatics and higher hydrogen consumption, thus degrading the oil quality, in addition to decline in catalyst lifespan. Moreover, huge capital investment is required for high-pressure processes.

The alternative processes such as adsorption and oxidative desulfurization and denitrogenation has been of interest to achieve ultra-low sulphur level in oil because of very mild operating conditions and no usage of hydrogen. Various adsorbents including activated carbon, ionic resins, metal organic frameworks, metal oxides and zeolites have been used to adsorb sulphur containing compounds from diesel oil [3–10]. Ganiyu et al. [6] utilized activated carbon doped with 1.0 wt.% boron to selectively adsorb 4,6-DMDBT from model fuel. Srivastav and Srivastava [7] carried out the adsorption of DBT dissolved in hexanes on commercial grade activated alumina and Li et al. [4] presented a study on the challenges associated with removal of aromatic sulphur compounds using metal-organic frameworks. McKinley and Angelici [9] used silver salts on SBA-15 for adsorptive removal of DBT from simulated hydrotreated petroleum feedstocks. They were able to lower the sulphur level from 411 ppm to 8 ppm.

Oxidative desulfurization (ODS) is one of the alternate routes for deep desulfurization. In ODS process (as shown in Figure 1), the sulphur present in compounds such as DBT is first oxidized to sulfoxides or sulfones in the presence of oxidizing agents such as hydrogen peroxide, organic peroxides, nitrogen oxide or air and catalysed by organic acids, heteropolyic acids or solid catalysts. The oxidation of sulphur leads to increase in polarity of sulphur containing compounds. Therefore, the sulfoxides and sulfones were easily extracted from the oil by using polar solvents or adsorbents, thus achieving deep desulfurization. Most commonly tested solvents include dimethylsulfoxide (DMSO), dimethylformamide (DMF), acetonitrile, methanol and acetone [11,12]. There are several well-known disadvantages with solvent exactions such as toxicity, reusability, disposal, explosiveness and cost. Therefore, selection of solvent is a challenge. DMSO poses challenges during recovery due to similar boiling point, whereas acetonitrile is highly polar and extracts lots of aromatics [11]. Methanol is a good solvent for extracting sulfones however, it has similar density as diesel and thus separation is difficult. The ease of oxidation of various sulphur containing compounds depends on the electron densities on the sulphur atom. Sulphur with higher electron densities are easier to oxidize, hence, follows the order 4,6-DMDBT > DBT > BT > Thiophene [13]. Bunthid et al. [14] have utilized formic acid and H_2O_2 as oxidizing agent to oxidize DBT. The corresponding sulfone from the solution was then extracted by adsorption on pyrolysis char. The small amount of water remaining was extracted by drying over anhydrous sodium sulphate. They reported 72% sulphur removal. In another study, Ahmad et al. [15], used acetic acid as a catalyst for sulphur oxidation with H_2O_2. They used Fuller's earth for the adsorption of sulfones and achieved 50% sulphur removal.

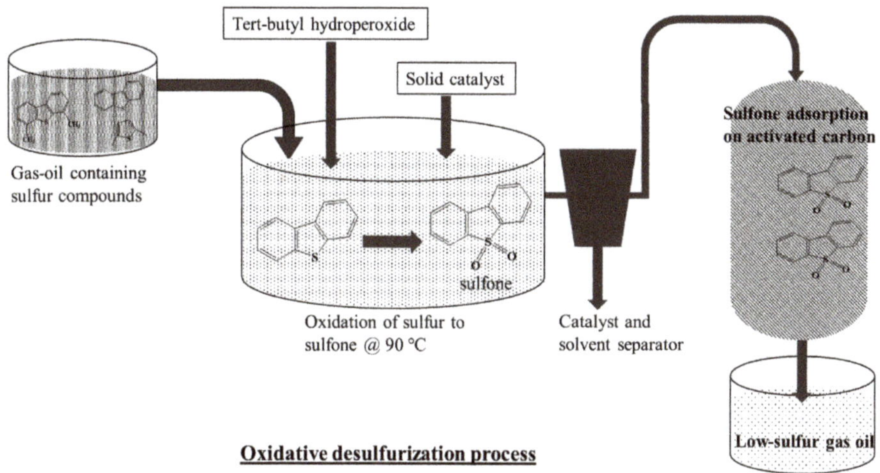

Figure 1. Schematic for Oxidative desulfurization process.

Palaic et al. [16] studied the oxidative desulfurization of diesel fuels for the removal of refractory sulphur compounds which are difficult to remove during conventional hydrotreating. They utilized hydrogen peroxide as oxidant and acetic acid as catalyst and performed reactions in a batch reactor. The effects of process conditions of ultrasound-assisted ODS with *N,N* dimethylformamide and methanol as extraction solvents were also reported. They successfully removed 98% sulphur from 4000 ppm DBT spiked diesel fuel. The usage of acid catalyst such as acetic acid or formic acid for oxidation of sulphur compounds requires recovery of these organic acids after treatment. It requires additional set up for recovery of such corrosive and toxic organic acids. Therefore, in view of this, solid catalysts were utilized. Fattahi et al. [17] synthesized $CoMo/\gamma-Al_2O_3$ catalyst with different Co/Mo ratio and utilized it for the oxidative desulfurization of DBT and benzothiophene (BT) using H_2O_2 as oxidizing agent. They reported 90% removal of DBT and 30% removal of BT. Chica et al. [18] studied the effect of catalyst on ODS in a continuous fixed bed reactor for model feed containing different types of sulphur compounds including thiophenes and alkyl substituted DBTs. Tert-butyl hydroperoxide was utilized as oxidizing agent. They reported that $MoOx/Al_2O_3$ was very active but had faster deactivation rate. However, the Ti-MCM-41 was stable and active for longer time. Gatan et al. [19] from UOP also proposed the oxidation with organic perpoxide in the presence of heterogeneous catalyst followed by separation via adsorption and/or extraction. They proposed ODS as complementary to hydrodesulfurization for the removal of refractory sulphur compounds to attain ultra-low sulphur diesel. Leng et al. [20] synthesized titanium doped hierarchical mordenite to catalyse ODS of DBT in octane. They used acetonitrile for the extraction of sulfones and were successful in lowering the sulphur content from 1000 ppm to 14 ppm. Lorencon et al. [21] utilized titanate nanotubes and H_2O_2 for the oxidation of DBT in model feed (500 ppm sulphur) and evidenced ~98% sulphur removal. Tian et al. [22] performed ODS using H_2O_2 and phosphomolybic acid supported on silica for the removal of DBT and BT from model oil (~400 ppm sulphur) and were successful in removing ~95% sulphur. García-Gutiérrez et al. [23] utilized heterogeneous tungsten catalyst for oxidation of sulphur compounds in diesel (~320 ppm sulphur) using H_2O_2 as oxidizing agent and achieved ~70% removal. Therefore, it has been seen that various catalyst systems have been tested for the oxidation of sulphur present in model and simulated diesel fuels. Moreover, the extraction of sulfones and sulfoxides from oil was carried out using both solvent extraction and adsorption.

The oxidation using peroxide in the presence of catalyst is not only selective to oxidize sulphur in heterocyclic aromatic compounds such as DBTs and alkyl substitute DMDBTs but it also oxidizes nitrogen containing aromatic compounds present in real feed. The oxidation of nitrogen containing aromatic compounds such as quinoline, indole and carbazole is a complex reaction. According to the literature [24–26], it was found that the peroxide group oxidizes that carbon in aromatic ring, which is having least electron density to form -oxy or -oxyl compounds. Further oxidation leads to ring opening and formation of various oxygenated products of ketone and carboxylic acid category. A study by Ogunlaja et al. [25] also reported the oxidation of nitrogen in quinoline to form of quinoline N-oxide. The oxidation follows the order indole > quinoline > acridine > carbazole. However, the oxidation of nitrogen compounds increases their polarity, which makes it easy to remove organonitrogen compounds from oil by adsorption or extraction, thus resulting in oxidative denitrogenation (ODN). The removal of nitrogen prevents the downstream catalysts from poisoning and hence ODN increases the quality of the treated HGO and LGO.

Therefore, to keep bitumen-based fuels competitive in the current market, the production of low sulphur and low nitrogen HGO and LGO is required, which still remains a challenge to industry. Thus, the potential of ODS and ODN to further lower the sulphur and nitrogen level in hydrotreated LGO and HGO using heterogeneous catalyst needs to be explored. In this work, NiMo/γ-Al$_2$O$_3$ catalyst was synthesized and utilized to hydrotreat the HGO and LGO at typical industrial conditions of 370–390 °C, 9 MPa, 1–1.5 h^{-1} LHSV and 1:600 oil to H$_2$ ratio, in fixed bed flow reactor, to generate hydrotreated gas oil for ODS and ODN process. Further, alumina and alumina-titania supported Mo, W, Mn and P catalysts were synthesized and tested for the oxidation of sulphur and nitrogen compounds present in real gas oil using tert-butylhydroperoxide (TBHP) as oxidizing agent. The extraction of oxidized sulphur and nitrogen compounds was carried out by adsorption on activated carbon and compared with liquid extraction using methanol. This study on integration of conventional hydrotreating technology with ODS and ODN for upgrading bitumen derived gas oil has resulted in lowering the sulphur levels to less than 500 ppm in both HGO and LGO, which can be further lowered down in existing refineries to meet the EPA regulations for diesel fuel. Additionally, the nitrogen levels for HGO were brought down to ~200 ppm, which enhances the quality of crude oil and eliminates the nitrogen removal process prior to refining of treated HGO. The catalysts were thoroughly characterized using BET, XRD, FTIR and XPS and their physico-chemical properties were related to their catalytic activities.

2. Results and Discussion

2.1. Material Characterization

2.1.1. N$_2$ Adsorption-Desorption Analysis

The textural properties of all the materials synthesized in this work were determined using N$_2$ adsorption-desorption analysis. Table 1 shows the surface area, pore volume and pore diameter of each material. The surface area of γ-Al$_2$O$_3$ decreased from 330 m^2/g to 300 m^2/g and pore volume decreased from 0.90 cm^3/g to 0.75 cm^3/g on addition of TiO$_2$. However, the pore structure stayed intact, as evidenced from the pore size distribution and adsorption-desorption profiles (figure not shown). The γ-Al$_2$O$_3$ and γ-Al$_2$O$_3$-TiO$_2$ exhibited the type IV isotherm with H1 type hysteresis loop confirming the mesoporous structure (Figure 2). On addition of Mo, W, P and Mn on γ-Al$_2$O$_3$-TiO$_2$, the resulting materials maintained the type IV isotherm. The decline in pore volume, surface area and pore diameter in catalysts Mo/γ-Al$_2$O$_3$, W/γ-Al$_2$O$_3$, Mo/γ-Al$_2$O$_3$-TiO$_2$, W/γ-Al$_2$O$_3$-TiO$_2$, PMo/γ-Al$_2$O$_3$-TiO$_2$ and MnPMo/γ-Al$_2$O$_3$-TiO$_2$ is proportional to the amounts of metals (Mo, W, P, Mn) loaded on alumina and alumina-titania supports.

Table 1. Textural properties of the catalysts used for oxidative desulfurization and denitrogenation. (error ±3.0%).

Catalyst	Surface Area (m^2/g)	Pore Volume (cc/g)	Pore Diameter (nm)
γ-Al$_2$O$_3$	330	0.90	7.6
γ-Al$_2$O$_3$-TiO$_2$	300	0.75	7.1
Mo/γ-Al$_2$O$_3$	242	0.60	7.2
Mo/γ-Al$_2$O$_3$-TiO$_2$	210	0.50	7.0
W/γ-Al$_2$O$_3$	240	0.62	7.3
W/γ-Al$_2$O$_3$-TiO$_2$	208	0.49	7.0
PMo/γ-Al$_2$O$_3$-TiO$_2$	181	0.46	6.9
MnPMo/γ-Al$_2$O$_3$-TiO$_2$	165	0.43	6.9

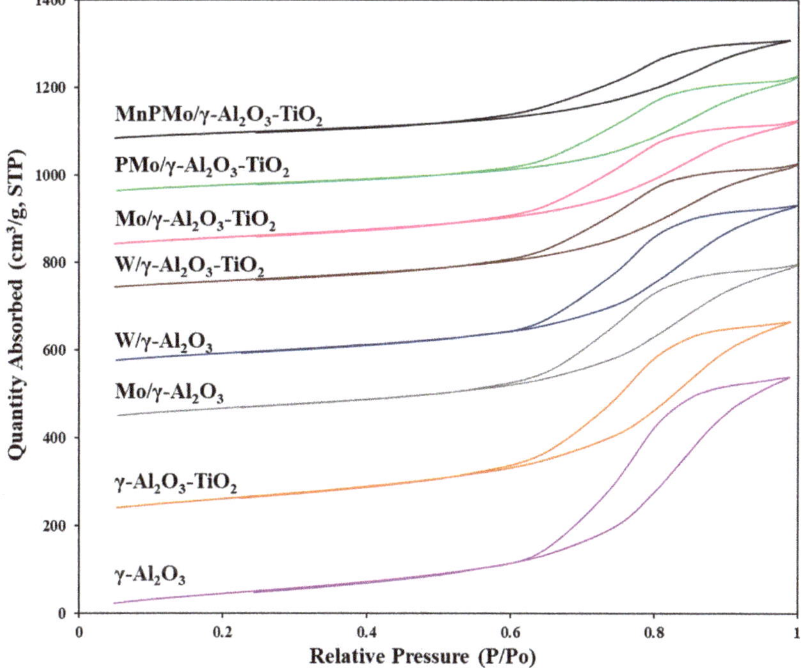

Figure 2. N$_2$ adsorption-desorption isotherms for all materials.

2.1.2. X-ray Diffraction Analysis

The X-ray diffraction analysis was performed to determine the presence of various crystalline phase of metals doped on the γ-Al$_2$O$_3$ and γ-Al$_2$O$_3$-TiO$_2$ support materials. The XRD profile for each catalyst is shown in Figure 3. The peaks at 2θ = 37.0°, 46.0° and 66.7° in the XRD profile for γ-Al$_2$O$_3$ corresponds to the planes of cubic Al$_2$O$_3$ [27]. The additional peaks at 2θ = 25.5, 37.9, 48.0, 54.0, 55.0 and 62.8 degrees in XRD profile for γ-Al$_2$O$_3$ -TiO$_2$ material is due to the presence of TiO$_2$ in anatase phase, which is the active phase of titania for the reaction [28]. The addition of W on alumina and alumina-titania, does not show any additional peak in XRD profile of W/γ-Al$_2$O$_3$ and W/ γ-Al$_2$O$_3$ -TiO$_2$ indicating the fine dispersion of tungsten oxide. However, the materials Mo/γ-Al$_2$O$_3$ and Mo/γ-Al$_2$O$_3$-TiO$_2$ showed small peaks at 2θ = 23.3° and 27.2° corresponding to the presence of MoO$_3$. There was no change observed in the crystal orientation on addition of phosphorous. The XRD profile for MnPMo/γ-Al$_2$O$_3$-TiO$_2$ further showed new peaks at 2θ = 25.9°, 26.9°, 27.9°, 31.3°, 32.3°, 33.2°, 35.8°, 37.9°, 39.2°, 40.5°, 44°, 51.5–53.0° and 57.2–59.5°, which are related to the presence of α-MnMoO$_4$ [29]. The intense peak at $2\theta = 25.9°$ may correspond to the 220 plane of MnMoO$_4$. These reflections are in accordance with JCPDS Card No: 01-72-0285 as reported by Veerasubramani et al. [30]. Additionally, the peaks at $2\theta = 21.0°$, $26.9°$, $27.9°$, $32.3°$, $33.2°$, $40.5°$, $44°$, 51.5–$53.0°$ and 57.2–$59.5°$ can also be attributed to the presence of Mo$_4$O$_{11}$ as described by Yang et al. [29]. Therefore, further details on the peak elucidation can be obtained from XPS analysis.

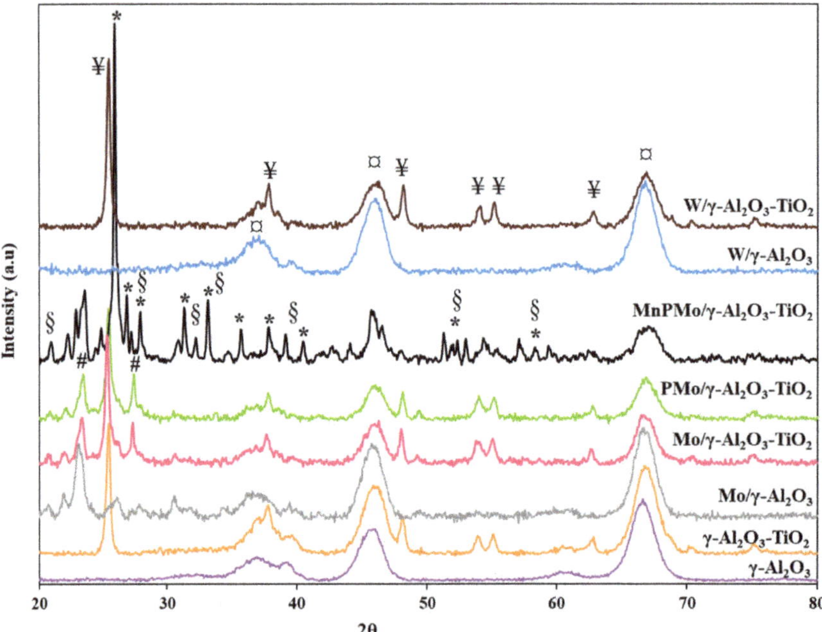

Figure 3. X-ray diffractograms of all catalysts. (§—Mo$_4$O$_{11}$, ¥—TiO$_2$ (anatase phase), *—α-MnMoO$_4$, ¤—planes of γ-Al$_2$O$_3$, #—MoO$_3$).

2.1.3. Fourier Transformed Infrared Spectroscopy

The FT-IR spectroscopy was performed to determine the presence of various vibrational bands corresponding to the functional groups and active metals. The FT-IR spectra for all the materials in this study are shown in Figure 4. The band in the range of 400–700 cm^{-1} with few spikes at 600–700 cm^{-1} in the spectra for γ-Al$_2$O$_3$ is due to the stretching of Al-O bonds in octahedral alumina. The band between 750–950 cm^{-1} represents the vibration of Al-O bond in tetrahedrally coordinated Al [31]. This type of broad band from 400–950 cm^{-1} with no sharp peaks is a typical for poorly ordered material such as γ-Al$_2$O$_3$. The IR bands corresponding to the vibration of Ti-O bonds appearing at 400–760 cm^{-1} are superimposed with Al-O vibrations and were not distinguished [32]. The IR active bonds corresponding to molybdenum and tungsten were not clearly identified in Mo and W supported γ-Al$_2$O$_3$ and γ-Al$_2$O$_3$-TiO$_2$ materials, as the broad peak of 400–700 cm^{-1} masks most of the peaks. The broad band at 1050–1250 cm^{-1} in IR spectra of PMo/γ-Al$_2$O$_3$-TiO$_2$ can be assigned to the superimposition of the bands corresponding symmetrical vibrations of PO$_4$ and asymmetrical stretching of P-O-P [33,34]. The catalyst MnPMo/γ-Al$_2$O$_3$-TiO$_2$ showed additional IR bands at 719 cm^{-1}, 725 cm^{-1}, 796 cm^{-1}, 865 cm^{-1} and 910–970 cm^{-1}, which are the characteristic peaks for tetrahedrally coordinated Mo in α-MnMoO$_4$ [35,36]. The presence of α-MnMoO4 was also detected in XRD analysis (Section 2.1.2) of the MnPMo/γ-Al$_2$O$_3$-TiO$_2$ catalyst. Additionally, the peaks at 865 cm^{-1} and 910 cm^{-1} could also be assigned to the molybdenum present in Mo$_4$O$_{11}$ species [37].

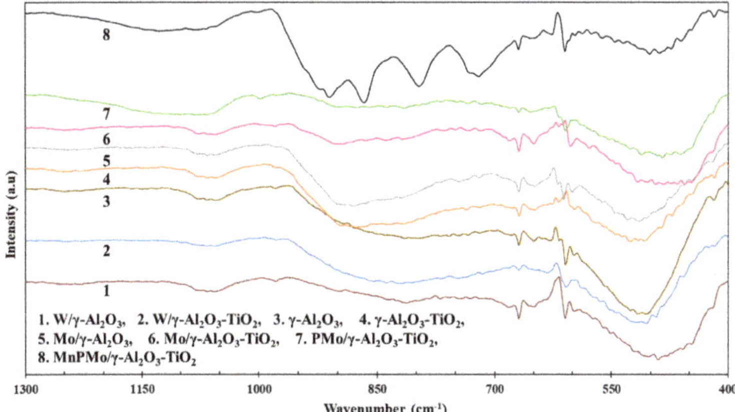

Figure 4. Fourier transform-infrared (FT-IR) spectra of supported catalysts.

2.1.4. X-ray Photoelectron Spectroscopy

The X-ray photoelectron spectroscopy (XPS) was employed to determine the oxidation state of Mo and Mn in TiO_2-Al_2O_3 supported catalysts. Both Mo and Mn tend to form various oxides with different oxidation states. The XPS spectra for Mn 2p and 3s region are shown in Figure 5. The Mn 2p spectrum (Figure 5b) shows the spin-orbit doublet at 641.6 eV and 652.9 eV representing 2p3/2 and 2p1/2 electronic states (ΔE of 11.3 eV). The Mn 2p spectrum also shows a satellite feature at 645–648 eV, which is a characteristic of Mn^{2+} [38,39]. The shake-up satellite feature appears when the x-ray ejected core electron excites a valance electron to a higher energy level, thus, reducing the energy of core electron leading to appearance of satellite at few eV lower than the core level binding energy. Therefore, the Mn 2p XPS spectra indicates the presence of +2 oxidation state of manganese. To further confirm the oxidation state of manganese in catalyst Mn/P/Mo/Al-Ti, the Mn 3s XPS spectra was analysed. The ΔE between the spilt 3s peaks indicates the oxidation state. The splitting of 3s peak is due to the interaction of 3s core hole (after photoemission) and 3d valance electrons. The energy difference between the parallel and antiparallel spin configuration of both 3s and 3d determines the split. Thus, the ΔE for 3s splitting in +2 oxidation sate of manganese is expected to be highest due to 3d5 high spin configuration [40]. The reported difference in binding energy (ΔE) for Mn^{2+}, Mn^{3+} and Mn^{4+} is ~6.0 eV, 5.3 eV and 4.7 eV, respectively [41]. The Mn 3s XPS spectra for catalyst MnPMo/γ-Al_2O_3 -TiO_2 shows peak split ΔE of 6.0 eV confirming the +2 oxidation state of manganese. This concludes that manganese is present in the form of and MnMoO4 as shown by XRD analysis.

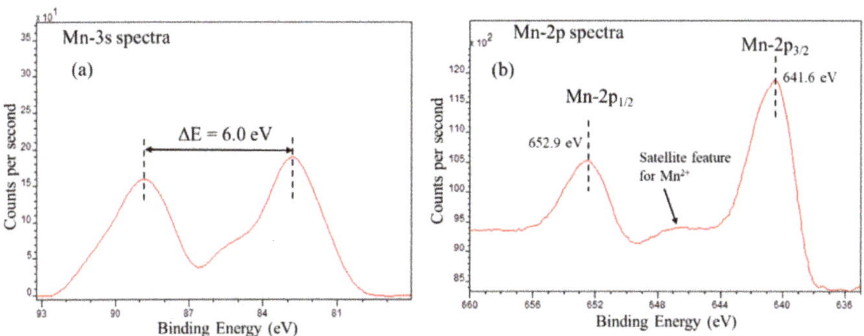

Figure 5. X-ray photoelectron (XPS) spectra for manganese in MnPMo/TiO_2-Al_2O_3 catalyst, (**a**) Mn-3s spectra and (**b**) Mn-2p spectra.

The Mo 3d XPS spectra for materials Mo/γ-Al$_2$O$_3$-TiO$_2$, PMo/γ-Al$_2$O$_3$-TiO$_2$ and MnPMo/γ-Al$_2$O$_3$-TiO$_2$ is shown is Figure 6. The Mo 3d region shows well separated spin-orbit component 3d5/2 and 3d3/2 with ΔE ≈ 3.2eV [38]. The binding energy of 3d5/2 peak indicates the oxidation state of molybdenum. For +4, +5 and +6 oxidation states of molybdenum, the 3d5/2 peak appears at binding energy ~229.7 eV, 231.5 eV and 232.5 eV, respectively [42–44]. The peak fitting was performed using the CasaXPS Demo version and results are shown in Figure 6. The area under 3d3/2 peak is kept constant to two-third of the 3d5/2 peak area. The ΔE was kept constant at 3.2 eV. It can be seen from Figure 6a that 86% molybdenum in catalyst Mo/γ-Al$_2$O$_3$-TiO$_2$ is in +6 oxidation state indicating the presence of MoO$_3$. On addition of phosphorus to this catalyst the increase in octahedral molybdenum was observed as shown in Mo 3d XPS in Figure 6b. It could be due the stronger interaction of phosphorus with surface hydroxyl groups present on the support material as explained in our previous work [45]. The review of influence of phosphorous on alumina supported hydrotreating catalyst by Iwamoto and Grimblot [46] also explained the formation of AlPO$_4$ on the surface of alumina using surface hydroxyl groups. Thus, the interaction of phosphorus with support reduces the number of surface hydroxyl group for molybdenum to anchor and thus decreases the metal support interactions, resulting in formation of octahedral molybdenum.

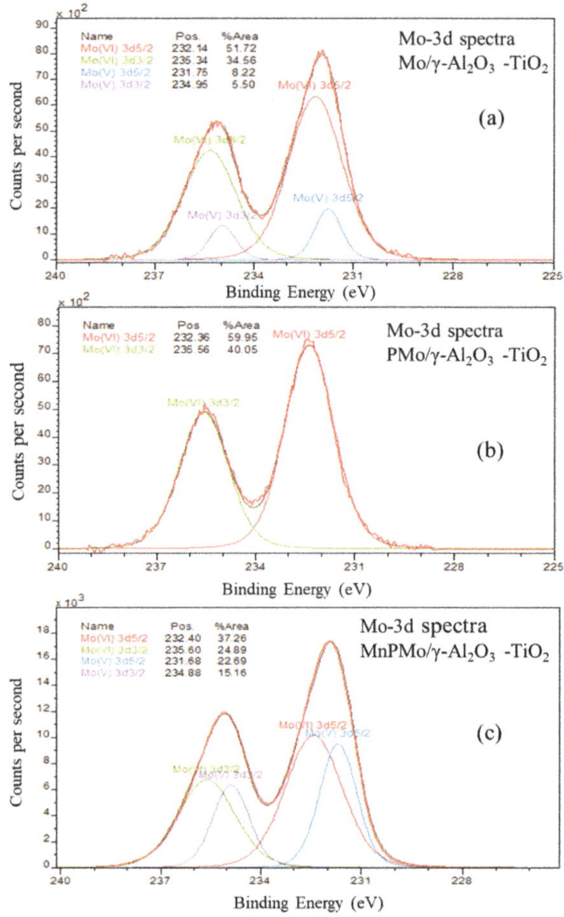

Figure 6. Mo 3d XPS spectra for (**a**) Mo/γ-Al$_2$O$_3$-TiO$_2$, (**b**) PMo/γ-Al$_2$O$_3$-TiO$_2$ and (**c**) MnPMo/γ-Al$_2$O$_3$-TiO$_2$.

The addition of manganese to the PMo/γ-Al$_2$O$_3$ -TiO$_2$ catalyst has significantly altered the oxidation state of the molybdenum as seen from the Mo 3d XPS of MnPMo/γ-Al$_2$O$_3$-TiO$_2$ (Figure 6c). Around 62% of the molybdenum is in +6 oxidation state and balance (~38%) is in +5 oxidation state. This indicates the presence of Mo$_4$O$_{11}$, which has a ratio of 1:1 for Mo^{+6} and Mo^{+5} and MnMoO$_4$ which contains polymolybdate (MoO4)$^{-2}$ with +6 oxidation state for Mo [47]. This confirms the finding of Mo$_4$O$_{11}$ and MnMoO$_4$ in MnPMo/γ-Al$_2$O$_3$-TiO$_2$ catalyst and supports the XRD conclusions.

The XPS analysis for determining the oxidation state of tungsten in W/γ-Al$_2$O$_3$ and W/γ-Al$_2$O$_3$ -TiO$_2$ was also performed. The 4f W XPS spectra (figure not shown) shows a 4f$_{7/2}$ and 4f$_{5/2}$ doublet with 4f7/2 peak at 36.1 eV and a loss feature at 42 eV indicating the presence of WO$_3$, where oxidation state of W is +6 [38]. The analysis of Ti 2p spectra for all titanium containing catalysts was also performed and it shows the doublet for 2p with 2p3/2 at 458.1eV confirming the +4 oxidation state of titanium.

2.2. Hydrotreating of Bitumen Derived Gas Oils

The hydrotreating of LGO and HGO facilitated by NiMo/γ-Al$_2$O$_3$ in fixed bed flow reactor operating at 9 MPa, 370 °C, LHSV 1.5 h^{-1}, 75 mL/min H$_2$ for LGO and 9 MPa, 390 °C, LHSV 1.0 h^{-1}, 50 mL/min H$_2$ for HGO, significantly lowered the sulphur and nitrogen content. For HGO the sulphur content was lowered from 41,000 ppm to 2100 ppm and nitrogen content dropped from 3900 ppm to 1750 ppm, resulting in 94.8 wt.% S and 55.0 wt.% N removal. The LGO hydrotreating succeeded in lowering the sulphur content from 24,000 ppm to 950 ppm and nitrogen content from 1400 ppm to 175 ppm, thus achieving 96.0 wt.% S and 87.5 wt.% N removal. The hydrotreated HGO and LGO were called as HDT-HGO and HDT-LGO, respectively and were further treated by oxidative desulfurization and denitrogenation process to achieve less than 500 ppm in treated LGO and HGO.

2.3. Oxidative Desulfurization and Denitrogenation

The synthesized catalysts were tested for the oxidation of sulphur and nitrogen containing aromatic compounds present in hydrotreated HGO and LGO. The oxidation increases the polarity of organosulfur and nitrogen compounds and therefore, the oxidized sulphur and nitrogen compounds were then removed by adsorption or extraction. The oxidation of sulphur and nitrogen compounds by tert-butyl hydroperoxide is facilitated by the supported metal oxide catalyst. TBHP nucleophilically attacks the Mo in molybdenum oxide (and W in tungsten oxide), to form peroxymolybdate (peroxymetallic) complex. This complex activates the peroxy group to oxidize the sulphur present in organosulfur compounds. Thus, forming sulfoxide and regenerating molybdenum oxide. The molybdenum oxide again undergoes reaction with another TBHP molecule to form peroxymolybdate complex, which further oxidizes sulfoxide to sulfone (see Figure 7) and regenerate molybdenum oxide. Simple mechanism shown in Figure 7 is inspired from previous studies in literature [48–50]. The sulfoxides and sulfones were then separated out using adsorption or extraction due to increase in their polarity. The oxidation of nitrogen containing organic compounds such as quinoline does not necessarily oxidizes the nitrogen, however it is a complex reaction. Studies in literature [25] showed that the carbon with least electron density in the ring of organonitrogen compound is oxidized by peroxides and then followed by ring opening. Further oxidation will result in formation of various oxygenates of organonitrogen compounds, which can be then removed from gas oil via adsorption or extraction due to increase in polarity of oxygenated nitrogen compounds.

Three adsorbents including Activated carbon, Amberlyst 15 ion exchange resin and Amberlite IRA-400 ion exchange resin were tested to determine a suitable adsorbent for this work. Equal amount of each adsorbent was mixed with HDT-LGO and HDT-HGO in separate flasks to determine the adsorption capacity of each adsorbent for adsorbing sulphur and nitrogen compounds from hydrotreated gas oils. The results shown in Table 2 indicates that all adsorbents can remove 4–9 wt.% sulphur. Activated carbon and Amberlite IRA-400 ion exchange resin were able to adsorb and remove 11–15 wt.% nitrogen. However, the Amberlyst 15 ion exchange resin removed 25 wt.% N from HDT-HGO and 33 wt.% N from HDT-LGO. This could be due the attractive forces between basic nitrogen compounds and highly acidic Amberlyst 15 resin. Further, the adsorption followed by oxidation of hydrotreated gas oil was also carried out using above mentioned three adsorbents. Results shown in Table 2 suggests the effectiveness of activated carbon as adsorbent for the removal of oxygenated sulphur and nitrogen compounds. Mo/γ-Al$_2$O$_3$ catalysed oxidation followed by adsorptive removal of oxygenates with activated carbon removed 32% and 44.9% sulphur from single stage oxidative desulfurization of HDT-HGO and HDT-LGO respectively, as against to less than 10% removal by ion exchange resins. Additionally, the nitrogen removal was also higher with activated carbon after single stage oxidative denitrogenation.

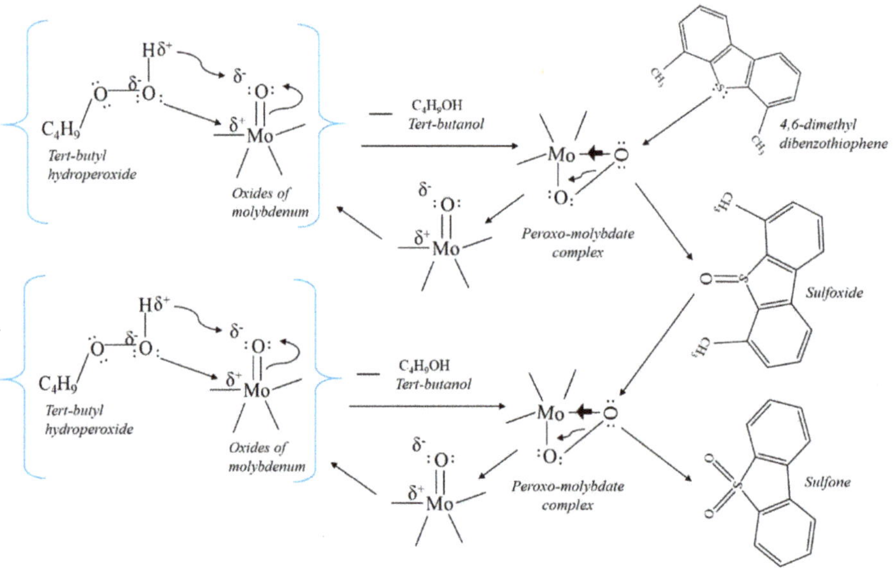

Figure 7. Simplistic mechanism for the oxidation of sulphur containing compounds via TPHP in presence of catalyst.

Table 2. Adsorption capacity of various adsorbents for sulphur and nitrogen compounds present in gas oil. (error $\pm$ 1%).

Adsorption Capacity (wt.%) of Following Adsorbents →			Activated Carbon	Amberlyst 15, Ion Exchange Resin	Amberlite IRA-400(CI), Ion Exchange Resin
HDT-HGO	For hydrotreated gas oil	% S removal	5.1	4.8	4.3
		% N removal	11.0	25.0	12.0
	After single state oxidative desulfurization (catalyst: Mo/γ-Al$_2$O$_3$)	% S removal	32.0	7.1	6.8
		% N removal	31.3	22.0	19.0
HDT-LGO	For hydrotreated gas oil	% S removal	8.1	9.0	6.3
		% N removal	15.0	33.0	14.0
	After single state oxidative desulfurization (catalyst: Mo/γ-Al$_2$O$_3$)	% S removal	44.9	9.9	8.1
		% N removal	49.7	28.0	20.0

Liquid-liquid extraction of oxidized sulphur and nitrogen compounds from oil phase was done with methanol. It was observed that methanol phase containing polar sulfoxides, sulfone and oxides of nitrogen compounds separates out easily from treated HDT-HGO. However, due to less difference between densities of HDT-LGO and methanol the separation was difficult and takes longer time. Nevertheless, for oxidized HDT-HGO and HDT-LGO the activated carbon could remove higher amount of oxidized sulphur and nitrogen compounds in comparison to extraction with methanol as shown in Tables 3 and 4 for oxidative desulfurization and denitrogenation of HDT-HGO and HDT-LGO. Therefore, in this this study activated carbon was used as adsorbent for further reactions with various catalysts.

Table 3. Oxidative desulfurization and denitrogenation of hydrotreated heavy gas oil. (error ± 4 ppm).

S conc. in HDT-HGO = 2100 ppm. N conc. in HDT-HGO = 1750 ppm. Single-Stage ODS and ODN Using Following Catalysts and TBHP		Mo/γ-Al$_2$O$_3$	Mo/γ-Al$_2$O$_3$-TiO$_2$	W/γ-Al$_2$O$_3$	W/γ-Al$_2$O$_3$-TiO$_2$	PMo/γ-Al$_2$O$_3$-TiO$_2$	MnPMo/γ-Al$_2$O$_3$-TiO$_2$
S conc. (ppm)	Adsorption using activated carbon	1427	1307	1612	1518	1244	1167
	Solvent Extraction using Methanol	1508	1417	1729	1740	-	-
N conc. (ppm)	Adsorption using activated carbon	1203	1115	1405	1294	1037	938
	Solvent Extraction using Methanol	1293	1208	1506	1412	-	-
% removal S	Adsorption using activated carbon	32.0	37.8	23.2	27.7	40.8	44.4
	Solvent Extraction using Methanol	28.2	32.5	17.7	17.1	-	-
% removal N	Adsorption using activated carbon	31.3	36.3	19.7	26.1	40.7	46.4
	Solvent Extraction using Methanol	26.1	31.0	13.9	19.3	-	-

The purified gas oil after adsorption is then tested for sulphur and nitrogen content and the percentage desulfurization and denitrogenation was used as a measure to define the catalytic activity. Tables 3 and 4 show the data for oxidative desulfurization and denitrogenation using various catalysts. It was observed that the Mo containing catalysts performs better than W containing catalysts for both desulfurization and denitrogenation for HDT-HGO and HDT-LGO. Mo/γ-Al$_2$O$_3$ catalyst showed 32% S and 31.3% N removal for HDT-HGO and 44.9% S and 49.7% N removal for HDT-LGO, whereas, the W/γ-Al$_2$O$_3$ catalyst showed 23.2% S and 19.7% N removal for HGO and 41.7% S and 47.4% N removal for LGO. It should be noted that although the weight percentage removal of nitrogen is higher than percentage removal of sulphur for HDT-LGO but in terms of ppm sulphur removal is higher. The higher activity of molybdenum supported catalyst in comparison to tungsten supported catalyst could be attributed to the higher rate of activation of TBHP by molybdenum, which results in the formation of peroxometallic species leading to the oxidation of sulphur to sulfoxide and sulfone. This could be due to the stronger bonding between Mo 4d orbital and peroxy oxygen 2p orbital during nucleophilic attack by TBHP on molybdenum oxide in comparison to bonding between W 5d and 2p orbital.

The increase in desulfurization and denitrogenation activity on incorporation of titanium as shown in Tables 3 and 4 is due to its Lewis acid character. XRD analysis has confirmed the presence of anatase TiO$_2$ and Ti 2p XPS spectra confirmed +4 oxidation state of titanium, which indicates the Lewis acid character of titanium. Therefore, the electron deficient titania will further provide the site for nucleophilic attack by TBHP for the formation of peroxo-titanium complex [51]. This complex activates the oxidation of sulphur and nitrogen compounds. Thus, Ti increases ODS and ODN activity by 8–12%.

Considering the promising activity of Mo/γ-Al$_2$O$_3$-TiO$_2$ catalyst further promotion by phosphorus (P) and manganese (Mn) was studied for Mo/γ-Al$_2$O$_3$-TiO$_2$ catalyst. The addition of P as PO$_4^{3-}$ makes molybdenum more electrophilic. Thereby boosting the rate of reaction of peroxides (TBHP) with Mo, which results in the creation of active species for oxidation of sulphur and nitrogen compounds. The results showing increase in desulfurization and denitrogenation from 37.8% to 40.8% and 36.3% to 40.7%, respectively, on addition of P are presented in Table 3 for HDT-HGO. Similar increase in activity was observed for HDT-LGO as shown in Table 4.

Table 4. Oxidative desulfurization and denitrogenation of hydrotreated light gas oil. (error $\pm$4 ppm).

S conc. in HDT-LGO = 950 ppm. N conc. in HDT-LGO = 175 ppm. Single-Stage ODS and ODN Using Following Catalysts and TBHP		Mo/γ-Al$_2$O$_3$	Mo/γ-Al$_2$O$_3$-TiO$_2$	W/γ-Al$_2$O$_3$	W/γ-Al$_2$O$_3$-TiO$_2$	PMo/γ-Al$_2$O$_3$-TiO$_2$	MnPMo/γ-Al$_2$O$_3$-TiO$_2$
S conc. (ppm)	Adsorption using activated carbon	523	487	554	508	462	428
	Solvent Extraction using Methanol	570	540	582	541	-	-
N conc. (ppm)	Adsorption using activated carbon	88	76	92	82	57	40
	Solvent Extraction using Methanol	103	85	115	101	-	-
% removal S	Adsorption using activated carbon	44.9	48.7	41.7	46.5	51.4	54.9
	Solvent Extraction using Methanol	40.0	43.2	38.7	43.1	-	-
% removal N	Adsorption using activated carbon	49.7	56.6	47.4	53.1	67.4	77.1
	Solvent Extraction using Methanol	41.1	51.4	34.3	42.3	-	-

The catalyst MnPMo/γ-Al$_2$O$_3$-TiO$_2$ further increased the desulfurization and denitrogenation activity from 40.8% to 44.4% and 40.7% to 46.4%, respectively, with reference to activity shown by PMo/γ-Al$_2$O$_3$-TiO$_2$ catalyst for HDT-HGO. The increase in activity could be assigned to the oxidizing or electron-accepting tendency of manganese. Mn 3s and Mn 2p XPS analysis has confirmed that manganese is in +2 oxidation state, indicating the presence of MnO and MnMoO$_4$. However, FTIR analysis and XRD analysis for MnPMo/γ-Al$_2$O$_3$-TiO$_2$ catalyst have confirmed the polymolybdate species present in MnMoO$_4$ form. Therefore, the manganese in MnMoO$_4$, will make Mo more electrophilic ($-$Mo$^{\delta+}$ = O). Hence, this increases the rate of reaction of TBHP with catalyst, which eventually increase the ODS and ODN activity. Similar increase in activity was observed for HDT-LGO as shown in Table 4.

Single stage oxidative desulfurization has lowered the sulphur content of HDT-HGO from 2100 ppm to 1167 ppm using MnPMo/γ-Al$_2$O$_3$-TiO$_2$ catalyst. However, to remove sulphur down to less than 500 ppm, multistage oxidation process was employed. To perform multistage oxidation, large amount of product from first stage ODS is required in order to account for the samples, which will be extracted after each stage for analysis. The spent catalyst from single stage process was utilized to oxidize another fresh batch of HDT-HGO to generate more product oil from first stage oxidative desulfurization process. It was observed that the sulphur removal efficiency decline from 44.4% to 40.1% on reusing the catalyst. The decrease in sulphur removal efficiency could be due to the active site inhibition by bulky heteroatomic molecules present in the HGO or might be due to the leaching of the fraction of Mo. However, 4.3% decline in sulphur removal or 9.7% decline in catalyst activity is not a clear affirmation of leaching due to other possible factors. Therefore, considering the scope of this paper, reusability, metal leaching and catalyst stability studies were not conducted and fresh catalyst was used to generate more product oil from stage 1 ODS.

Table 5 shows that after 3 stage ODS the sulphur in treated HGO was lowered to 478 ppm, accounting to 77.2% sulphur removal. It was also observed that between stage 2 and stage 3 of ODS process only 87 ppm (~4%) sulphur was removed. Therefore, increasing the number of ODS stages beyond 3rd stage will not be able to further lower the sulphur concentration significantly. Interestingly, the denitrogenation with multistage oxidation process was more effective than desulfurization. The nitrogen content was lowered from 1750 to 206 ppm (~88% nitrogen removal) for HDT-HGO. Nitrogen is a well-known poison to many downstream processing catalysts, therefore, the decrease in nitrogen content will benefit the downstream processing of treated HGO and LGO.

Table 5. Multistage ODS and ODN of hydrotreated gas oils. (error ±4 ppm).

ODS and ODN Stages with MnPMo/γ-Al$_2$O$_3$-TiO$_2$ Catalyst, TBHP and Activated Carbon		Stage 1	Stage 2	Stage 3
HDT-HGO (S = 2100 ppm; N = 1750 ppm)	S conc. (ppm)	1167	565	478
	Cumulative % S removal	44.4	73.1	77.2
	N conc. (ppm)	938	261	206
	Cumulative % N removal	46.4	85.1	88.2
HDT-LGO (S = 950 ppm; N = 175 ppm)	S conc. (ppm)	428	394	354
	Cumulative % S removal	54.9	58.5	62.7
	N conc. (ppm)	40	30	30
	Cumulative % N removal	77.1	82.9	82.9

Three-stage oxidative desulfurization and denitrogenation was also carried out for HDT-LGO and the results are presented in Table 5. After first stage the sulphur and nitrogen contents were lowered to 428 ppm and 40 ppm respectively. It was observed that the increase in percentage sulphur and nitrogen removal was not significant after two-stage and three-stage processes. The total S and N removal after third stage is 62.7% and 82.9% in comparison to 54.9% and 77.1% after single stage ODS and ODN, respectively. Therefore, the combination of hydrotreating and oxidative desulfurization-denitrogenation can lower the sulphur content from 41,000 ppm to 480 ppm and nitrogen content from 3900 ppm to 210 ppm for HGO. In future, this study can serve as a basis to design the process at commercial scale, which can enable oil-sands derived gas oils to stay competitive to conventional crude oil.

3. Experimental

3.1. Materials

Titanium isopropoxide, ammonium dihydrogen phosphate, Ammonium heptamolybdate tertrahydrate, Manganese nitrate hexahydrate, tert-butylhydroperoxide (TBHP) (70 *v/v*%), Ammonium metatungstate hydrate, were purchased from Sigma Aldrich, Edmonton, Canada. Activated carbon was provided by Norit, Canada. Amberlyst 15 ion exchange resin and Amberlite IRA-400 ion exchange resin were purchased from Alfa Aesar, USA. γ-Al$_2$O$_3$ and methanol was purchased from Fischer-Scientific, Toronto, Canada.

3.2. Catalyst Synthesis

All metals were impregnated using incipient wetness method. Al$_2$O$_3$-TiO$_2$ material was synthesized by impregnating the solution of titanium isopropoxide and ethanol on γ-Al$_2$O$_3$ to obtain 10 wt.% TiO$_2$ in Al$_2$O$_3$-TiO$_2$. 15 wt.% W and 15 wt.% Mo were impregnated on γ-Al$_2$O$_3$ and Al$_2$O$_3$-TiO$_2$ materials using ammonium metatungstate hydrate and ammonium heptamolybdate tertrahydrate as precursors for W and Mo, respectively. The materials were dried overnight at 100 °C and then calcined at 550 °C for 6 h. The resulting materials were named as Mo/γ-Al$_2$O$_3$, Mo/γ-Al$_2$O$_3$-TiO$_2$, W/γ-Al$_2$O$_3$ and W/γ-Al$_2$O$_3$-TiO$_2$. Further, 2 wt.% phosphorus was impregnated to Mo/γ-Al$_2$O$_3$-TiO$_2$ using ammonium dihydrogen phosphate as precursor. This was followed by drying and calcination at 400 °C for 4 h to obtain PMo/γ-Al$_2$O$_3$-TiO$_2$ catalyst. Manganese nitrate hexahydrate was used as a precursor for Mn and it was wet impregnated on PMo/γ-Al$_2$O$_3$-TiO$_2$ to obtain 2 wt.% Mn in MnPMo/γ-Al$_2$O$_3$-TiO$_2$ catalyst. The catalyst containing 13 wt.% Mo and 3 wt.% Ni supported on γ-Al$_2$O$_3$ was also synthesized using sequential wetness impregnation method and named as NiMo/γ-Al$_2$O$_3$.

3.3. Material Characterization

All materials were characterized in oxide state. Micrometrics ASAP2020 instrument was used to determine the textural properties. The Brunauer-Emmett-Teller (BET) method was used to determine the surface area and Barrett-Joyner-Halenda (BJH) method was used to calculate the pore diameter and pore volume. The wide angle (20° to 80°) X-ray diffractograms for powder samples were obtained using Bruker Advance D8 series II equipment having Cu Kα radiation. The FTIR analysis for the materials was performed using PerkinElmer Spectrum GX instrument. The details on the sample preparation and method of analysis for XRD, FTIR and BET are mentioned in our previous work [52]. The X-ray photoelectron spectroscopy (XPS) analysis was performed to determine the oxidation state of active metals. The XPS spectra for catalysts were collected using Kratos Axis SUPRA XPS instrument equipped with monochromatic Al Kα X-ray source. This instrument is located at Saskatchewan Structural Sciences Centre, University of Saskatchewan, Canada. The XPS data were fitted using CasaXPS software.

3.4. Hydrotreating Experiment

The fixed bed flow reactor system was used to perform hydrotreating reactions for HGO and LGO. The physical properties of HGO and LGO are mentioned in Table 6. The reaction setup consists of a liquid feed pump, H_2 and He gas inlet lines via mass flow controllers, tube type fixed bed reactor heated by furnace, back pressure regulator, NH_3 scrubber and gas-liquid separator to collect the hydrotreated product followed by nitrogen stripping. The schematic and details of the setup are mentioned in our previous work [52]. 5g of NiMo/γ-Al_2O_3 catalyst was diluted with SiC (90 mesh size) and loaded in the reactor. The catalyst was sulphided by pumping 5 mL/h of 2.9 vol% butanethiol solution (in transformer oil) through the catalyst bed for 48 hrs. The first 24 h sulphidation was performed at 190 °C and then the temperature was raised to 340 °C and maintained for next 24 h. The reactor was pressurized to 9 MPa with helium prior to start of sulphidation. The entire sulphidation process was carried out at 50 mL/min of hydrogen flow.

Table 6. Physical properties of heavy gas oil and light gas oil.

Properties	Heavy Gas Oil	Light Gas Oil
Sulphur (wt.%)	4.1	2.4
Nitrogen (wt.%)	0.39	0.14
Density (g/mL)	0.96	0.89
Aromatic content (%)	42.0	30
Boiling range (°C)	wt.%	wt.%
Initial boiling point–300	3.0	46.4
301–400	20.8	42.7
401–500	54.0	10.9
501–600	20.2	-
600–Final boiling point	2.0	-

The liquid feed was switched to gas oil (HGO or LGO) after sulphidation and the reaction temperature was increased to 370 °C for LGO and 390 °C for HGO. The gas oil feed rate was maintained at 5 mL/h and 7.5 mL/h for HGO and LGO, respectively. The H_2 feed rate was maintained at 50 mL/h for HGO and 75 mL/h for LGO. The hydrotreated oil was collected every 24 h and analysed for sulphur and nitrogen content. The constant conversion was observed from day 5 to day 12 and then the reactor was shut down. Therefore, the hydrotreated oil product from day 6–12 was mixed to obtain a big batch of hydrotreated gas oil, hereafter called as HDT-HGO and HDT-LGO. Hydrotreating of LGO and HGO was done with fresh catalyst in the same experimental set-up. The Antek 9000 N/S analyser was used to determine the nitrogen and sulphur content in liquid samples. ASTM D4629 method using combustion/chemiluminescence technique was adopted to determine the total nitrogen content of the liquid product and the ASTM D5463 method using combustion/fluorescence technique was deployed for measuring sulphur content. The details on catalyst loading procedure and sulphidation procedure are mentioned in our previous works [52].

3.5. Experimental Procedure for Oxidative Desulfurization and Denitrogenation

20 ml of HDT-HGO was taken in a round bottom flask and 1 g catalyst was added to it. To this mixture 20 ml of TBHP (70% *v/v*) was added as an oxidizer and stirred at 400 rpm at 90 °C under reflux for 15 h. The 1:1 volume ratio of TBHP and HDT-HGO was used for screening experiments considering the complex nature of bitumen derived heavy gas oil. This makes molar ratio of TBHP to sulphur equal to 110 and TBHP to nitrogen molar ratio equal to 58, which is quite high. The Metal (Mo) to sulphur molar ratio is 1.2. Further, handling equal amount of TBHP and gas oil in existing refineries will be a concern. However, this lab scale experiment was performed for the proof of concept and further optimization studies on the amount of oxidant, catalyst, oil, stirring speed and temperature can help to determine the techno-economical and logistical viability of this process integration with existing refineries. The reaction was cooled down to room temperature and then filtered to recover the solid catalyst. The catalysts tested for this reaction are: $Mo/\gamma\text{-}Al_2O_3$, $Mo/\gamma\text{-}Al_2O_3\text{-}TiO_2$, $W/\gamma\text{-}Al_2O_3$ and $W/\gamma\text{-}Al_2O_3\text{-}TiO_2$. $PMo/\gamma\text{-}Al_2O_3\text{-}TiO_2$ and $MnPMo/\gamma\text{-}Al_2O_3\text{-}TiO_2$. The reaction products were phase separated due to the difference in the density of oil phase and water phase. The oil phase was collected and water washed to remove dissolved butanol, which is the by-product of oxidation of sulphur and nitrogen compounds by TBHP.

The oxidized sulphur compounds known as sulfones and sulfoxides and oxides of aromatic nitrogen compounds were than extracted from the oil phase using adsorption or solvent extraction. Activated carbon was used as an adsorbent. In a typical experiment, 1 g of activated carbon was mixed with 20 mL of oil for 12 h at room temperature. The mixture was then filtered to separate the solids and the resultant liquid was analysed to determine total S and N content. The type and amount of oxidized sulphur and nitrogen compounds were not determined in this work due to the complex composition of heavy gas oil. Therefore, the total S and N content in product was used as a basis to define the catalyst activity. The liquid was again treated with TBHP in presence of fresh catalyst followed by adsorption with activated carbon to perform double stage adsorptive extraction, when required. Similar procedure was followed to perform multi-stage ODS and ODN.

The solvent extraction process was also tested to extract the sulfones, sulfoxides and oxides of organonitrogen compounds from the oil phase collected after oxidation with TBHP. 10 mL methanol was mixed with 20 mL oil and allowed to settle for 4 h at room temperature. The mixture was phase separated to obtain desulfurized and denitrogenated oil, which was then tested to determine the sulphur and nitrogen content. Similar procedure was followed to perform oxidative desulfurization and denitrogenation on HDT-LGO.

4. Conclusions

Highly efficient upgrading of oil-sands bitumen derived heavy gas oil (~ 41,000 ppm sulphur, 3900 ppm nitrogen) and light gas oil (~24,000 ppm sulphur, 1400 ppm nitrogen) is required to generate synthetic crude, which can compete with conventional crude oil. Therefore, in this work the combination of hydrotreating, oxidative desulfurization (ODS) and oxidative denitrogenation (ODN) was performed to achieve less than 500 ppm sulphur in LGO and HGO. The synthetic crude produced using treated HGO and LGO will be highly competitive and easily processed in existing refineries to produce various petroleum fractions including diesel fuel with less than 15 ppm sulphur. The hydrotreating of gas oils was carried out in a fixed bed flow reactor operating at typical industrial conditions of 370–390 °C, 9 MPa, 1–1.5 h^{-1} space velocity and 600:1 H_2 to oil ratio. $NiMo/\gamma\text{-}Al_2O_3$ was used as a catalyst. Hydrotreating resulted in lowering the sulphur and nitrogen content to 2100 ppm and 1750 ppm for HGO and 950 ppm and 175 ppm for LGO, respectively.

Various catalysts including $Mo/\gamma\text{-}Al_2O_3$, $Mo/\gamma\text{-}Al_2O_3\text{-}TiO_2$, $W/\gamma\text{-}Al_2O_3$, $W/\gamma\text{-}Al_2O_3\text{-}TiO_2$, $PMo/\gamma\text{-}Al_2O_3\text{-}TiO_2$ and $MnPMo/\gamma\text{-}Al_2O_3\text{-}TiO_2$ were synthesized and characterized using X-ray diffractions, N_2 adsorption-desorption, FTIR and XPS. All catalysts were tested for the oxidation of sulphur and nitrogen compounds present in gas oil using tert-butyl hydroperoxide (TBHP). The removal of oxides of sulphur and nitrogen compounds was carried out using adsorption and extraction. Methanol was used as solvent for liquid-liquid extraction, however, the minimal difference in densities of hydrotreated LGO and methanol possess challenge in separation and longer settling time was required. Among activated carbon and ion-exchange resins, the higher adsorption capacity for polar oxidized sulphur and nitrogen aromatic compounds was shown by activated carbon. The sulphur and nitrogen removal were higher with activated carbon in comparison to methanol. The catalytic activity measured in terms of percent sulphur and nitrogen removal was related to the catalyst characterization. XPS analysis has confirmed the oxidation state of metals such as Mo, Ti, Mn, P and W, which helped in identifying the catalyst structure and relate it to the metal oxide structures as indicated by XRD and FTIR analysis. Mo supported catalyst outperformed the W supported catalyst due to stronger interaction between 2p O (TBHP) and 4d Mo orbitals in contrast to bonding between 5d W and 2p O orbitals. The Ti acts as an additional site for TBHP activation, which caused oxidation of sulphur and nitrogen containing compounds. P and Mn in $MnPMo/\gamma$-Al_2O_3-TiO_2 catalyst makes molybdenum more electrophilic, thereby promoting the nucleophilic attack by TBHP, thus facilitating oxidation. Hence, the catalyst $MnPMo/\gamma$-Al_2O_3-TiO_2 performed best among the series tested and removed 44.4 wt.% sulphur and 46.4 wt.% nitrogen from hydrotreated HGO and 54.9 wt.% sulphur and 77.1 wt.% nitrogen from hydrotreated LGO. Further three stage ODS and ODN process was performed using $MnPMo/\gamma$-Al_2O_3-TiO_2 catalyst. Thus, the integration of hydrotreating, oxidative desulfurization and oxidative denitrogenation lowered the sulphur and nitrogen content to 478 ppm (~98.8% removal) and 206 ppm (~94.7% removal), respectively, in HGO and 354 ppm (~98.5% removal) and 30 ppm (~97.8% removal), respectively in LGO. The decrease in nitrogen content is beneficial for the downstream processing of gas oils because nitrogen acts as a poison to various catalysts. Therefore, the combination of oxidative desulfurization and denitrogenation with hydrotreating is a promising methodology to improve the quality of synthetic crude derived from oil sands bitumen and keeping them competitive with respect to conventional petroleum crude oil.

Author Contributions: Conceptualization, S.B. and A.K.D.; Methodology, S.B. and G.K.; Validation, A.K.D. and Y.Z.; Formal Analysis, S.B., P.M. and G.K.; Writing-Original Draft Preparation, S.B.; Writing-Review & Editing, P.M.; Supervision, Y.Z. and A.K.D.; Project Administration, A.K.D.; Funding Acquisition, Y.Z.

Funding: This research received no private funding.

Acknowledgments: The authors acknowledge financial assistance from NSERC and are thankful to Syncrude Canada ltd. for providing heavy and light gas oils. The authors are also thankful to Saskatchewan Structural Sciences Centre for the XPS analysis.

Conflicts of Interest: The authors declare no conflict of interest.

References

1. Sahoo, K. Sustainable Design and Simulation of Multi-Feedstock Bioenergy Supply Chain. Ph.D. Thesis, University of Georgia, Athens, Georgia, USA 2017.
2. Badoga, S.; Ganesan, A.; Dalai, A.K.; Chand, S. Effect of synthesis technique on the activity of CoNiMo tri-metallic catalyst for hydrotreating of heavy gas oil. *Catal. Today* **2017**, *291*, 160–171. [CrossRef]
3. Li, H.; Han, X.; Huang, H.; Wang, Y.; Zhao, L.; Cao, L.; Shen, B.; Gao, J.; Xu, C. Competitive adsorption desulfurization performance over K-Doped NiY zeolite. *J. Colloid Interface Sci.* **2016**, *483*, 102–108. [CrossRef]
4. Li, Y.X.; Jiang, W.J.; Tan, P.; Liu, X.Q.; Zhang, D.Y.; Sun, L.B. What matters to the adsorptive desulfurization performance of metal-Organic frameworks? *J. Phys. Chem. C* **2015**, *119*, 21969–21977. [CrossRef]
5. Saleh, T.A.; Sulaiman, K.O.; AL-Hammadi, S.A.; Dafalla, H.; Danmaliki, G.I. Adsorptive desulfurization of thiophene, benzothiophene and dibenzothiophene over activated carbon manganese oxide nanocomposite: with column system evaluation. *J. Clean. Prod.* **2017**, *154*, 401–412. [CrossRef]

6. Ganiyu, S.A.; Ajumobi, O.O.; Lateef, S.A.; Sulaiman, K.O.; Bakare, I.A.; Qamaruddin, M.; Alhooshani, K. Boron-doped activated carbon as efficient and selective adsorbent for ultra-deep desulfurization of 4,6-dimethyldibenzothiophene. *Chem. Eng. J.* **2017**, *321*, 651–661. [CrossRef]

7. Srivastav, A.; Srivastava, V.C. Adsorptive desulfurization by activated alumina. *J. Hazard. Mater.* **2009**, *170*, 1133–1140. [CrossRef]

8. Seredych, M.; Bandosz, T.J. Adsorption of dibenzothiophenes on nanoporous carbons: Identification of specific adsorption sites governing capacity and selectivity. *Energy Fuels* **2010**, *24*, 3352–3360. [CrossRef]

9. McKinley, S.G.; Angelici, R.J. Deep desulfurization by selective adsorption of dibenzothiophenes on Ag$^+$/SBA-15 and Ag$^+$/SiO$_2$. *Chem. Commun.* **2003**, 2620–2621. [CrossRef]

10. Zubaidi, I.A.; Darwish, N.N.; Sayed, Y.E.; Shareefdeen, Z.; Sara, Z. Adsorptive Desulfurization of Commercial Diesel oil Using Granular Activated Charcoal. *Int. J. Adv. Chem. Eng. Biol. Sci.* **2015**, *2*, 15–18.

11. Campos-Martin, J.M.; Capel-Sanchez, M.C.; Perez-Presas, P.; Fierro, J.L.G. Oxidative processes of desulfurization of liquid fuels. *J. Chem. Technol. Biotechnol.* **2010**, *85*, 879–890. [CrossRef]

12. Khalfalla, H.A. Modelling and Optimization of oxidative Desulphurization Process for Model Sulphur Compounds and Heavy Has Oil. Ph.D. Thesis, University of Bradford, Bradford, UK, 2009.

13. Jiang, Z.; Lu, H.; Zhang, Y.; Li, C. Oxidative Desulfurization of Fuel Oils. *Chin. J. Catal.* **2011**, *32*, 707–715. [CrossRef]

14. Bunthid, D.; Prasassarakich, P.; Hinchiranan, N. Oxidative desulfurization of tire pyrolysis naphtha in formic acid/H 2O2/pyrolysis char system. *Fuel* **2010**, *89*, 2617–2622. [CrossRef]

15. Ahmad, S.; Ahmad, M.; Naeem, K.; Humayun, M.; Sebt-E-Zaeem, S.; Faheem, F. Oxidative desulfurization of tire pyrolysis oil. *Chem. Ind. Chem. Eng. Q.* **2016**, *22*, 249–254. [CrossRef]

16. Palaić, N.; Sertić-bionda, K.; Margeta, D.; Podolski, Š. Oxidative Desulphurization of Diesel Fuels. *Chem. Biochem. Eng. Q.* **2015**, *29*, 323–327. [CrossRef]

17. Fattahi, A.M.; Omidkhah, M.R.; Moghaddam, A.Z.; Akbari, A. Synthesis and Characterization of Co-Mo/γ-Al2O3 New Catalyst for Oxidative Desulfurization (ODS) of Model Diesel Fuel. *Pet. Coal* **2014**, *56*, 442–447.

18. Chica, A.; Corma, A.; Dómine, M.E. Catalytic oxidative desulfurization (ODS) of diesel fuel on a continuous fixed-bed reactor. *J. Catal.* **2006**, *242*, 299–308. [CrossRef]

19. Gatan, R.; Barger, P.; Gembicki, V.; Cavanna, A.; Molinari, D. Oxidative desulfurization: A new technology for ULSD. *ACS Div. Fuel Chem. Prepr.* **2004**, *49*, 577–579.

20. Leng, K.; Sun, Y.; Zhang, X.; Yu, M.; Xu, W. Ti-modified hierarchical mordenite as highly active catalyst for oxidative desulfurization of dibenzothiophene. *Fuel* **2016**, *174*, 9–16. [CrossRef]

21. Lorençon, E.; Alves, D.C.B.; Krambrock, K.; Ávila, E.S.; Resende, R.R.; Ferlauto, A.S.; Lago, R.M. Oxidative desulfurization of dibenzothiophene over titanate nanotubes. *Fuel* **2014**, *132*, 53–61. [CrossRef]

22. Tian, Y.; Wang, G.; Long, J.; Cui, J.; Jin, W.; Zeng, D. Ultra-deep oxidative desulfurization of fuel with H2O2 catalyzed by phosphomolybdic acid supported on silica. *Chin. J. Catal.* **2016**, *37*, 2098–2105. [CrossRef]

23. García-Gutiérrez, J.L.; Laredo, G.C.; García-Gutiérrez, P.; Jiménez-Cruz, F. Oxidative desulfurization of diesel using promising heterogeneous tungsten catalysts and hydrogen peroxide. *Fuel* **2014**, *138*, 118–125. [CrossRef]

24. Shiraishi, Y.; Tachibana, K.; Hirai, T.; Komasawa, I. Desulfurization and denitrogenation process for light oils based on chemical oxidation followed by liquid-liquid extraction. *Ind. Eng. Chem. Res.* **2002**, *41*, 4362–4375. [CrossRef]

25. Ogunlaja, A.S.; Abdul-quadir, M.S.; Kleyi, P.E.; Ferg, E.E.; Watts, P.; Tshentu, Z.R. Towards oxidative denitrogenation of fuel oils: Vanadium oxide-catalysed oxidation of quinoline and adsorptive removal of quinoline-N-oxide using 2,6-pyridine-polybenzimidazole nanofibers. *Arab. J. Chem.* **2017**, in press. [CrossRef]

26. Ishihara, A.; Wang, D.; Dumeignil, F.; Amano, H.; Qian, E.W.; Kabe, T. Oxidative desulfurization and denitrogenation of a light gas oil using an oxidation/adsorption continuous flow process. *Appl. Catal. A Gen.* **2005**, *279*, 279–287. [CrossRef]

27. Badoga, S.; Sharma, R.V.; Dalai, A.K.; Adjaye, J. Synthesis and characterization of mesoporous aluminas with different pore sizes: Application in NiMo supported catalyst for hydrotreating of heavy gas oil. *Appl. Catal. A Gen.* **2015**, *489*, 86–97. [CrossRef]

28. Filippo, E.; Carlucci, C.; Capodilupo, A.L.; Perulli, P.; Conciauro, F.; Corrente, G.A.; Gigli, G.; Ciccarella, G.; Street, M.; Street, A.; et al. Enhanced Photocatalytic Activity of Pure Anatase TiO_2 and Pt-TiO_2 Nanoparticles Synthesized by Green Microwave Assisted Route. Experimental Section. *Mater. Res.* **2015**, *18*, 473–481. [CrossRef]

29. Senthilkumar, B.; Vijaya Sankar, K.; Kalai Selvan, R.; Danielle, M.; Manickam, M. Nano α-$NiMoO$ 4 as a new electrode for electrochemical supercapacitors. *RSC Adv.* **2013**, *3*, 352–357. [CrossRef]

30. Veerasubramani, G.K.; Krishnamoorthy, K.; Sivaprakasam, R.; Kim, S.J. Sonochemical synthesis, characterization, and electrochemical properties of $MnMoO_4$ nanorods for supercapacitor applications. *Mater. Chem. Phys.* **2014**, *147*, 836–842. [CrossRef]

31. Heuer, A.H.; Nakagawa, T.; Azar, M.Z.; Hovis, D.B.; Smialek, J.L.; Gleeson, B. SXPS, FTIR, EDX, and XRD Analysis of Al_2O_3 Scales Grown on PM2000 Alloy. *J. Spectrosc.* **2015**, *2015*, 1–16.

32. Adamczyk, A.; Długoń, E. The FTIR studies of gels and thin films of Al_2O_3-TiO_2 and Al_2O_3-TiO_2-SiO_2 systems. *Spectrochim. Acta—Part A Mol. Biomol. Spectrosc.* **2012**, *89*, 11–17. [CrossRef]

33. Ng, E.-P.; Ghoy, J.-P.; Awala, H.; Vicente, A.; Adnan, R.; Ling, T.C.; Mintova, S. Ionothermal synthesis of FeAPO-5 in the presence of phosphorous acid. *CrystEngComm* **2016**, *18*, 257–265. [CrossRef]

34. Klähn, M.; Mathias, G.; Kötting, C.; Nonella, M.; Schlitter, J.; Gerwert, K.; Tavan, P. IR spectra of phosphate ions in aqueous solution: Predictions of a DFT/MM approach compared with observations. *J. Phys. Chem. A* **2004**, *108*, 6186–6194. [CrossRef]

35. Ghosh, D.; Giri, S.; Moniruzzaman, M.; Basu, T.; Mandal, M.; Das, C.K. α $MnMoO_4$/graphene hybrid composite: high energy density supercapacitor electrode material. *Dalt. Trans.* **2014**, *43*, 11067–11076. [CrossRef]

36. Yan, X.; Tian, L.; Murowchick, J.; Chen, X. Partially amorphized $MnMoO_4$ for highly efficient energy storage and the hydrogen evolution reaction. *J. Mater. Chem. A* **2016**, *4*, 3683–3688. [CrossRef]

37. Kurumada, M.; Kaito, C. Change in IR spectra of molybdenum oxide nanoparticles due to particles size or phase change. *J. Phys. Soc. Japan* **2006**, *75*, 2–6. [CrossRef]

38. XPS Data for Elements. Available online: https://xpssimplified.com/periodictable.php (accessed on 10 April 2018).

39. Di Castro, V.; Polzonetti, G. XPS study of MnO oxidation. *J. Electron. Spectrosc. Relat. Phenom.* **1989**, *48*, 117–123. [CrossRef]

40. Cerrato, J.M.; Hochella, M.F.J.; Knocke, W.R.; Dietrich, A.M.; Cromer, T.F. Use of XPS to Identify the Oxidation State of Mn in Solid Surfaces of Filtration Media Oxide Samples from Drinking Water Treatment Plants. *Environ. Sci. Technol.* **2010**, *44*, 5881–5886. [CrossRef]

41. Majumdar, S.; Elovaara, T.; Huhtinen, H.; Granroth, S.; Paturi, P. Crystal asymmetry and low-angle grain boundary governed persistent photoinduced magnetization in small bandwidth manganites. *J. Appl. Phys.* **2013**, *113*, 063906. [CrossRef]

42. Nikolova, D.; Edreva-Kardjieva, R.; Gouliev, G.; Grozeva, T.; Tzvetkov, P. The state of $(K)(Ni)Mo/\gamma$-Al_2O_3 catalysts after water-gas shift reaction in the presence of sulfur in the feed: XPS and EPR study. *Appl. Catal. A Gen.* **2006**, *297*, 135–144. [CrossRef]

43. Castillo, C.; Buono-Core, G.; Manzur, C.; Yutronic, N.; Sierpe, R.; Cabello, G.; Chornik, B. Molybdenum trioxide thin films doped with gold nanoparticles grown by a sequential methodology: Photochemical Metal-Organic Deposition (PMOD) and DC-magnetron sputtering. *J. Chil. Chem. Soc.* **2016**, *61*, 2816–2820. [CrossRef]

44. Naumkin, A.V.; Kraut-Vass, A.; Gaarenstroom, S.W.; Powell, C.J. NIST X-ray Photoelectron Spectroscopy Database. Available online: https://srdata.nist.gov/xps/ElmSpectralSrch.aspx?selEnergy=PE (accessed on 10 April 2018).

45. Badoga, S.; Dalai, A.K.; Adjaye, J.; Hu, Y. Insights into individual and combined effects of phosphorus and EDTA on performance of $NiMo/MesoAl_2O_3$ catalyst for hydrotreating of heavy gas oil. *Fuel Process. Technol.* **2017**, *159*, 232–246. [CrossRef]

46. Iwamoto, R.; Grimblot, J. Influence of Phosphorus on the Properties of Alumina-Based Hydrotreating Catalysts. *Adv. Catal.* **1999**, *44*, 417–503. [CrossRef]

47. Deng, X.; Quek, S.Y.; Biener, M.M.; Biener, J.; Kang, D.H.; Schalek, R.; Kaxiras, E.; Firend, C.M. Selective Thermal Reduction of Single-layer MoO_3 nanostructures on Au (111). *Surf. Sci.* **2007**, *602*, 1166–1174. [CrossRef]

48. Abdullah, W.N.W.; Ali, R.; Bakar, W.A.W.A. In depth investigation of Fe/MoO$_3$-PO$_4$/Al$_2$O$_3$ catalyst in oxidative desulfurization of Malaysian diesel with TBHP-DMF system. *J. Taiwan Inst. Chem. Eng.* **2016**, *58*, 344–350. [CrossRef]

49. García-Gutiérrez, J.L.; Fuentes, G.A.; Hernández-Terán, M.E.; García, P.; Murrieta-Guevara, F.; Jiménez-Cruz, F. Ultra-deep oxidative desulfurization of diesel fuel by the Mo/Al$_2$O$_3$-H$_2$O$_2$ system: The effect of system parameters on catalytic activity. *Appl. Catal. A Gen.* **2008**, *334*, 366–373. [CrossRef]

50. García-Gutiérrez, J.L.; Fuentes, G.A.; Hernández-Terán, M.E.; Murrieta, F.; Navarrete, J.; Jiménez-Cruz, F. Ultra-deep oxidative desulfurization of diesel fuel with H2O2 catalyzed under mild conditions by polymolybdates supported on Al$_2$O$_3$. *Appl. Catal. A Gen.* **2006**, *305*, 15–20. [CrossRef]

51. Shen, C.; Wang, Y.J.; Xu, J.H.; Luo, G.S. Oxidative desulfurization of DBT with H$_2$O$_2$ catalysed by TiO$_2$/porous glass. *Green Chem.* **2016**, *18*, 771–781. [CrossRef]

52. Badoga, S. Synthesis and Characterization of NiMo Supported Mesoporous Materials with EDTA and Phosphorus for Hydrotreating of Heavy Gas Oil. Ph.D. Thesis, University of Saskatchewan, Saskatoon, SK, Canada, 2015.

 *catalysts*

Article

Hydrotreated Vegetable Oil as a Fuel from Waste Materials

Petr Zeman [1], Vladimír Hönig [1,*], Martin Kotek [2], Jan Táborský [1], Michal Obergruber [1], Jakub Mařík [2], Veronika Hartová [2] and Martin Pechout [2]

[1] Faculty of Agrobiology, Food and Natural Resources, Department of Chemistry, Czech University of Life Sciences Prague, Kamýcká 129, Prague 6, 169 21, Czech Republic; pzeman@af.czu.cz (P.Z.); taborsky@af.czu.cz (J.T.); obergruber@af.czu.cz (M.O.)

[2] Faculty of Engineering, Department of Vehicles and Ground Transport, Czech University of Life Sciences Prague, Kamýcká 129, Prague 6, 169 21, Czech Republic; kotekm@oikt.czu.cz (M.K.); marikj@tf.czu.cz (J.M.); nidlova@tf.czu.cz (V.H.); pechout@tf.czu.cz (M.P.)

* Correspondence: honig@af.czu.cz; Tel.: +420-22438-2722

Received: 26 February 2019; Accepted: 3 April 2019; Published: 4 April 2019

Abstract: Biofuels have become an integral part of everyday life in modern society. Bioethanol and fatty acid methyl esters are a common part of both the production of gasoline and diesel fuels. Also, pressure on replacing fossil fuels with bio-components is constantly growing. Waste vegetable fats can replace biodiesel. Hydrotreated vegetable oil (HVO) seems to be a better alternative. This fuel has a higher oxidation stability for storage purposes, a lower temperature of loss of filterability for the winter time, a lower boiling point for cold starts, and more. Viscosity, density, cold filter plugging point of fuel blend, and flash point have been measured to confirm that a fuel from HVO is so close to a fuel standard that it is possible to use it in engines without modification. The objective of this article is to show the properties of different fuels with and without HVO admixtures and to prove the suitability of using HVO compared to FAME. HVO can also be prepared from waste materials, and no major modifications of existing refinery facilities are required. No technology in either investment or engine adaptation of fuel oils is needed in fuel processing.

Keywords: biofuel; biodiesel; hydrotreating; hydrocarbon; waste

---

1. Introduction

A long-term European strategy is an effort for a so-called "recycling society". With the growing volume of waste, expanding industries are dealing with waste management and recycling. Despite noticeable progress, there is still great potential in previously underutilized sources of waste. The main obstacles are the legislative problems and the low application of approved rules, the differentiation of regulations in different countries, and generally the low awareness of the professional and lay public about new possibilities and prospects. The current EU waste policy is based on the concept of the so-called waste hierarchy, which states that it is primarily necessary to prevent the generation of waste itself, and if this is not possible, it must be recycled or otherwise exploited, under condition of minimal dump disposal. Anything which can be reused in some way may be considered waste, even materials like grey water, wood chips, old clothing, kitchen scraps or diseased fruit and vegetables [1–7].

Legislative requirements are higher for double-counting materials, in order to meet the 10% share of biofuels. This double counting applies to biofuels made from waste and residues, as well as to biofuels made from raw materials that have been grown on so-called degraded areas, and it is thus another supportive step towards meeting the sustainability criteria [8–10].

Sustainability criteria in the EU are determined by Directive 2009/28/ES. Among these, we include reducing GHG (greenhouse gas) emissions, optimizing land use and carbon stocks, biodiversity, and environmental requirements for crop production. To achieve 10% CO_2 savings between 2010 and 2020, a minimum biofuel share of 15% must be reached.

Hydrotreated vegetable oils are one of the possible ways of using the increased biofuel content in diesel fuel. This would provide another option for meeting the CO_2 reduction target for the year 2020. At the same time, it will be necessary to include these advanced biofuels in legislation, and to establish clear rules for their use. In addition to these changes, changes in the composition of diesel fuel can also be expected in the future. This mainly concerns requirements for increasing the cetane number and cetane index, adjusting the course of the distillation curve (reduction of temperature by 95% of the pre-distilled volume), further reducing the content of polycyclic aromatic hydrocarbon, introducing a limit for aromatics similar to automotive gasoline and tightening requirements for lubrication and mechanical impurities for fuels for diesel engines. Introducing changes that have a positive effect on reducing harmful emissions and particulate matter pollution will entail an increase in production costs and therefore the speed of their deployment will depend on the economic situation and legislative changes adopted within the EU [11–13]. Compared to fossil fuels, biofuels are renewable. As far as their technological development is concerned, the issue of biofuels is only at an early stage. The most commonly declared "first generation" of biofuels is bioethanol produced from starch and sugar, biodiesel produced from vegetable oils (rape, soy, etc.) and animal fats without chemical treatment, or produced due to the transesterification process to fatty acid methyl esters (FAME—rapeseed oil) or similar non-edible feedstocks like soursop seed oil [14,15]. These are sophisticated technologies, and above all, commercially available [16–18].

Figure 1 shows a simplified difference between the production of hydrotreated oils and fatty acid methyl esters (FAME).

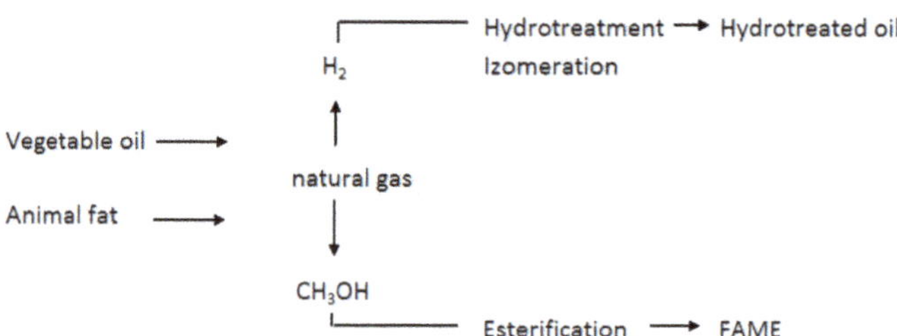

Figure 1. Simplified diagram for the production of hydrotreated oils and fatty acid methyl esters. FAME: fatty acid methyl esters.

The production of hydrotreated vegetable oils is based on introducing hydrogen molecules into the raw fat or oil molecule. This process is associated with the reduction of the carbon compound. When hydrogen is used to react with triglycerides, different types of reactions can occur, and different resultant products are combined.

The original oil obtained by hydrotreatment achieves higher oxidation stability, which is desirable for frying oils. Partial fat stiffening is used for raw margarine production. For fuel purposes, such a final product is not suitable. In hydrotreated fuels, therefore, partial hydrotreatment is mostly omitted and overall hydrotreatment continues, often with free fatty acids (Figure 2).

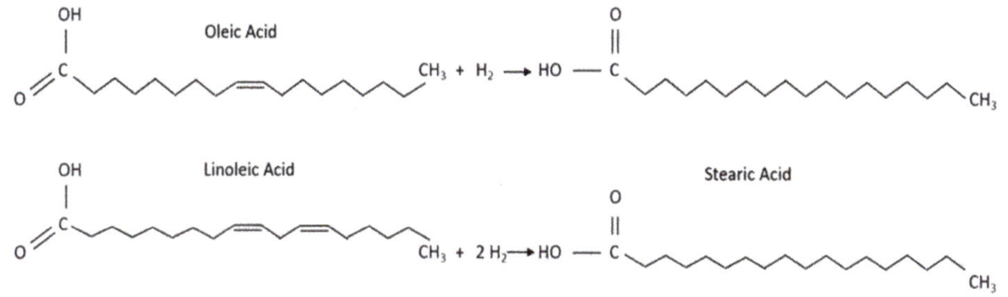

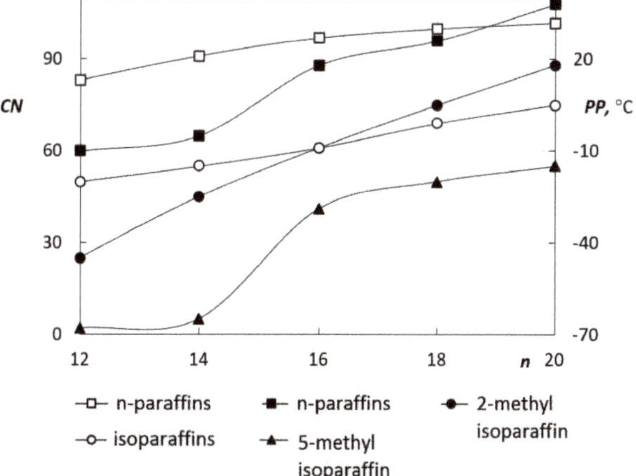

Figure 2. Hydrotreatment of fatty acids.

Another method of converting triglycerides by hydrogen is the cleavage of the ester to hydrocarbon and glycerol-derived propane and free fatty acids. These fatty acids (n is the number of carbons) are either:

1. reduced to hydrocarbons (n) and water by hydro deoxygenation,
2. subject to decarboxylation, i.e. carbon dioxide CO_2 is cleaved to give n-1 hydrocarbons,
3. or decarboxylation is carried out by removing the carbon monoxide (CO) and water molecule to produce an n-1 hydrocarbon.

For the reaction of hydrogen with vegetable oils and vacuum gas oil, the same catalysts and the same types of reactors and equipment were used as for the oil processing [19,20]. In recent decades, efforts have been made to find the best catalysts, to optimize hydrogen reaction processes and to find suitable sources of vegetable oil or fats. A lack of resource availability and high hydrogen consumption are increasing production costs, but these shortcomings are being gradually managed commercially [21].

A key process for obtaining hydrocarbons is hydroisomerization. This is a radical reaction where branching of hydrocarbon molecules is achieved by the use of form-selective catalysts, such as zeolites or other acid catalysts. N-paraffins having a boiling point corresponding to diesel fuel generally have a higher cetane number than their branched isomers. In contrast, isoparaffins have lower solidification points than n-paraffins. Therefore, there is a compromise in the quality of the paraffin-rich fuel; the fuel has either good combustion properties, or good low-temperature properties. The result of hydroisomerization is therefore a fuel with a lower solidification point and a lower cetane number. The relationship of these two properties is illustrated in Figure 3 [22].

Figure 3. Cetane number and paraffin solidification points depending on carbon number.

Hydrotreated oils are characterized by very good low temperature properties. The cloud point also occurs below −40 °C. Therefore, these fuels are suitable for the preparation of premium fuel with a high cetane number and excellent low temperature properties. The cold filter plugging point (CFPP) virtually corresponds to the cloud point value, which is why the value of the cloud point is significant in the case of hydrotreated oils.

The production process of hydrotreated oils can also produce fuel with appropriate low temperature properties from palm oils and other oils, including animal fats, whose methyl esters have a very poor applicability at lower temperatures [23]. This results in the use of hydrotreated oils throughout the year, without risking loss of serviceability or fuel logistics problems. Hydrotreated vegetable oils thus find their potential usability as aviation fuels.

Hydrotreated vegetable oils meet EN 15940:2014 for paraffinic diesel fuels from synthesis or hydrotreatment, formerly TS 15940:2012 for paraffinic diesel fuels [24]. This specification also applies to Fischer-Tropsch synthesis products: GTL, BTL, and CTL. Specification TS 15940:2012 was preceded by CWA 15940:2009 CEN Workshop Agreement, which was created in cooperation between car and fuel manufacturers. HVO is usually supplied without FAME, however, it is allowed to add up to 7% vol. under specification EN 15940, which the earlier CWA 15940 did not allow. EN 14214 for FAME for HVO does not apply, as HVO is composed only of hydrocarbons. However, HVO meets the requirements of EN 590 for diesel fuel, except for the density below the lower limit of this standard [25]. This also applies to the US ASTM D975 [26]. Table 1 shows the differences between these standards.

Table 1. Requirements of EN 15940, EN 590 and ASTM D975 [27].

Parameter	EN 15940	EN 590:2013	ASTM D975
Cetane number	≥70.0	≥51.0	≥40
Density at 15 °C (kg·m^{-3})	765–800	820–845	
Viscosity at 40 °C (mm^2·s^{-1})	2.00–4.50	2.00–4.50	1.9–4.1
Hydrocarbons (% m/m)			≤35
Polyaromatic	-	≤8	-
Aromatic	≤1.0	-	-
Olefin	≤0.1	-	-
Sulfur content (mg·kg^{-1})	≤5.0	≤10.0	≤15
Flash point (°C)	>55	>55	>52
Lubricity HFRR at 60 °C (μm)	≤460 *	≤460	≤520
95% by volume distils at (°C)	<360	<360	282–338
CFPP (°C)	≥−34	≥−34	-
Ash content (% m/m)	≤0.01	≤0.01	≤0.01
Total impurity content (mg·kg^{-1})	≤24	≤24	-

* Including lubricating additives for use in vehicles approved for driving on the fuel according to the standard.
CFPP: cold filter plugging point; HFRR: high frequency reciprocating rig.

2. Results

The density of the hydrotreated vegetable oil (about 780 kg·m^{-3}) is lower than the density of fossil diesel (800 to 845 kg·m^{-3}) because of its paraffinic character, and also a lower temperature distillation end. The density of the fuel has traditionally been an important factor in terms of volume of fuel consumption and maximum performance, and we can say that the reduction of volume calorific value is only a function of density. At a lower calorific value, the engine generates less energy and needs more fuel to ensure the same power output at a partial load.

In the case of hydrotreated oils, the effect is different, as the calorific value compensates for the lower density effect (Figures 4 and 5). The higher calorific value of the hydrotreated oils is due to the fact that the amount of hydrogen in the hydrotreated oils is about 15.2% (m/m), as opposed to 13.5% (m/m) in standard diesel.

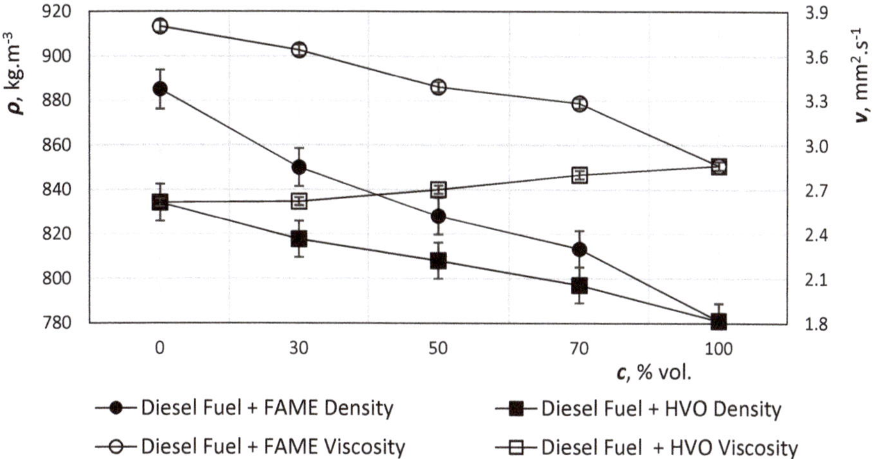

Figure 4. Density and kinematic viscosity of diesel fuel with FAME and hydrotreated vegetable oil (HVO).

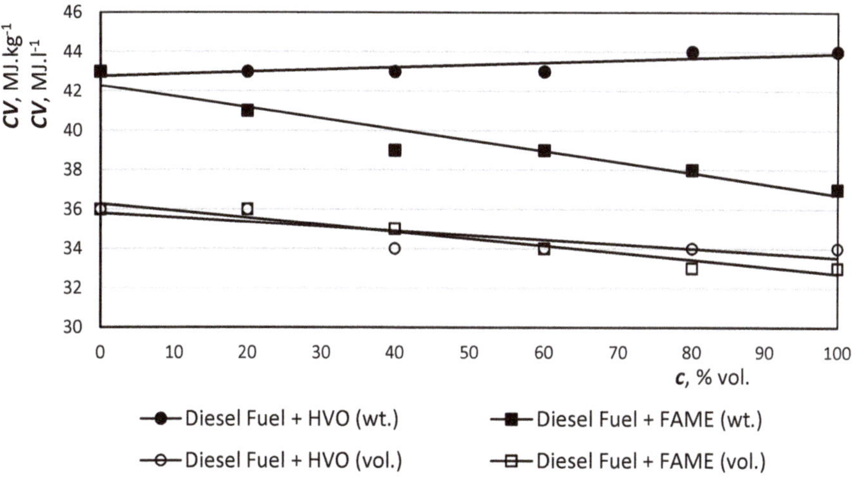

Figure 5. Calorific value of mixed fuels.

Figure 4 shows that the density with increasing HVO concentration in the mixture is expected to decrease. This is due to the lower permissible water content and paraffinic character with a higher hydrogen content than diesel, which results in a higher energy content per kg. Mixture of diesel fuel with HVO > 30% vol. exceeds the EN 590 (Table 1) limit for diesel (820–845 kg·m^{-3}). High proportion mixtures did not meet the standard limits. On the contrary, low density offers the possibility of blending HVO into diesel fuels with higher contents of heavier fractions, or their incorporation into less profitable products, such as fuel oil. Conversely, all HVO mixtures meet standardized kinematic viscosity limits.

On the y-axis, lubricity (WSD—wear scar diameter) is in μm. HVO has very low lubricity, therefore up to 80% of the volume may be added as maximum. Once the concentration reaches 80% or more, the mixture of the fuel does not match the EN standard—see Figure 6. It could lead to seizure of the fuel system of the machine.

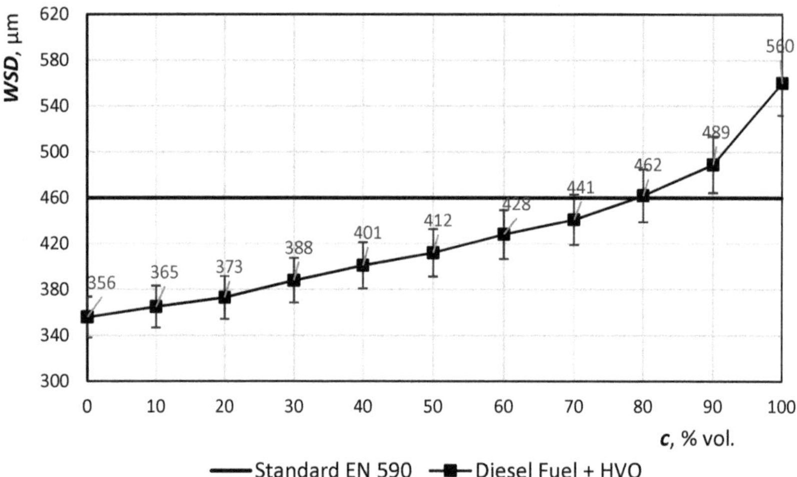

Figure 6. Lubricity of diesel fuel with addition of HVO.

An aromatic-free hydrocarbon composition results in a lower lubricity of the fuel. Lubrication of hydrotreated oils corresponds to sulfur-free winter grade diesel or GTL. It is essential that lubricant additives are added to these fuels to meet the requirements of EN 12156-1 (HFRR high frequency reciprocating rig, corrected abrasion area diameter at 60 °C < 460 mm). It is possible to apply commonly used lubricants for diesel with a similar dosage. When using this fuel at higher concentrations, it is assumed that a further test for lubricity verification will be added.

Hydrotreated oil can also be supplied with lubricating additives for use in pure form, or as an additive. It is common, however, to supply, for example, HVO without these additives if the fuel is designated as a component of the mixture. The lubricity of the resulting fuel must then be controlled and must be increased, to cover the HVO in the mixture. From the point of view of lubricity, this is the only parameter where FAME is better and in itself can replace the additive in all hydrocarbon fuels. The lubricating ability can be improved by adding the appropriate additive or a small amount of FAME content. The expanded uncertainty of the result is ± 5%.

Also, the distillation curve is different from diesel fuel and FAME. Distillation properties show how the fuel evaporates when injected into the combustion chamber of the diesel engine. Low-boiling fractions are important for the engine's start-up, and the heavier fractions with higher boiling points can cause problems with fuel being incompletely burned and increasing the proportion of harmful emissions in the engine exhaust gases. Standard diesel fuel has a boiling range approximately from 180 °C to 360 °C.

A distillation test to determine the distillation curve is a test that must always be carried out when assessing the quality of fuel. The distillation curve expresses the volume percentage of the fuel that is distilled to a certain distillation temperature. T50 is the temperature at which 50% of the fuel is distilled. At this point the amount of air is bound for perfect combustion. For HVO and diesel, there is no need to worry about combustion air as the temperature in T50 has not increased significantly; the value is shown in Figure 7.

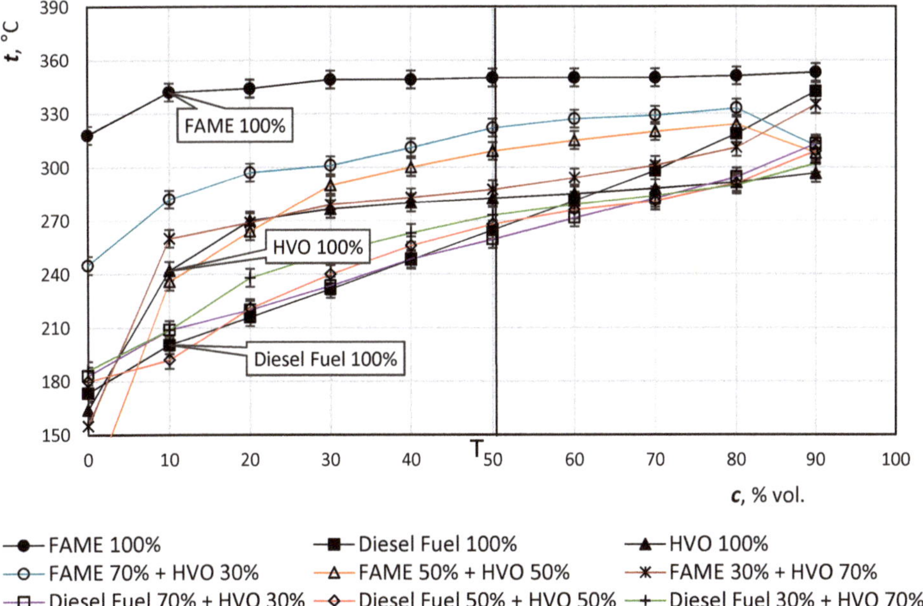

Figure 7. Distillation curves of FAME, diesel fuel, HVO and their mixtures.

It is clear from Figure 7 that HVO does not affect the beginning of fuel distillation. The presence of "light components" is not compromised, so the moving parts of the fuel system cannot be damaged. The addition of HVO results in a flattening of the distillation curve. Distillation indicates a lower proportion of high boiling heavy fractions, thereby reducing carbonization shares and reducing exhaust emissions. Higher concentrations of HVO can be expected to affect engine performance.

Figure 8 shows the difference of values due to the HVO admixtures. The addition of HVO to diesel fuel positively affects the loss of filterability (CFPP—cold filter plugging point). Standard EN 590 sets the temperature −20 °C as the maximum value for F-class winter diesel, 14 is marked with a thick horizontal line.

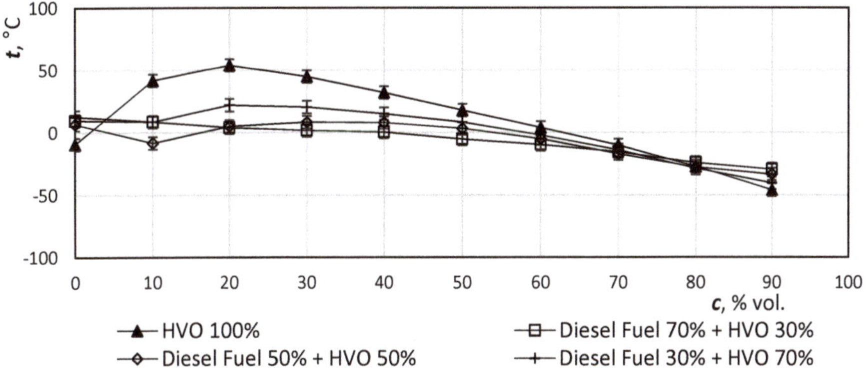

Figure 8. Changes in the mixed fuel distillation curve as compared to pure mineral diesel on the *x*-axis.

The cloud point (CP) or wax appearance temperature (WAT) is the temperature at which n-paraffins begin to precipitate in fuel, but generally, it is not a mandatory indicator.

Figure 9 illustrates the decrease of temperature with increasing addition of HVO. In the case of 100% HVO, the CP value is practically the same as the CFPP value. HVO addition to diesel fuel favors low temperature properties, which, even in the case of an HVO 30 mixture, are 6 degrees below the winter diesel fuel EN Class F. EN 590 also sets the CFPP for an arctic climate (for 1st class it is −26 °C, while for 2nd class it is −32 °C), both of which exceed FAME + HVO 70–100.

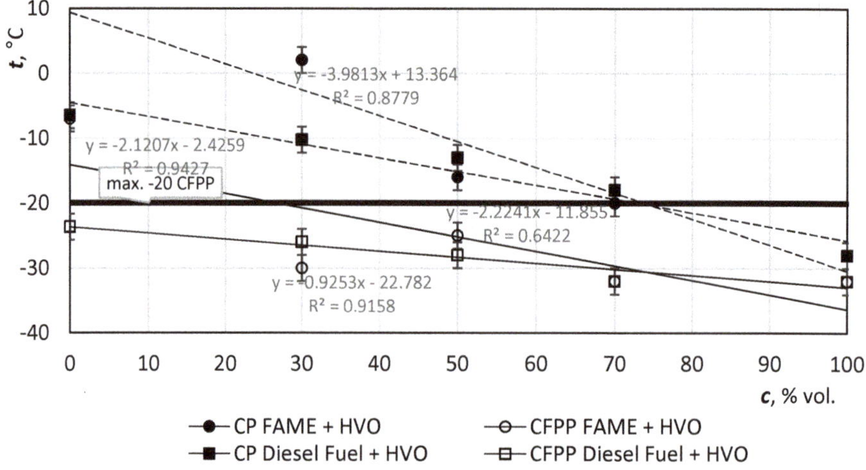

Figure 9. CFPP and cloud point (CP) of mineral diesel fuel with FAME and HVO.

At temperatures above the cloud point, the hydrotreated oil is colorless, clear as water. It has no characteristic aroma typical for other fuels. It does not contain any visible dirt at temperatures above the cloud point. The cloud point causes the creation of scum characteristic for diesel.

The flash point is the lowest temperature at which the flammable substance produces so many flammable vapors at atmospheric pressure that they will briefly ignite being in contact with open flame but they do not continue to burn.

Figure 10 shows the temperature increase of the flash point with increasing HVO content. Temperature range III of the hazard class, in which the measured temperatures of all HVO mixtures are located, is defined here. However, in the case of a flash point value, these values do not affect the work of the diesel engine.

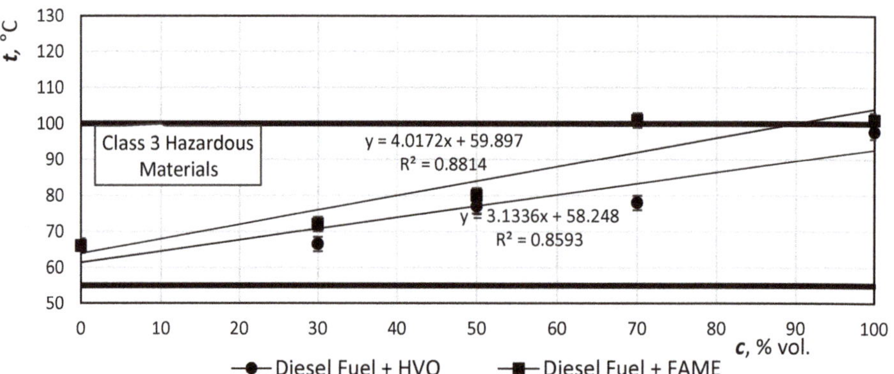

Figure 10. Flash points of diesel fuel with HVO and FAME.

The cetane number indicates the reactivity of diesel fuel in terms of its diesel characteristics. The higher the fuel gets, the better it is, the higher the cetin number is, the more regular and better is its combustion, as well as the engine running and noise. Because of the relative difficulty of the cetane number test, the cetane index, which can be determined based on a calculation from the results of the laboratory density and distillation tests, has been introduced as a characteristic of ignition capability. According to EN 590, the cetane number is at least 51 units, the cetane index is 46 units.

Figure 11 shows the increasing trend lines of the cetane number and cetane index with increasing HVO content. The cetane number of hydrotreated vegetable oils ranges from 75 to 95 units due to the composition (n-paraffins and isoparaffins). In mixed fuels, there is a linear increase in the cetane number, corresponding to the proportion of components. Hydrotreated oil is a suitable additive for increasing the cetane number due to the nature of the fuel, where its effect is greater than the use of conventional additives. For measuring the cetane number on the test engine, the hydrotreated oil must be mixed with a fuel with a known and low cetane number, such that the cetane number of the resulting mixture is below 70 units within the measuring range. Then, the cetane number of the hydrotreated vegetable oil is determined by linear extrapolation. The calculation of the cetane index is suitable for standard diesel fuels (with FAME) and its use for hydrotreated oils is not appropriate.

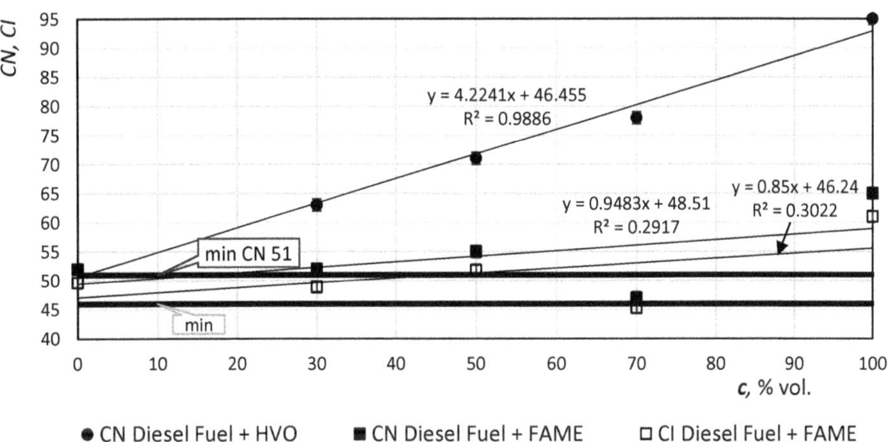

Figure 11. Cetane numbers of diesel fuel with HVO and FAME, cetane index of diesel fuel with FAME.

The limit value for the cetane number according to EN 590 is highlighted in the graph with a horizontal line; for the cetane index, 46 is also highlighted by the horizontal line. The extrapolated cetane number is high because of the very high content of n- and isoparaffins in HVO; the value of the HVO 30 mixture already means a much higher cetane number than the minimum value according to EN 590.

The measurement in this work compared mineral diesel fuel without bio-components, 100% pure HVO and their mixtures. Prepared ratios were based on the possibility of comparison of the results of our own measurements and results of measurements already published in the literature. The individual measurement procedures were performed according to applicable valid standards and were repeated three times to avoid any measurement error. The density measurement results of mixtures with increasing HVO concentration had an expected decreasing trend, which should have a negative effect on the calorific value (energy content per kg). According to the HVO manual, this drop is due to the lower permitted water content and the pure paraffinic character of HVO fuel, which mineral diesel does not have, but due to its higher hydrogen content, HVO has a higher calorific value at lower density.

Since hydrotreated oil consists only of hydrocarbons, traditional methods for fossil diesel, but not FAME, are also suitable for determining fuel stability. This especially applies to "Rancimat" methods according to EN 15751, which is intended for pure diesel fuel and FAME containing 2–7% vol. of FAME. This method is not suitable for pure hydrotreated oil, even as an additive in diesel fuel. The stability of hydrotreated oils is at the standard level of diesel fuel and there should be no risks except for long-term vehicle shutdown or storage.

The sulfur content of hydrotreated oil is based on the production process and is <1 mg·kg^{-1}. As the standard oil logistics system is used for hydrotreated oil, the sulfur content due to contamination may be higher, and then the normalized value is ≤5.0 mg·kg^{-1}. Addition of hydrotreated oils can also positively reduce the sulfur content, for example in diesel, where the value exceeds the relevant standard EN 590.

The ash content in hydrotreated oils is very low (<0.001%). Also, the content of P, Ca and Mg is well below the detection limits of analytical methods (<1 mg·kg^{-1}).

Hydrotreated oil is, like fossil hydrocarbons, nonpolar, while water is polar. Water solubility is thus similar to traditional diesel fuel, or even lower. Therefore, the issue of water requires no further action in the field of logistics (as well as for diesel we use).

The issue of microbial growth is primarily about FAME, which also promotes microbial growth in diesel fuel blends. FAME is biodegradable and tends to increase the water content of diesel fuel. Unlike FAME, the presence of hydrotreated oil mixed with diesel fuel does not require any further action. However, monitoring the quality indicator is useful because microbes can proliferate even in pure fossil fuel during long term storage in the presence of free water. Higher temperatures, especially in the summer, can increase microbial growth, mainly if mineral salts are present in the water phase. At lower temperatures, growth of microorganisms slows down.

The hydrocarbon composition of HVO corresponds approximately to the hydrocarbons of which diesel fuel is composed. The composition of HVO is composition closer to diesel oil than to a FAME mixture, which is an advantage for the use of HVO as a substitute for FAME.

R is detector response in Volts; t_R is retention time in minutes. Figures 12–14 are chromatograms of fuel samples (diesel fuel, HVO, and FAME). In Figure 12, we can see the composition of analyzed diesel fuel with labeled n-alkanes, which represent major constituents of the sample. Similarly, Figure 13 shows the composition of hydrocarbon compounds in HVO. The chromatogram in Figure 14 allowed us to specify the presence of major components (methyl esters) in FAME; the peaks are well separated. All components labeled in the chromatograms were identified according to retention times of analytical standards (mixture of n-alkanes), which are specified in the section Materials and Methods.

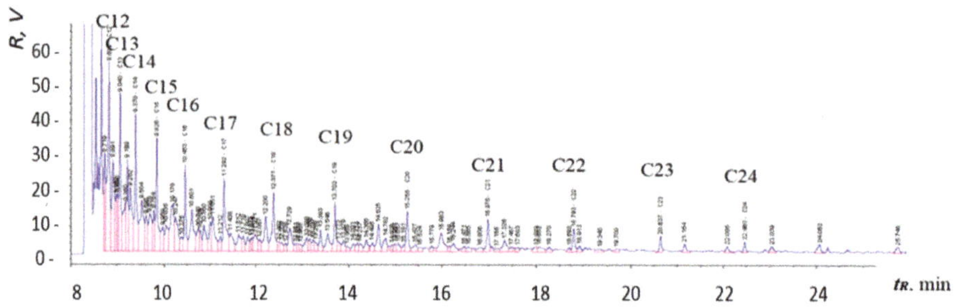

Figure 12. Chromatogram of 100% diesel fuel with identified (labeled) n-alkanes.

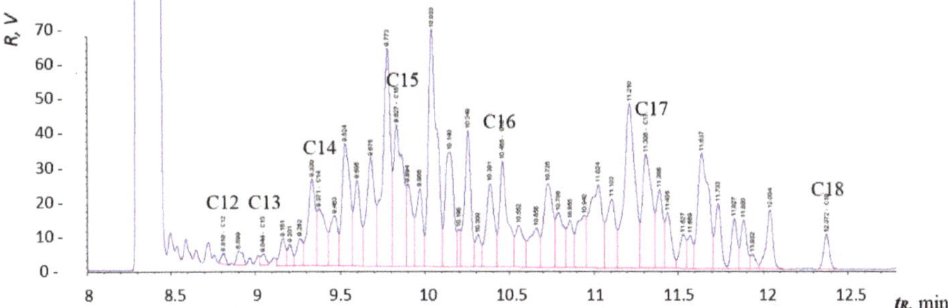

Figure 13. Chromatogram of 100% HVO with identified (labeled) n-alkanes.

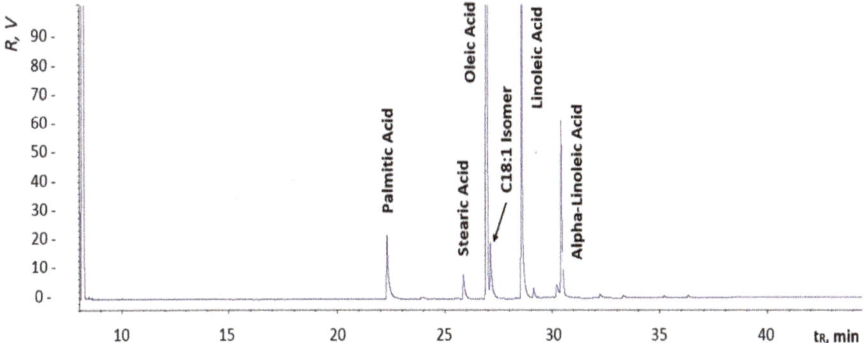

Figure 14. Chromatogram of 100% FAME with identified main components (methyl esters).

Chromatographic results amongst individual samples may vary, depending on the refinery they come from. Generally, the final composition of the fuel depends on the season, the country of origin, the class of fuel, and more.

A statistical analysis was then conducted to obtain a general equation of density and viscosity for independent variable concentration. An analysis was done for a mixture of diesel and HVO, and a mixture of FAME and HVO. For the needs of the article, statistical tool R, and the built-in library lm() were used.

As a first analysis, a linear regression analysis was calculated to predict the density of a diesel and HVO mixture based on concentration. A significant regression equation was found ($F_{(1, 3)} = 1411$, $p < 0.01$, N = 20), with R^2 of 0.9972. The model predicted density in the form of equation (1). Assumptions of linear regression were verified by gvlma library. All assumptions were accepted: Global Stats ($p = 0.5704$), Heteroscedasticity ($p = 0.9905$), Skewness ($p = 0.2229$), Kurtosis ($p = 0.9911$), and Link Function ($p = 0.2302$).

$$\rho = 0.52147c + 780.97672 \tag{1}$$

Complementary linear regression was calculated to predict the viscosity of a diesel and HVO mixture based on concentration. A significant regression equation was found ($F_{(1, 3)} = 49.97$, $p < 0.01$, N = 20), with R^2 of 0.9245. The model predicted viscosity in the form of Equation (2). All assumptions were accepted: Global Stats ($p = 0.5884$), Heteroscedasticity ($p = 0.8937$), Skewness ($p = 0.7208$), Kurtosis ($p = 0.4098$), and Link Function ($p = 0.1579$).

$$v = -0.003875c + 2.89085 \tag{2}$$

As a second analysis, a linear regression analysis was calculated to predict the density of a FAME and HVO mixture based on concentration. A significant regression equation was found ($F_{(1, 3)} = 94.09$,

$p < 0.01$, N = 20), with R^2 of 0.9588. The model predicted density in the form of equation (3). Assumptions of linear regression were verified by gvlma library. All assumptions were accepted: Global Stats ($p = 0.5035$), Heteroscedasticity ($p = 0.8260$), Skewness ($p = 0.4299$), Kurtosis ($p = 0.1330$), and Link Function ($p = 0.5241$).

$$\rho = 0.7478c + 785.6795 \tag{3}$$

Complementary linear regression was calculated to predict the viscosity of a FAME and HVO mixture based on concentration. A significant regression equation was found (F(1, 3) = 88.94, $p < 0.01$, N = 20), with R^2 of 0.9674. The model predicted viscosity in the form of equation (4). All assumptions were accepted: Global Stats ($p = 0.8565$), Heteroscedasticity ($p = 0.6979$), Skewness ($p = 0.5792$), Kurtosis ($p = 0.4987$) and Link Function ($p = 0.5205$).

$$v = 0.011987c + 2.887347 \tag{4}$$

3. Discussion

The experiment shows that the hydrotreatment process is an alternative to the production of biofuels for the esterification process to eliminate the undesirable effects, as described in [16] and [28]. These include, in particular, increased NO_x content, emissions, fuel storage problems, engine oil wear, and so on. HVO are also characterized by high cetane numbers, as confirmed by the measured results [16]. The characteristics for diesel compared to HVO were practically the same, as illustrated by Sugiyama [17]. Experiments and proven measurements show that HVO impurities have a positive effect on the characteristics of diesel engines.

Characteristics to be monitored include, above all, the lubricity to provide the lubricating ability of the moving parts of the fuel system and the cetane number. The graphs show, in accordance with [28], that the recommended HVO ratio (addition) should be about 50% in order to be consistent with the diesel fuel characteristics.

According to Šimáček et al. [29], low density and low sulfur content have an effect on lower lubricity, which can be improved by the application of conventional lubricating additives as is the case with today's low-grade mineral diesel fuel. The kinematic viscosity of all HVO mixtures meets the standard parameters. The distillation curve determines that, by addition of HVO to mineral diesel fuel, its process is flattened. According to Hönig et al. [23], this has a positive effect on the reduction of carbon deposits and exhaust emissions. The HVO Manual [27] indicates CFPP up to −40 °C. This value has not been confirmed by its own measurement. The lowest measured CFPP was −36 °C in 100% HVO. This is even 11 °C less than that found in Aatola et al. [16]. Even with this mismatch, all blends have a positive effect on the CFPP drop and are well below the F-class for diesel fuels, the CFPP reported by EN 590 max −20 °C. Results of the flash point measurement have an increasing tendency, which corresponds to all other articles, the measured values were compared with. This has a positive effect on reducing the risk of a fuel explosion during handling and storage under standard conditions. The measured high values of the cetane number and the calculated cetane index value increase with the HVO content in the mineral oil mixture. The values of the cetane number and cetane index set out in this work correspond to the already high values of these figures in Aatola et al. [16], Šimáček et al. [29], Hönig et al. [30], Váchová and Vozka [31], and the HVO guidelines [27].

The properties of hydrotreated oils are much more similar to high quality sulfur-free diesel or synthetic GTL diesel fuel than to FAME.

In the production of fuels, components with n-hydrocarbons and branched hydrocarbons are suitably combined to achieve suitable fuel properties (cetane number, pour point). Biodiesel produced by the hydrotreating of vegetable oil consists mainly of C_{17} and C_{18} n-hydrocarbons with a high cetane number but with poor low temperature properties due to the melting point between 20 and 28 °C. Improvement of these parameters can be achieved by adding a second proportion of highly isomeric hydrocarbons into the fuel blend.

During long-term storage, pure hydrotreated oils as well as mixtures containing them, behave like traditional diesel fuels. Hydrotreated oils do not contain any hazardous impurities, such as saturated monoglycerides present in FAME. There is therefore no risk of clotting above the cloud point. However, as with standard diesel, this phenomenon may occur due to the presence of paraffins in the fossil fuel or hydrotreated oil during long-term storage at temperatures below the cloud point.

4. Materials and Methods

The sample of tested hydrotreated vegetable oil was received from the Neste Oil Company (Espoo, Finland). Simultaneously, diesel fuel free of fatty acid methyl esters compliant with EN 590 and a FAME mixture compliant with EN 14 214 were used for laboratory tests.

The following tests of blends were carried out:

1. Density at 15 °C according to EN ISO 3675
2. Kinematic viscosity at 40 °C according to EN ISO 3104
3. Cold filter plugging point (CFPP) according to EN 116
4. Flash point according to EN 2719
5. Oxidation stability of vegetable oil according to EN 15 751
6. Cetane number according to EN ISO 5165
7. Lubricity according to EN 12156-1 (HVO lubrication 460–650 μm)
8. Calorific value according to ISO 1928 on IKA C200 Calorimeter
9. Gas Chromatography—Flame Ionization Detector GC-FID

The following samples and their mixtures were analyzed:

1. 100% FAME
2. 100% HVO (from Neste Oil Company)
3. 100% Diesel Fuel (without any FAME)

For the GC measurements, the samples were diluted 1/50 (20 μL sample + 980 μL hexane). Analytical Standards:

1. Mixed standard: n-alkanes C10 to C30 in hexane
2. Mixed standard: Supelco 37 Components FAME Mix

For sample analysis, an Agilent Technologies 7890A gas chromatograph (Santa Clara, CA, USA) equipped with an autosampler, a fused-silica capillary column SPB-2560 and a flame ionization detector (FID) was used. The basic instrument parameters and GC analysis conditions are shown in Table 2.

Table 2. Device parameters and GC analysis conditions.

Device	Type and Settings
Gas Chromatograph	Agilent Technologies 7890A
Autosampler	G4513A (16 positions) with a syringe Agilent Gold Standard 10 μL
Analytical column	SP-2560, 100 m $\times$ 0.25 mm i.d., film thickness 0.2 μm
Temperature program	140 °C (5 min), increase 4 °C·min^{-1}, 245 °C (20 min) $\rightarrow$ 51.25 min
Carrier gas	Helium 5.6, const. inlet pressure 50 psi (flow rate 1.58 mL·min^{-1} at 140 °C)
Injection chamber	Temperature 280 °C, injection volume 1 μL, split ratio 1:100
Detector	Temperature 280 °C, gas flow: hydrogen (6.0) 30 mL·min^{-1}, air (5.0) 400 mL·min^{-1}, makeup = nitrogen (6.0) 25 mL·min^{-1}
Data collection software	Agilent ChemStation (Revision B.04.02 SP1)

Lubrication was measured on a PCS instrument: HFRR (high frequency reciprocating rig). The PCS instrument uses an electromagnetically vibrating moving body with low amplitude, while simultaneously compressing it against a solid body. The instrument measures the frictional forces between the bodies and the electrical contact resistance between them. Settings of the instrument are in Table 3.

Table 3. Technical parameters of HFRR.

Parameter	Value
Frequency	10–200 Hz
Shift	20 μm–2 mm
Load	0.1–1 kg with supplied weights
Maximum fractional force	by amplitude, max 10 N
Temperature	From room temperature to 150 °C
Standard upper test body	Ball ∅ 6 mm
Standard lower test body	Disk ∅ 10 mm and thickness 3 mm
Power supply	100–230 V
Heating	Two heating cartridges 24 kW, 15 kW

5. Conclusions

Due to the pressure of the European Union to reduce the total amount of greenhouse gases released into the atmosphere, there is also a need to reduce greenhouse gas emissions in transport. One way is to increase the share of biofuels in mineral diesel over 7%.

Biodiesel from FAME is not very suitable as a higher percentage mixture because of the low oxidation stability, higher temperature of cold filter plugging point (CFPP), carbonization tendency and possible microbial contamination in the presence of water. A much better biofuel is HVO, whose hydrocarbon character can be compared to high quality mineral diesel with a very high cetane number and a very low temperature of cold filter plugging point (CFPP). As confirmed by the actual measurement, HVO does not have the above-mentioned drawbacks as fatty acid methyl esters do. HVO can be mixed into mineral diesel fuel without limitation. Its presence in mineral oil blends improves engine performance and reduces fuel consumption, exhaust emissions, and cold filter plugging point (CFPP), so it can also be used in aviation turbine engines. The properties of hydrotreated oils are much more similar to high quality sulfur-free diesel or synthetic GTL diesel fuel than to FAME.

Adding HVO can achieve a lowering of NO_x and particulate matter emission, which has a positive impact on environment. This is a suitable way of accomplishing emissions below the limits of a newly introduced ban for highly-polluting, older diesel vehicles, for example in Germany. HVO has a naturally high cetane number, which is very useful for increasing lower cetane fuels, which could also be conducive to other alternative fuels for lowering emissions, as well as having sufficient fuel properties.

The production technology of this biofuel is intergenerational in its own way, as raw materials, both food and waste, can be used for its production without significantly changing the hydrotreatment conditions. It is also beneficial for this technology that it can be operated, after minor modifications, directly in existing refineries.

Author Contributions: P.Z., V.H., M.K., J.T., M.O., J.M., V.H., and M.P. designed methodology; P.Z., V.H., M.K., J.T., M.O., J.M., V.H., and M.P. performed the calculation and analyzed the data; P.Z., V.H., M.K., J.T., M.O., J.M., V.H., and M.P. wrote the paper.

Acknowledgments: This work was supported by grants of the Grant Agency of Czech University of Life Sciences Prague IGA 2018: 31150/1312/3113—Analysis of the influence of biofuels on the operating parameters of combustion engines and IGA 2017: 31150/1312/3116—The Examination of the Influence of Blended Biofuels on Operating Parameters of CI Engines.

Conflicts of Interest: The authors declare no conflict of interest.

References

1. European Commission. *Being Wise with Waste: The EU's Approach to Waste Management*; Publications Office of the European Union: Luxembourg, 2010; ISBN 978-92-79-14297-0.

2. Masi, F.; Bresciani, R.; Rizzo, A.; Edathoot, A.; Patwardhan, N.; Panse, D.; Langergraber, G. Green walls for greywater treatment and recycling in dense urban areas: a case-study in Pune. *J. Water Sanit Hyg. Dev.* **2016**, *6*, 342–347. [CrossRef]

3. Rizzo, A.; Boano, F.; Revelli, R.; Ridolfi, L. Role of water flow in modeling methane emissions from flooded paddy soils. *Adv. Water Resour.* **2013**, *52*, 261–274. [CrossRef]

4. Kučerová, V.; Lagaňa, R.; Výbohová, E.; Hýrošová, T. The Effect of Chemical Changes during Heat Treatment on the Color and Mechanical Properties of Fir Wood. *BioResources* **2016**, *11*. [CrossRef]

5. Výbohová, E.; Kučerová, V.; Andor, T.; Balážová, Ž.; Veľková, V. The Effect of Heat Treatment on the Chemical Composition of Ash Wood. *BioResources* **2018**, *13*. [CrossRef]

6. Clark, J.S.; Prochazka, P.; Yiridoe, E.K.; Prochazkova, K. PVYn and Potato Wart Disease Outbreaks in Prince Edward Island: Policy Response and Analysis. *Can. J. Agr. Sci. Econ.* **2007**, *55*, 527–534. [CrossRef]

7. Hrcka, R.; Kučerová, V.; Hýrošová, T. Correlations between Oak Wood Properties. *BioResources* **2018**, *13*. [CrossRef]

8. Hönig, V.; Linhart, Z.; Procházka, P. Biobutanol from local bio-wastes. In *Agrarian Perspectives XXVI. Competitiveness of European Agriculture and Food Sectors, Proceedings of the 26th International Conference, Prague, Czech Republic, 13–15 September 2017*; Czech University of Life Sciences Prague, Faculty of Economics and Management: Prague, Czech Republic, 2017; pp. 102–108.

9. Corral Bobadilla, M.; Lostado Lorza, R.; Escribano García, R.; Somovilla Gómez, F.; Vergara González, E. An Improvement in Biodiesel Production from Waste Cooking Oil by Applying Thought Multi-Response Surface Methodology Using Desirability Functions. *Energies* **2017**, *10*, 130. [CrossRef]

10. Corral Bobadilla, M.; Fernández Martínez, R.; Lostado Lorza, R.; Somovilla Gómez, F.; Vergara González, E. Optimizing Biodiesel Production from Waste Cooking Oil Using Genetic Algorithm-Based Support Vector Machines. *Energies* **2018**, *11*, 2995. [CrossRef]

11. Pražnikar, J. Particulate matter time-series and Köppen-Geiger climate classes in North America and Europe. *Atmos. Environ.* **2017**, *150*, 136–145. [CrossRef]

12. Žibert, J.; Cedilnik, J.; Pražnikar, J. Particulate matter (PM10) patterns in Europe: An exploratory data analysis using non-negative matrix factorization. *Atmos. Environ.* **2016**, *132*, 217–228. [CrossRef]

13. Krause, J.; Špička, J. Economic Analysis of Chemical Industry. *Chemické listy* **2013**, *107*, 573–578.

14. Procházka, P.; Hönig, V. Economic Analysis of Diesel-Fuel Replacement by Crude Palm Oil in Indonesian Power Plants. *Energies* **2018**, *11*, 504. [CrossRef]

15. Su, C.-H.; Nguyen, H.; Pham, U.; Nguyen, M.; Juan, H.-Y. Biodiesel Production from a Novel Nonedible Feedstock, Soursop (Annona muricata L.) Seed Oil. *Energies* **2018**, *11*, 2562. [CrossRef]

16. Aatola, H.; Larmi, M.; Sarjovaara, T.; Mikkonen, S. Hydrotreated Vegetable Oil (HVO) as a Renewable Diesel Fuel: Trade-off between NOx, Particulate Emission, and Fuel Consumption of a Heavy Duty Engine. *SAE Int. J. Engines* **2008**, *1*, 1251–1262. [CrossRef]

17. Sugiyama, K.; Goto, I.; Kitano, K.; Mogi, K.; Honkanen, M. Effects of Hydrotreated Vegetable Oil (HVO) as Renewable Diesel Fuel on Combustion and Exhaust Emissions in Diesel Engine. *SAE Int. J. Engines* **2011**, *5*, 205–217. [CrossRef]

18. Pešek, M.; Samková, E.; Špička, J. Fatty acids and composition of their important groups in milk fat of Czech Pied cattle. *Czech J. Anim. Sci.* **2011**, *51*, 181–188. [CrossRef]

19. Ali, M.F.; El Ali, B.M.; Speight, J.G. *Handbook of Industrial Chemistry: Organic Chemicals*; McGraw-Hill handbooks; McGraw-Hill: New York, NY, USA, 2005; ISBN 978-0-07-141037-3.

20. Ancheyta, J.; Trejo, F.; Rana, M. *Asphaltenes: Chemical Transformation during Hydroprocessing of Heavy Oils*; Chemical Industries; CRC Press: Boca Raton, FL, USA, 2010; Volume 20101657, ISBN 978-1-4200-6630-2.

21. Sotelo-Boyas, R.; Trejo-Zarraga, F.; de Jesus Hernandez-Loyo, F. Hydroconversion of Triglycerides into Green Liquid Fuels. In *Hydrogenation*; Karam, I., Ed.; InTech: London, UK, 2012; ISBN 978-953-51-0785-9.

22. Kovács, S.; Kasza, T.; Thernesz, A.; Horváth, I.W.; Hancsók, J. Fuel production by hydrotreating of triglycerides on NiMo/Al2O3/F catalyst. *Chem. Eng. J.* **2011**, *176–177*, 237–243. [CrossRef]

23. Hönig, V.; Linhart, Z.; Procházka, P. Hydrotreated Vegetable Oil (HVO) for local bio-wastes. In *Agrarian Perspectives XXVI. Competitiveness of European Agriculture and Food Sectors, Proceedings of the 26th International Conference, Prague, Czech Republic, 13–15 September 2017*; Czech University of Life Sciences Prague, Faculty of Economics and Management: Prague, Czech Republic, 2017; pp. 109–115.

24. EN 15940: Automotive Fuels—Paraffinic Diesel Fuel from Synthesis or Hydrotreatment—Requirements and Test Methods. CEN. Available online: https://standards.globalspec.com/std/13065707/en-15940 (accessed on 26 February 2019).

25. EN 590. 2013: Automotive fuels—Diesel—Requirements and Test Methods. CEN. Available online: https://standards.globalspec.com/std/10257634/EN%20590 (accessed on 26 February 2019).

26. *ASTM D975: Standard Specification for Diesel Fuel Oils*; ASTM: West Conshohocken, PA, USA, 2019.

27. Neste Corporation Neste Renewable Diesel Handbook. Available online: https://www.neste.com/sites/default/files/attachments/neste_renewable_diesel_handbook.pdf (accessed on 27 March 2017).

28. Nasikin, M.; Susanto, B.H.; Hirsaman, M.A.; Wijanarko, A. Biogasoline from palm oil by simultaneous cracking and hydrogenation reaction over nimo/zeolite catalyst. *World Appl. Sci. J.* **2008**, *5*, 74–79.

29. Šimáček, P.; Vrtiška, D.; Mužíková, Z.; Pospíšil, M. Motor Fuels Produced by Hydrotreating of Vegetable Oils and Animal Fats (in Czech). *Chem. Listy* **2017**, *111*, 206–212.

30. Hönig, V.; Táborský, J.; Linhart, Z. Use of blend of hydrotreated vegetable oil with biobutanol for applications in diesel engines. In Proceedings of the Engineering for Rural Development, Jelgava, Latvia, 20–22 May 2015; pp. 324–329.

31. Váchová, V.; Vozka, P. Processing of vegetable oils to diesel fuel. *Paliva* **2015**, *7*, 66–73.

 *catalysts*

Article

The Promoting Effect of Ni on Glycerol Hydrogenolysis to 1,2-Propanediol with In Situ Hydrogen from Methanol Steam Reforming Using a Cu/ZnO/Al$_2$O$_3$ Catalyst

Yuanqing Liu [1], Xiaoming Guo [2], Garry L. Rempel [1,†] and Flora T. T. Ng [1,*]

[1]   Department of Chemical Engineering, University of Waterloo, 200 University Ave. W., Waterloo, ON N2L3G1, Canada; y289liu@uwaterloo.ca

[2]   Research Institute of Applied Catalysis, School of Chemical and Environmental Engineering, Shanghai Institute of Technology, Shanghai 201418, China; guoxiaoming@sit.edu.cn

*   Correspondence: fttng@uwaterloo.ca; Tel.: +1-519-888-4567 (ext. 33979)

†   Deceased November 2, 2018, this paper is dedicated to the memory of Professor Garry L. Rempel.

Received: 26 February 2019; Accepted: 18 April 2019; Published: 1 May 2019

Abstract: Production of green chemicals using a biomass derived feedstock is of current interest. Among the processes, the hydrogenolysis of glycerol to 1,2-propanediol (1,2-PD) using externally supplied molecular hydrogen has been studied quite extensively. The utilization of methanol present in crude glycerol from biodiesel production can avoid the additional cost for molecular hydrogen storage and transportation, as well as reduce the safety risks due to the high hydrogen pressure operation. Recently the hydrogenolysis of glycerol with a Cu/ZnO/Al$_2$O$_3$ catalyst using in situ hydrogen generated from methanol steam reforming in a liquid phase reaction has been reported. This paper focusses on the effect of added Ni on the activity of a Cu/ZnO/Al$_2$O$_3$ catalyst prepared by an oxalate gel-co-precipitation method for the hydrogenolysis of glycerol using methanol as a hydrogen source. It is found that Ni reduces the conversion of glycerol but improves the selectivity to 1,2-PD, while a higher conversion of methanol is observed. The promoting effect of Ni on the selectivity to 1,2-PD is attributed to the slower dehydration of glycerol to acetol coupled with a higher availability of in situ hydrogen produced from methanol steam reforming and the higher hydrogenation activity of Ni towards the intermediate acetol to produce 1,2-PD.

Keywords: glycerol hydrogenolysis; in situ hydrogen; methanol steam reforming; Ni/Cu/ZnO/Al$_2$O$_3$ catalysts

1. Introduction

Fossil based fuels such as diesel, gasoline and jetfuel have been the most important energy resources affecting human life and modern society in the past century. However, fossil fuel is a non-renewable resource and energy demand has rapidly increased. Currently many researchers are working on alternative sources of renewable energy to reduce the dependence on fossil fuel, especially in view of the emission of greenhouse gases and climate change. Biodiesel has been used to supplement fossil diesel [1] and has already been commercialized in the world. Glycerol, the major by-product from the biodiesel production process, can be utilized to produce a number of value added chemicals such as 1,2-propanediol, 1,3-propanediol, acrolein, acrylic acid and some other special chemicals. Approximately 1 kg of glycerol can be formed for every 9 kg of biodiesel produced via a transesterification reaction using vegetable oil or animal fat as the feedstocks. Adding value to glycerol will not only lower the production cost of biodiesel but also avoid the chemical waste and environmental hazards caused by the large amount of surplus crude glycerol [2,3]. Among all of these applications, the production of 1,2-propanediol (1,2-PD), also known as propylene glycol, has been extensively researched because high 1,2-PD selectivity can be obtained under relatively mild reaction conditions compared with other routes of glycerol upgrading [4–7]. Xia et al. in 2012 reported a Cu/Zn/Mg/Al mixed oxide catalyst where 1,2-PD selectivity can reach 99.7% under very mild reaction conditions with molecular hydrogen [8]. It has been most frequently reported that 1,2-PD is produced via a glycerol dehydration to form acetol followed by a hydrogenation of acetol [5,6,9]. The conventional glycerol hydrogenolysis reaction is carried out in a batch reactor with heterogeneous catalysts under hydrogen pressure up to 10 MPa [10,11]. However, the high hydrogen pressure will incur a significant cost issue related to molecular hydrogen transportation and storage [12]. Additionally, the high pressure hydrogen in the storage tanks and reactors can cause safety problems due to the potential leak and explosion on contact with air.

To overcome the drawbacks of using high pressure molecular hydrogen, the process without adding external hydrogen has received a lot of interest from researchers. The most frequently reported approach is that the in situ hydrogen is produced via liquid phase glycerol steam reforming (known as APR—aqueous phase reforming) in a batch reactor and used for the glycerol hydrogenolysis process. D'Hondt et al. was the first group who reported this process in 2008 using a Pt/NaY catalyst to convert glycerol to 1,2-PD in the absence of added hydrogen [13]. Under inert atmosphere using 20 wt% aqueous glycerol at 230 °C, after 15 h reaction time, the glycerol conversion was reported to be 85.4% with 64% 1,2-PD selectivity. The advantage of this process is that glycerol itself is the raw material for both hydrogen and 1,2-PD production, no other hydrogen donor is needed avoiding the additional downstream separation steps. However, in order to get sufficient hydrogen from the glycerol reforming process at relatively low temperature (≤250 °C), a Pt based catalyst is usually used [14,15]. Pt is an expensive metal resulting in a high production cost for this process. Barbelli et al. in 2012 investigated a supported Pt catalyst with a lower Pt loading (1 wt%) as well as the promoting effect of Sn for the APR process [16]. The experimental results showed that using monometallic 1 wt% Pt on SiO_2 catalyst and 10 wt% aqueous glycerol feedstock, at 200 °C after 2 h reaction, the glycerol conversion is only 1%; when 0.4 wt% Sn was added, the glycerol conversion and 1,2-PD selectivity were improved to 49% and 63% respectively. Roy et al. in 2010 [17] and Pendem et al. in 2012 [18] also reported the hydrogenolysis of glycerol using various Pt based catalysts, the 1,2-PD selectivities were all quite low compared with the process using externally supplied hydrogen.

The other route of the glycerol hydrogenolysis process without external hydrogen added is to use a hydrogen donor such as iso-propanol. Musolino et al. reported a glycerol hydrogenolysis process under an inert atmosphere using 10 wt%Pd supported on Fe_2O_3 [19,20]. Using glycerol iso-propanol solution (12 wt% with respect to glycerol) as feedstock, at 180 °C for 24 h, the glycerol conversion and 1,2-PD selectivity were reported to be 100% and 94% respectively. Gandarias et al. in 2011 reported a process with iso-propanol as a hydrogen donor using a Ni-Cu/Al_2O_3 catalyst prepared by a so-gel method [21] which is relatively cheaper than a Pd based catalyst as mentioned previously. Using a 4 wt% glycerol solution as the feedstock and an equal molar of isopropanol with respect to glycerol as hydrogen donor, at 220 °C, after 24 h reaction time, the glycerol conversion was 41.2% and the selectivity of 1,2-PD was only 48.3%. Another active hydrogen donor for this process is formic acid, which can be obtained from non-food biomass sources. Gandarias et al. in 2012 developed a semi-batch system using a Ni-Cu/Al_2O_3 catalyst with the hydrogen donor being pumped into the reactor continuously [22]. Three different hydrogen donors were investigated, i.e., methanol, iso-propanol and formic acid. The glycerol conversion and 1,2-propanediol selectivity using formic acid were the highest among those three sources which were 33.5% and 85.9% respectively at 220 °C after 10 h reaction time. Recently, Gandarias et al. modified the catalyst and optimized the reaction conditions and developed a kinetic model for this process [23,24]. The optimum glycerol conversion and 1,2-PD selectivity were 55.2% and 84.6% respectively. The advantage of adding another hydrogen donor is that hydrogen can be produced under milder conditions compared with glycerol aqueous phase reforming resulting in a higher 1,2-PD selectivity. However, other downstream separation steps are needed to separate the impurities, such as acetone, unreacted formic acid and iso-propanol, causing a higher production cost.

We have been working on the upgrading of glycerol using in situ hydrogen generated from steam reforming of methanol [25,26]. Methanol has the highest H/C ratio (4:1) compared with formic acid or iso-propanol and is widely used for transesterification reaction for biodiesel production. Stoichiometrically, one mole of triglyceride requires three moles of methanol to produce three moles of methyl ester and one mole of glycerol. An excess amount of methanol is usually added to drive the transesterification reaction towards methyl ester, the ratio of methanol to triglyceride usually ranges from 6:1 to 12:1. In a conventional biodiesel production plant, the unreacted methanol is recovered before sending the crude biodiesel and crude glycerol mixture into a decanter for separation [27]. It has been studied that methanol can be more preferably dissolved in glycerol phase suggested by very small distribution coefficients of methanol in biodiesel to glycerol being less than 0.2 [28]. Therefore, if the crude glycerol is separated from the crude biodiesel before a methanol recovery process, a large amount of un-reacted methanol will be presented in the crude glycerol phase. The amount of methanol present in the crude glycerol before methanol recovery was estimated to range from 29% to 62% depending on the methanol to oil feed molar ratio. The real industrial data provided by Shandong Dingyu Bio-energy Co. Ltd. (Laiwu, China), which is one of the largest biodiesel manufacturing plants in China, using refined palm stearin oil as the feedstock meets a good agreement with our estimation (see the Supplementary Document A1). The excess methanol present in glycerol can be utilized to provide hydrogen in situ for glycerol hydrogenolysis, and the methanol stored in the biodiesel plant can always ensure that the desired methanol content in crude glycerol can be obtained. Biodiesel plants normally do not have H_2 plants on site, utilization of the methanol for hydrogen production on site could increase the overall economics for the production of 1,2-PD.

It is interesting to note that professor Lemonidou's group has recently published a few papers on glycerol hydrogenolysis to produce 1,2-PD using in situ hydrogen produced from methanol steam reforming [29–33]. Their initial work compared the activities of $Cu/ZnO/Al_2O_3$ and Pt/SiO_2 catalysts [29]. Under the conditions of 3.5 MPa N_2, 250 °C, 7.2 wt% methanol, 11.4 wt% glycerol, the glycerol conversion and 1,2-PD selectivity using a $Cu/ZnO/Al_2O_3$ catalyst prepared via oxalate gel-co-precipitation method (88.8% and 39.2% respectively) were significantly higher than that using a Pt/SiO_2 catalyst (58.9% and 36.2% respectively). By comparison, using a Cu based catalyst, the selectivities of propanol and ethylene glycol were significantly lower than those using a Pt/SiO_2 catalyst revealing that the $Cu/ZnO/Al_2O_3$ catalyst has less promotion effect on the C-C cleavage reaction and sequential 1,2-PD hydrogenolysis. It is generally accepted that the dehydration of glycerol to acetol is the rate determining step in the hydrogenolysis of glycerol to produce 1,2-PD. In order to achieve high selectivity to 1,2-PD, rapid hydrogenation of the acetol intermediate is required as acetol is known to be active to produce other un-desired products via reactions between acetol and alcohols [9,31,34,35]. Thus, a major challenge of the glycerol hydrogenolysis process without adding molecular hydrogen is to ensure a fast hydrogenation of acetol compared to the other side reactions caused by acetol to produce undesired by-products. Hence a catalyst with high activity for dehydration, hydrogenation and methanol steam reforming would be desirable for a high yield and selectivity to 1,2-PD. We have previously reported that the activity of a $Cu/ZnO/Al_2O_3$ catalyst to produce 1,2-PD in the hydrogenolysis of glycerol using molecular hydrogen is dependent on the catalyst preparation method. Among the three preparation methods, namely, alkaline co-precipitation, impregnation and oxalate gel-co-precipitation, the catalyst prepared by the gel-co-precipitation method is the most active and selective [9,36]. Furthermore, $Cu/ZnO/Al_2O_3$ prepared by the gel-co-precipitation method has also been reported to be more active for steam reforming of methanol [37,38]. The main focus of this paper is to investigate the effect of Ni on the activity of a $Cu/ZnO/Al_2O_3$ catalyst prepared via oxalate gel-coprecipitation for the hydrogenolysis of glycerol using the in situ hydrogen produced via methanol steam reforming as illustrated in Figure 1. Ni was chosen because Ni has been reported to be active for both glycerol hydrogenolysis and methanol steam reforming [39–41] and it is less expensive than precious metals such as Pt and less active for C-C bond cleavage [42,43]. The characterization and activity of $Cu/ZnO/Al_2O_3$ catalysts prepared by two different methods, namely, oxalate gel-co-precipitation and sodium carbonate co-precipitation were also reported as this provided a rational for adding Ni as a promoter to the $Cu/ZnO/Al_2O_3$ catalyst prepared by the oxalate gel-co-precipitation method.

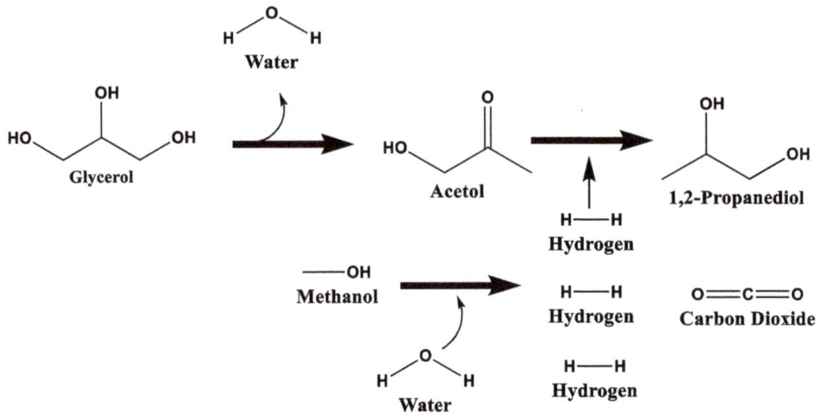

Figure 1. Reaction pathway of glycerol hydrogenolysis process using in situ hydrogen produced from methanol steam reforming.

2. Results and Discussion

2.1. Catalyst Characterization

2.1.1. Acidity of the Catalysts

The NH_3 temperature programmed desorption (TPD) technique was used to investigate the acidity of the catalysts. The $Cu/ZnO/Al_2O_3$ catalysts prepared via oxalate gel-co-precipitation and sodium carbonate co-precipitation referred to as $Cu/ZnO/Al_2O_3$-OA and $Cu/ZnO/Al_2O_3$-Na respectively. The NH_3 desorption profiles for the $Cu/ZnO/Al_2O_3$-OA and $Cu/ZnO/Al_2O_3$-Na have been presented in our previous work [9] and the desorption data are listed in Table 1. The NH_3 TPD data for $Cu/ZnO/Al_2O_3$-OA catalysts with added Ni are listed in Table 1 and Figure 2. The $Cu/ZnO/Al_2O_3$-Na catalyst has mainly acid sites with weak acidity ranging from 107 to 392 °C and very low total number of acidic sites, while $Cu/ZnO/Al_2O_3$-OA possesses both moderate and strong acidic sites corresponding to the desorption peaks ranging from 290 to 470 °C and from 590 to 800 °C respectively. It is noted that the strong acidic sites are a majority, which can facilitate the glycerol dehydration step. More acidic sites are being provided by $Cu/ZnO/Al_2O_3$-OA compared to $Cu/ZnO/Al_2O_3$-Na due to the smaller particle size, which will be investigated in the later section. Therefore, the oxalate gel-co-precipitation method can significantly enhance the number of acidic sites and the strength of acidity. Since dehydration requires acidic sites, the $Cu/ZnO/Al_2O_3$-OA catalyst was chosen to study the promoting effect of Ni on the hydrogenolysis of glycerol with methanol steam reforming. The effect of Ni loading on the acidity of a $Cu/ZnO/Al_2O_3$-OA catalyst is presented in Figure 2. With different amounts of Ni loading, three distinct peaks representing weak, moderate and strong acidic sites were observed for all the catalysts with three different Ni loadings (0%, 1% and 5%) indicating that the acidic strength of the catalysts was not changed as no new desorption peak was generated by adding Ni. As Ni loading increased, the peak of the strong acidic sites ranging from 590 to 800 °C with the maximum desorption peak at 683 °C shrunk, suggesting a smaller amount of strong acidic sites as shown in Table 1. When 1% Ni was loaded, the number of strong acidic sites slightly decreased from 0.075 to 0.072 $mmolNH_3$/g-cat; when the Ni loading was increased to 5%, the number of strong acidic sites decreased significantly to

0.030 mmolNH$_3$/g-cat. It is possible that when Ni was added, some strong acidic sites were blocked by Ni. This negative effect of Ni on the acidity of the catalyst behavior has been reported previously [44]. The weak and moderate acidic sites were not significantly affected by the Ni loading, even though the adsorption peak ranging from 290 to 470 °C without Ni loading was observed to be slightly lower than those with 1% and 5% Ni loading.

Table 1. Effect of Ni on the acidity of Cu/ZnO/Al$_2$O$_3$-OA catalysts.

Catalysts	Number of Acidic Sites	Total Acidic Sites
	mmolNH$_3$/g-cat	mmolNH$_3$/g-cat
Cu/ZnO/Al$_2$O$_3$-Na [1]	0.04 (107.0–392.0 °C)	0.040
Cu/ZnO/Al$_2$O$_3$-OA [1]	0.216 (50.0–290.0 °C)	0.347
	0.056 (290.0–470.0 °C)	
	0.075 (590.0–800.0 °C)	
1% (molar) Ni/Cu/ZnO/Al$_2$O$_3$-OA [2]	0.198 (50.0–290.0 °C)	0.342
	0.072 (290.0–470.0 °C)	
	0.072 (590.0–800.0 °C)	
5% (molar) Ni/Cu/ZnO/Al$_2$O$_3$-OA [3]	0.197 (50.0–590.0 °C)	0.298
	0.071 (290.0–470.0 °C)	
	0.030 (590.0–800.0 °C)	

[1] Cu/Zn/Al (molar) = 35/35/30; [2] Ni/Cu/Zn/Al (molar) = 1/34.5/34.5/30; [3] Ni/Cu/Zn/Al (molar) = 5.0/32.5/32.5/30.0.

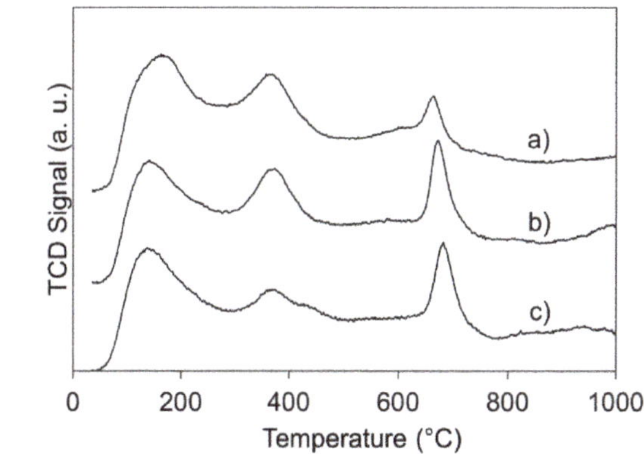

Figure 2. NH$_3$ temperature programmed desorption (TPD) profiles for Ni/Cu/ZnO/Al$_2$O$_3$-OA with different amounts of Ni loading (molar). (**a**) Ni/Cu/Zn/Al (molar) = 5/32.5/32.5/30, (**b**) Ni/Cu/Zn/Al (molar) = 1/34.5/34.5/30, (**c**) Cu/Zn/Al (molar) = 35/35/30. Conditions: 5%NH$_3$ balanced by Ar, 120 mg catalyst, temperature ramp 5 °C/min, flow rate 30 mL/min. All traces have been displaced for clarity.

2.1.2. Temperature Programmed Reduction

The reducibility of the catalysts was characterized by a temperature programmed reduction (TPR) technique. The TPR profiles for the calcined CuO/ZnO/Al$_2$O$_3$-OA and the CuO/ZnO/Al$_2$O$_3$-Na catalysts have been already reported [9]. All the profiles indicated that the reduction of the catalysts could be completed before 300 °C suggesting that reduction at 300 °C is sufficient to reduce the CuO/ZnO/Al$_2$O$_3$ catalysts. Figure 3 shows the TPR profiles of NiO/CuO/ZnO/Al$_2$O$_3$-OA, CuO/ZnO/Al$_2$O$_3$-OA and NiO only. The characteristic p values were calculated in the Supplementary Document A2 to verify that the experiments were carried out with absence of significant reducing agent concentration gradients along the catalyst bed. From the graph, it can be seen that both NiO/CuO/ZnO/Al$_2$O$_3$-OA and CuO/ZnO/Al$_2$O$_3$-OA catalysts show the reduction peaks between 180 and 330 °C with the peak maxima at around 250 °C. The reduction peak for NiO starts at 250 °C and ends at 370 °C with the peak maxima at 310 °C. No distinctive peak is observed in the NiO/CuO/ZnO/Al$_2$O$_3$-OA profile between

300 and 400 °C compared with the profile for $CuO/ZnO/Al_2O_3$-OA. This suggests that NiO and CuO are well mixed and both oxides can be effectively reduced at 300 °C [45]. Two shoulder peaks were observed for both $NiO/CuO/ZnO/Al_2O_3$-OA and $CuO/ZnO/Al_2O_3$-OA. The peak between 185 and 210 °C is due to the formation of bulk CuO and the broad shoulder peak between 270 and 300 °C is possibly due to the reduction of Cu^+ to Cu^0 [9]. A very broad peak after 300 °C was possibly due to the reduction of ZnO.

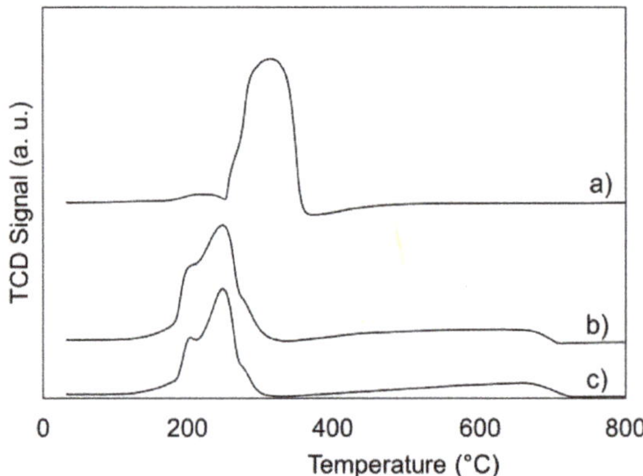

Figure 3. H_2 TPR profiles for $Ni/Cu/ZnO/Al_2O_3$-OA with different amounts of Ni loading (molar). (**a**) NiO, (**b**) Ni/Cu/Zn/Al (molar) = 5/32.5/32.5/30, (**c**) Cu/Zn/Al (molar) = 35/35/30. Conditions: 5%H_2 balanced by Ar, catalyst amount: (**a**) 20 mg, (**b**,**c**) 50 mg, temperature ramp 5 °C/min, flow rate 30 mL/min. *p* value 18–20 K. All traces have been displaced for clarity.

2.1.3. Transmission Electron Microscopy

The $CuO/ZnO/Al_2O_3$-OA and $CuO/ZnO/Al_2O_3$-Na catalysts were characterized by transmission electron microscopy (TEM) to study the effect of preparation method on the catalyst morphology as shown in Figure 4. From Figure 4a,b, it can be observed that for the catalyst prepared by oxalate gel-co-precipitation method, the particle shape is spherical and the particles are very uniformly distributed suggested by a smaller calculated standard deviation (SD). Figure 5 compared the histograms of the particle size distributions for the catalysts prepared by these two methods. The calculated mean particle size for the $Cu/ZnO/Al_2O_3$-OA catalyst is 10.41 nm with SD of 2.04. For the $Cu/ZnO/Al_2O_3$-Na catalyst, the particles are more elliptical and the average size of the major axis is 18.83 nm and SD is 4.05, which is much larger than the $Cu/ZnO/Al_2O_3$-OA catalyst and the size distribution is much wider as illustrated in Figure 5. This observation is in a good agreement with the previous literature reports using these preparation methods [9,46].

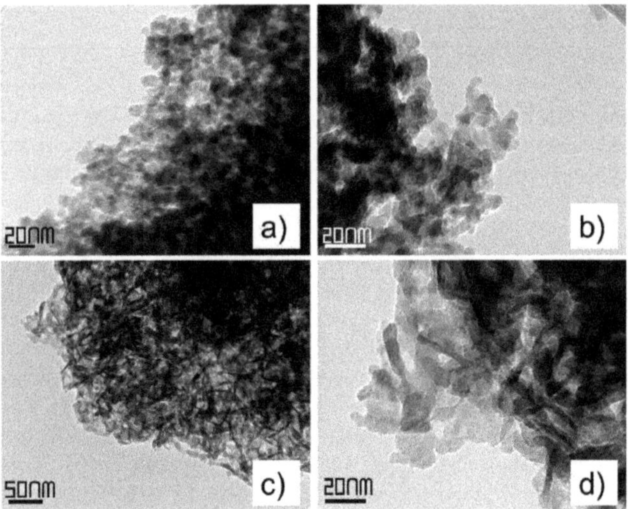

Figure 4. Transmission electron microscopy (TEM) images of: (**a,b**) CuO/ZnO/Al$_2$O$_3$ catalyst prepared via oxalate gel-co-precipitation; (**c,d**) CuO/ZnO/Al$_2$O$_3$ catalysts prepared via Na$_2$CO$_3$ co-precipitation. Cu/Zn/Al (molar) = 35/35/30.

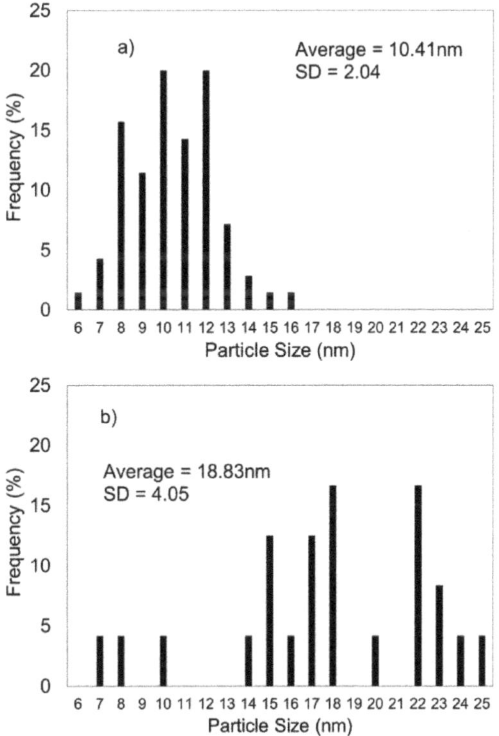

Figure 5. Histograms of the particle size distribution determined by TEM. (**a**) CuO/ZnO/Al$_2$O$_3$ catalyst prepared via oxalate gel-co-precipitation; (**b**) CuO/ZnO/Al$_2$O$_3$ catalysts prepared via Na$_2$CO$_3$ co-precipitation. Cu/Zn/Al (molar) = 35/35/30.

2.1.4. Thermal Gravimetric Analysis

In our previous work, the thermogravimetric analysis (TGA) results for the $Cu/ZnO/Al_2O_3$-OA and $Cu/ZnO/Al_2O_3$-Na catalyst have been presented [9]. The $Cu/ZnO/Al_2O_3$-OA and $Cu/ZnO/Al_2O_3$-Na were completely decomposed at 325 and 630 °C respectively. The oxalate gel-coprecipitation method favors a chemical homogenous phase of Cu and Zn, and the lower calcination temperature compared with Na_2CO_3 co-precipitation method can help to avoid the sintering of the particles during the calcination process. Figure 6 illustrates the TGA results for the oxalates of $Cu/ZnO/Al_2O_3$-OA and $Ni/Cu/ZnO/Al_2O_3$-OA. The temperature difference was used to demonstrate the heat flow during the thermal decomposition of metal oxalates process. For a $Cu/ZnO/Al_2O_3$-OA catalyst, the weight loss completes at 330 °C and a small weight loss peak is observed at around 210 °C indicating a thermal decomposition of a mixed metal oxalate with a higher Cu content as shown in Figure 6a, which is in agreement with the results we previously published [9]. When Ni was added, a similar trend was observed from Figure 6b that the weight loss started at 165 °C and completed at 330 °C with the highest rate of weight lost at 310 °C suggested by the temperature difference profile. A positive temperature difference reveals that the thermal decomposition of metal oxalate is an exothermic reaction. As the decomposition rate increased, the temperature difference was also increased and reached its maxima at 310 °C due to the maximum heat released associated with the metal oxalate decomposition. No separate peak was found for the decomposition of nickel oxalate suggesting that Ni was well mixed with other metals; this is also supported by the X-ray diffraction (XRD) data in the latter discussion. Therefore, the calcination temperature of 360 °C would be sufficient for the decomposition of all the catalyst precursors.

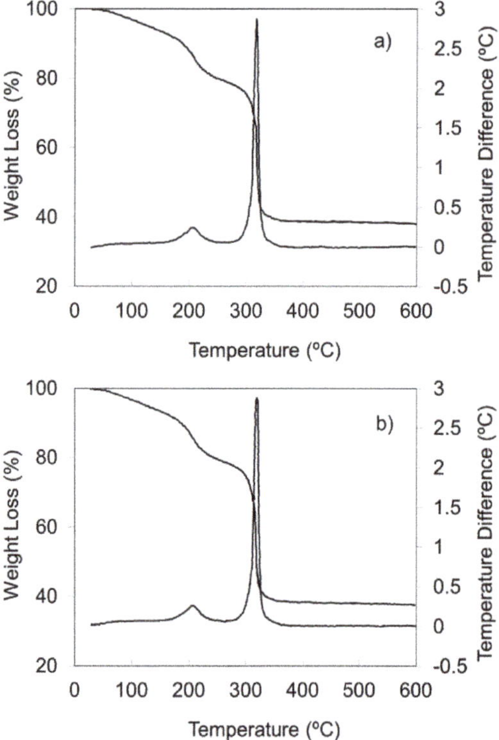

Figure 6. Thermogravimetric analysis (TGA) profiles for: (**a**) $Cu/ZnO/Al_2O_3$-OA oxalate, Cu/Zn/Al (molar) = 35/35/30; (**b**) $Ni/Cu/ZnO/Al_2O_3$-OA oxalate, Ni/Cu/Zn/Al (molar) = 5/32.5/32.5/30.

2.1.5. Diffuse Reflectance Infrared Fourier Transform

Figure 7 shows the Diffuse Reflectance Infrared Fourier Transform (DRIFT) spectra of adsorbed CO on the reduced catalysts at ambient temperature. At a CO equilibrium pressure of 50 Pa, two bands were observed for the sample of Cu/ZnO/Al$_2$O$_3$-OA catalyst. The band appears at 2106 cm^{-1}, which decreases in intensity with decreasing equilibrium CO pressure and disappears after evacuation, corresponding to the absorbance of Cu^0–CO species [47,48]. The band at 2025 cm^{-1} still remains even after evacuation at ambient temperature, and it is associated with the carbonyls linearly adsorbed on the Cu^0 atoms with lower coordination numbers [48]. It is worth noting that the band at 2025 cm^{-1} shifts to 2005 cm^{-1} after evacuation. This is related to the decrease in the coverage of CO on the Cu surface during evacuation, which leads to a decrease in the dipole–dipole coupling of the adsorbed CO molecules and further a band shift toward lower wavenumbers [49]. For the Ni-doped Cu/ZnO/Al$_2$O$_3$-OA catalysts, a new band at 2057 cm^{-1} was observed, which can be assigned to the stretching vibration of the linear-bonded CO on the reduced Ni. This band vanished after evacuation. In general, the linear-bonded CO on Ni appears in the region of 2020–2080 cm^{-1}, and the position of the band is a reflection of the crystallinity and dispersion of Ni. According to the literature [50,51], the band at 2057 cm^{-1} (>2050 cm^{-1}) implies that Ni exists in a high dispersion and a low crystallinity, which is in a good agreement with the results of XRD. Another band corresponding to bridged-bonded CO on Ni usually can be observed in the region of 1990–1940 cm^{-1} [50]. In this case, however, the band cannot be distinguished because it overlaps with the signal of the carbonyls linearly adsorbed on the Cu^0 atoms.

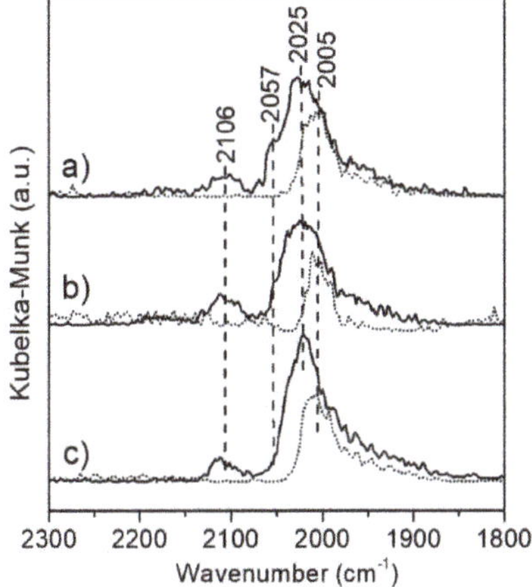

Figure 7. Diffuse Reflectance Infrared Fourier Transform (DRIFT) spectra of CO adsorbed on the reduced catalysts at ambient temperature. Solid: equilibrium pressure of 50 Pa CO; dash: after evacuation. Catalysts: Ni/Cu/ZnO/Al$_2$O$_3$-OA, (**a**) Ni/Cu/Zn/Al (molar) = 5/32.5/32.5/30, (**b**) Ni/Cu/Zn/Al (molar) = 1/34.5/34.5/30, (**c**) Cu/Zn/Al (molar) = 35/35/30. All traces have been displaced for clarity.

2.1.6. X-ray Diffraction and Other Physicochemical Properties

The crystalline phases for all the catalysts were investigated by XRD. The XRD pattern for CuO/ZnO/Al$_2$O$_3$-OA and CuO/ZnO/Al$_2$O$_3$-Na have been previously reported. It showed that for the catalyst prepared via the gel-co-precipitation method the particles were very well mixed and homogeneously distributed through the catalyst as suggested by much broader and low intensity peaks for CuO and ZnO [9,38,46]. Figure 8 illustrates the XRD patterns for the CuO/ZnO/Al$_2$O$_3$-OA catalysts with different Ni loadings. It can be observed that the crystal structures of the catalysts are not significantly affected by Ni loadings, the 2θ peaks located at 35.6° and 38.8° represent CuO and the 2θ peak located at 31.9° represents ZnO [6,9,38]; the particle sizes of CuO and ZnO calculated by Scherer's equation are also not significantly affected as shown in Table 2. The peak for NiO is not observed from the XRD patterns for the catalysts suggesting that NiO particles are in a low crystallinity and highly dispersed in the catalyst or the Ni loadings are so small that the peaks for NiO cannot be clearly observed which has been reported [52,53]. The copper surface area of the catalyst with different amounts of Ni loading are listed in Table 2, where addition of 1% Ni into the $Cu/ZnO/Al_2O_3$-OA catalyst causes a 3.7% reduction of Cu surface area, while a 5% Ni loading reduces the Cu surface area by 16.0%. This reduction of Cu surface area can significantly affect the activity of the catalyst which will be discussed later. Table 2 also lists the actual metal molar content measured by an Inductively coupled plasma (ICP) technique, the actual values of the metal content for the $Cu/ZnO/Al_2O_3$-OA catalysts with different Ni loading are not significantly changed compared with the feed composition during the preparation.

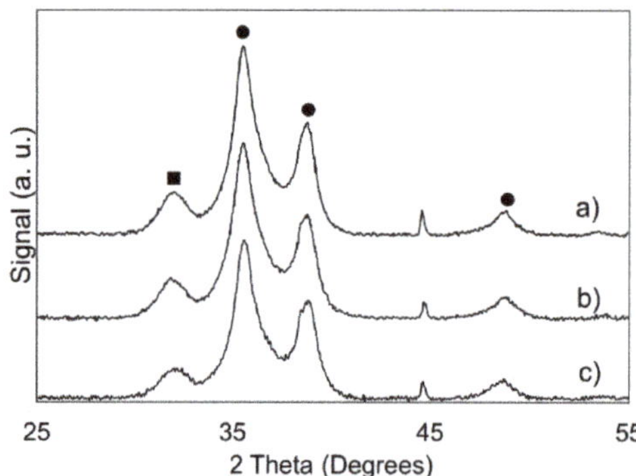

Figure 8. X-ray diffraction (XRD) patterns for $Ni/Cu/ZnO/Al_2O_3$-OA with different amounts of Ni loading (molar). (**a**) Ni/Cu/Zn/Al (molar) = 5/32.5/32.5/30, (**b**) Ni/Cu/Zn/Al (molar) = 1/34.5/34.5/30, (**c**) Cu/Zn/Al (molar) = 35/35/30. ● CuO, ■ ZnO. All traces have been displaced for clarity.

Table 2. Some physicochemical properties of the $Ni/Cu/ZnO/Al_2O_3$-OA.

	Crystal Size of CuO [1] nm	Crystal Size of ZnO [1] nm	Cu Surface Area m^2/g-cat	Metal Composition [2] (Theoretical Value)
0% Ni	14.7	10.2	18.8	37.1/36.4/26.5 (35.0/35.0/30.0)
1% Ni	14.2	11.6	18.1	0.9/36.2/35.6/27.3 (1.0/34.5/34.5/30.0)
5% Ni	15.0	10.9	15.8	4.7/33.1/34.4/27.8 (5.0/32.5/32.5/30.0)

[1] Calculated by Scherrer's equation, [2] Molar ratio of Ni/Cu/Zn/Al measured by ICP.

2.2. Glycerol Hydrogenolysis with In Situ Hydrogen Produced from Methanol Steam Reforming

2.2.1. Effect of Preparation Method for $Cu/ZnO/Al_2O_3$ Catalysts

The $Cu/ZnO/Al_2O_3$-OA catalyst has been reported to be more active than the $Cu/ZnO/Al_2O_3$-Na catalyst on both glycerol hydrogenolysis with external molecular hydrogen added [9] and methanol steam reforming [37,38]. The conversion of glycerol, selectivity of 1,2-PD and the yields of different products were calculated through the equations listed below, where "i" stands for each product/by-product formed from glycerol. The experimental results on glycerol hydrogenolysis with in situ hydrogen generated from methanol steam reforming are listed in Table 3. The space-time yields are provided from Tables S1–S4 in the Supplementary Material. It can be clearly observed in Table 3 that the glycerol conversion and 1,2-PD selectivity using a $Cu/ZnO/Al_2O_3$-OA catalyst are significantly higher than those using a $Cu/ZnO/Al_2O_3$-Na catalyst; the space-time yield of 1,2-PD using a $Cu/ZnO/Al_2O_3$-OA catalyst is about 10 times higher than that using a $Cu/ZnO/Al_2O_3$-Na catalyst as listed in Tables S1 and S2. When Cu/ZnO/Al$_2$O$_3$-Na was used, the selectivity of 1,2-PD was only 29.1% and the major by-products were acetol and some other higher molecular weight compounds. These higher molecular weight compounds were believed to be formed via condensation reactions of alcohols with acetol [9,35]. A significantly higher 1,2-PD selectivity obtained using Cu/ZnO/Al$_2$O$_3$-OA (70.7%) compared with the Cu/ZnO/Al$_2$O$_3$-Na catalyst (29.1%) attributed to its superior activity on methanol steam reforming to provide more hydrogen for acetol hydrogenation, as revealed in Table 3, since the acetol yield and other undesired by-products yields in the product mixture were significantly lower. The higher conversion obtained with the Cu/ZnO/Al$_2$O$_3$-OA catalyst is attributed to the higher copper surface area and higher number of acidic sites and higher acidic strength. Professor Lemonidou's group also reported that a Cu/ZnO/Al$_2$O$_3$-OA catalyst was more active than a Cu/ZnO/Al$_2$O$_3$-Na catalyst in the hydrogenolysis of glycerol using hydrogen derived from steam reforming of methanol [29]. In their optimization study, the space-time yield for 1,2-PD was reported to be 12.1 mmol/h.g-cat at 250 °C after 1 h reaction [30]; a very similar space-time yield for 1,2-PD was found using the Cu/ZnO/Al$_2$O$_3$-OA catalyst at 220 °C, as listed in Table S2, which was calculated to be 11.8 mmol/h.g-cat. The space-time yield of 1,2-PD was found to be slightly lower possibly due to a slightly lower reaction temperature. It was also reported that at 220 °C a maximum yield of 50.6% was obtained at 220 °C after 4 h reaction time; a similar 1,2-PD yield was also obtained, as listed in Table S2, that at the same reaction temperature, the yield of 1,2-PD at 4h reaction time was found to be 46.1%. Therefore, it is believed that the Cu/ZnO/Al$_2$O$_3$-OA catalyst is feasible for this reaction system. It is clear that the 1,2-PD selectivity strongly depends on the hydrogen supply, which is derived from methanol steam reforming in this reaction system. Higher ethylene glycol yield using a Cu/ZnO/Al$_2$O$_3$-OA catalyst (Table 3) is possibly due to the higher acidity of the catalyst and a higher hydrogen concentration generated by methanol steam reforming promoting the C-C cleavage to produce more ethylene glycol. The Cu:Zn:Al molar ratio of 35:35:30 was chosen for this work after a metal composition study, as listed in Table 3. The composition study for Cu/ZnO/Al$_2$O$_3$ catalysts have been extensively reported and the optimum molar ratio of Cu/Zn has been mostly reported to be 1 for both methanol steam reforming [37,38,54] and glycerol hdyrogenolysis [6,55,56]. In this work, the molar ratio for Cu/Zn of 1 was chosen and the molar content of aluminum was varied. The experimental results are shown in Table 3. A significant improvement for glycerol conversion and 1,2-PD selectivity was noticed when the aluminum molar content was increased from 10% to 30%. If the aluminum molar content was further increased from 30% to 50%, the glycerol conversion and 1,2-PD yield slightly dropped. Therefore, the catalyst with a Cu/Zn/Al molar ratio of 35/35/30 was the optimum ratio and was used for further study. The absence of external liquid to solid mass transfer limitation and intraparticle diffusion limitation were also verified via theoretical calculations provided in the Supplementary Document A3. It has been mathematically proven that the experiments were carried out in the reaction controlled regime.

$$Conversion_{Glycerol} = 100\% - \frac{n_{Glycerol}}{n_{Glycerol} + n_{1,2-PD} + n_{Acetol} + n_{EG} + n_{Propanol} + n_{Others}} \times 100\%,$$

$$Yield_i = \frac{n_i}{n_{Glycerol} + n_{1,2-PD} + n_{Acetol} + n_{EG} + n_{Propanol} + n_{Others}} \times 100\%,$$

$$Selectivity_i = \frac{Yield_i}{Conversion_{Glycerol}} \times 100\%.$$

Table 3. Products distribution for glycerol hydrogenolysis with in situ H_2 from methanol steam reforming over $Cu/ZnO/Al_2O_3$-OA and $Cu/ZnO/Al_2O_3$-Na catalysts [1].

Catalysts	Glycerol Conversion	1,2-PD Selectivity	1,2-PD Yield	Acetol Yield	EG [5] Yield	Propanol Yield	Others Yield
$Cu/ZnO/Al_2O_3$-Na [2]	60.3	29.1	17.5	16.1	0.0	0.0	26.7
$Cu/ZnO/Al_2O_3$-OA [2]	87.1	70.7	61.6	5.2	2.9	0.7	16.7
$Cu/ZnO/Al_2O_3$-OA [3]	80.2	65.7	52.7	9.5	2.6	0.5	14.9
$Cu/ZnO/Al_2O_3$-OA [4]	82.7	69.9	57.8	9.2	2.0	0.9	12.7

[1] Reaction Conditions: 220 °C, 1.5 MPa N_2, 100 g feedstock mixture, 20 wt% glycerol, 32.2 wt% of water and 47.8 wt% methanol (water/methanol molar ratio = 1.2), 3 g catalyst, 500 RPM, 8 h; [2] Cu/Zn/Al (molar) = 35/35/30; [3] Cu/Zn/Al (molar) = 45/45/10; [4] Cu/Zn/Al (molar) = 25/25/50; [5] EG: ethylene glycol.

2.2.2. Effect of Ni as a Promoter for the $Cu/ZnO/Al_2O_3$-OA Catalysts

Ni based catalysts are active in various hydrocarbon reforming [39,40] and glycerol hydrogenolysis processes [57,58] and Ni is less costly compared to some precious hydrogenation metals such as Pt, Pd and Ru. Since the $Cu/ZnO/Al_2O_3$-OA catalyst is more active and selective to 1,2-PD in the glycerol hydrogenolysis process than $Cu/ZnO/Al_2O_3$-Na, Ni was added to $Cu/ZnO/Al_2O_3$-OA to further improve the selectivity to 1,2-PD. The $Ni/Cu/ZnO/Al_2O_3$-OA catalysts with three different molar contents of Ni (1%, 3% and 5%) were used to investigate the promoting effect of Ni on the catalytic activity. The products distributions were listed in Table 4, the glycerol conversion, 1,2-PD selectivity, other by-products yields over the reaction time are illustrated in Figure 9. The space-time yields were provided from Tables S5–S7 in the Supplementary Material. From Table 4 and Figure 9a, it can be seen that as Ni content increased from 0% to 5%, the glycerol conversion dropped over the reaction time and the final glycerol conversion decreased from 87.1% to 70.0%. The lower glycerol conversion as more Ni added is attributed to the loss of Cu surface area and a reduction of acidity (Tables 1 and 2). The reaction rate of glycerol hydrogenolysis has been reported to be strongly dependent on the Cu surface area [59,60]. Sato et al. in 2008 reported a mechanism for glycerol dehydration to form acetol catalyzed by a Cu surface [61]. More experimental evidence will be shown later that Cu is the primary active site for glycerol dehydration in this reaction system. Since the addition of Ni reduces the number of strong acidic sites (Figure 2, Table 1), a slower reaction rate for the acid catalyzed glycerol dehydration is expected. Therefore, the glycerol conversion was lower with added Ni. Thus, the reaction rate of glycerol dehydration can be affected by two factors, namely the strong acidic sites and Cu surface area.

Table 4. Products distribution for glycerol hydrogenolysis with in situ H_2 from methanol steam reforming over $Cu/ZnO/Al_2O_3$-OA catalysts with different Ni loading [1].

Ni Molar Content	Glycerol Conversion	1,2-PD Selectivity	1,2-PD Yield	Acetol Yield	EG Yield	Propanol Yield	Others Yield
0% Ni [2]	87.1	70.7	61.6	5.2	2.9	0.7	16.7
1% Ni [3]	85.5	76.7	65.6	4.9	3.4	0.6	11.1
3% Ni [4]	77.5	82.8	64.2	3.2	3.6	0.5	5.9
5% Ni [5]	70.0	85.5	59.9	3.5	3.5	0.6	2.5
5% Ni [5,6]	80.8	81.2	64.8	2.5	6.8	1.1	5.6
5%Ni [5,7]	87.0	82.9	72.1	1.3	7.7	1.2	4.7

[1] Reaction Conditions: 220 °C, 1.5 MPa N_2, 100 g feedstock mixture, 20 wt% glycerol, 32.2 wt% of water and 47.8 wt% methanol (water/methanol molar ratio = 1.2), 3 g catalyst, 500 RPM, 8 h; [2] Cu/Zn/Al (molar) = 35/35/30; [3] Ni/Cu/Zn/Al (molar) = 1/34.5/34.5/30; [4] Ni/Cu/Zn/Al (molar) = 3/33.5/33.5/30; [5] Ni/Cu/Zn/Al (molar) = 5/32.5/32.5/30; [6] 6 g catalyst; [7] 9 g catalyst.

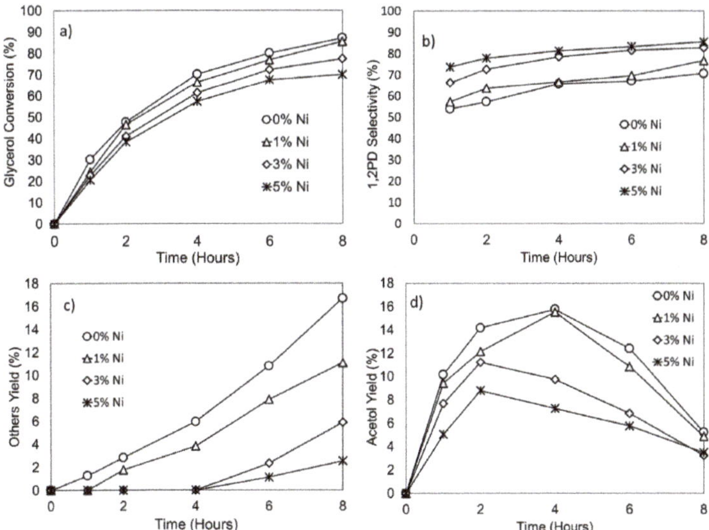

Figure 9. Effect of Ni molar contents on: (**a**) glycerol conversion; (**b**) 1,2-propanediol (1,2-PD) selectivity; (**c**) others yield; (**d**) acetol yield. Conditions: 220 °C, 1.5 MPa N_2, 100 g feedstock mixture, 20 wt% glycerol, 32.2 wt% of water and 47.8 wt% methanol (water/methanol molar ratio = 1.2), 3 g catalyst, 500 RPM. Catalysts: $Ni/Cu/ZnO/Al_2O_3$-OA with different Ni molar contents. 0% Ni: Cu/Zn/Al (molar) = 35/35/30; 1% Ni: Ni/Cu/Zn/Al (molar) = 1/34.5/34.5/30; 3% Ni: Ni/Cu/Zn/Al (molar) = 3/33.5/33.5/30; 5% Ni:Ni/Cu/Zn/Al (molar) = 5/32.5/32.5/30.

However, it is interesting to note that as Ni loading is increased from 0% to 5%, the 1,2-PD selectivity is increased from 70.7% to 85.5%. In Figure 9b, it can be seen that the 1,2-PD selectivity is always higher with a higher amount of Ni loaded over the reaction time. One reason that a higher 1,2-PD selectivity can be obtained as the Ni content is increased to 5% is because the acetol concentration in the reaction mixture is lower due to the lower dehydration rate suppressing the formation of other by-products caused by the condensation reactions between acetol and alcohols [31]. As illustrated in Figure 9d, for all types of catalysts, the acetol yields always increase at the early stage of the reaction and then decrease afterward. This is expected as acetol is the intermediate in the reaction system. As more Ni was added into $Cu/ZnO/Al_2O_3$-OA, the acetol yields over the reaction time were always lower, revealing a lower acetol concentration in the reaction mixture when more Ni was present in the catalyst. For the $Cu/ZnO/Al_2O_3$-OA catalyst and the catalyst with 1% Ni added, the acetol yields increase and reach their maximum at the fourth hour and then decrease thereafter. When the Ni content is further increased from 3% to 5%, the acetol concentrations reach the maximum on the second hour, which is earlier than that when lower Ni is present in the catalyst and then decreases. This trend reveals that at the beginning of the reaction, the major product is acetol when the hydrogen being produced from methanol steam reforming is insufficient to effectively hydrogenate the acetol formed via glycerol dehydration. When the Ni is added into a $Cu/ZnO/Al_2O_3$-OA catalyst, the loss of strong acidic sites and Cu surface area cause a slower glycerol dehydration rate resulting in a lower yield of acetol. This can cause slower rates for the side reactions with acetol to produce lower amounts of undesired by-products as shown in Figure 9c. Ethylene glycol yield at the end of the reaction was also increased from 2.9% to 3.4% when 1% Ni was added onto the catalyst. When the Ni loading was further increased from 1% to 5%, the ethylene glycol yield was not significantly changed. The increment of ethylene glycol due to the addition of Ni is possibly due to the promoting effect of the Ni on C-C cleavage. As shown in Figure 10, both Cu surface area and the number of strong acidic sites have positive correlations with the glycerol conversion; but they all have negative effects on 1,2-PD selectivity since the higher acetol formation rate by glycerol dehydration can cause a higher formation rate of other un-desired by-products when the hydrogen availability is not sufficient for acetol hydrogenation. Therefore, even a slower glycerol conversion rate was obtained using a Ni promoted catalyst, the space-time yield of 1,2-PD was not significantly changed as listed in Table S5, which was found to be 11.0 mmol/h.g-cat, this was very close to the value obtained using a Cu/ZnO/Al$_2$O$_3$-OA catalyst in the previous discussion and the reported literature value [30]. It is important to obtain a high 1,2-PD selectivity, since the un-reacted glycerol can be recycled back to the reactor and hence it can reduce the production cost. Otherwise, the high yield of undesired by-products will require separation of other by-products prior to the recycling of glycerol. Even though the glycerol conversion is lower at higher Ni loading, the yield of 1,2-PD is not significantly lowered due to the high 1,2-PD selectivity, as listed in Table 4.

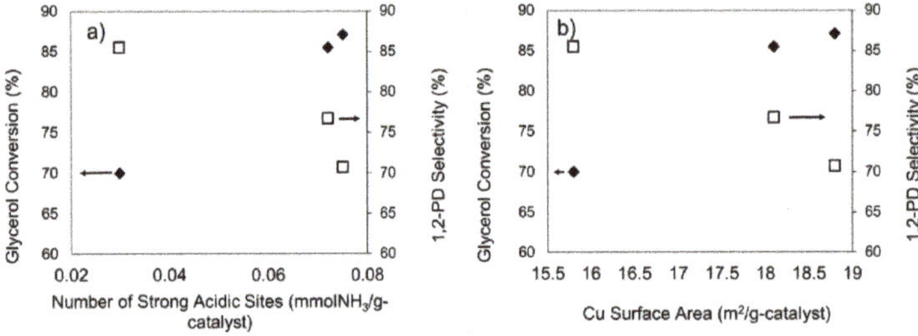

Figure 10. Effect of: (**a**) Number of strong acidic sites; (**b**) Cu surface area on glycerol conversion and 1,2-propanediol selectivity. Conditions: 220 °C, 1.5 MPa N$_2$, 100 g feedstock mixture, 20 wt% glycerol, 32.2 wt% of water and 47.8 wt% methanol (water/methanol molar ratio = 1.2), 3 g catalyst, 500 RPM, 8 h.

The effect of catalyst loading was also investigated by varying the catalyst loading amount from 3 to 9 wt% with respect to the total weight of feedstock mixture using a Ni/Cu/ZnO/Al$_2$O$_3$-OA catalyst with 5% Ni loading as listed in Table 4. When the catalyst loading was increased from 3 to 9 wt%, the glycerol conversion was increased from 70.0% to 87.0% with the 1,2-PD selectivity slightly reduced from 85.5% to 82.9%. The loss of 1,2-PD selectivity by increasing the catalyst loading is mainly due to the increments of ethylene glycol, propanol and other by-products yields. It is possibly because when the catalyst loading is increased, more active sites can be provided for C-C cleavage, the sequential hydrogenolysis of 1,2-PD and the condensation reactions between acetol and other alcohols resulting in higher yields of ethylene glycol, propanol and other by-products. In addition, when the catalyst loading is higher, more active sites are provided for methanol steam reforming generating more hydrogen, which can also favor the C-C cleavage and 1,2-PD hydrogenolysis to propanol.

As discussed previously, the 1,2-PD selectivity strongly depends on the acetol concentration in the reaction mixture. The un-desired by-products are mainly due to the side reactions of acetol present in the mixture when the hydrogen being produced from methanol steam reforming is not sufficient for the acetol hydrogenation. Therefore, the methanol steam reforming reaction is another key factor improving the 1,2-PD selectivity. The experiments were carried out without taking any sample during the reaction time to investigate the methanol conversion and mole balance of glycerol. In this case, the methanol conversion, glycerol conversion, 1,2-PD selectivity and yields of products were calculated using the equations listed below, where "i" stands for each product/byproduct formed from glycerol. Table 5 lists the effect of Ni loading on methanol conversion. To investigate the roles that Cu and Ni play in the reaction system, a Ni/ZnO/Al$_2$O$_3$-OA catalyst without Cu present was also used. The mole balances of glycerol were all close to 100% revealing that glycerol steam reforming did not occur to any significant extent. When Ni was added to Cu/ZnO/Al$_2$O$_3$-OA, the methanol conversion was higher. As more hydrogen was produced, acetol could be more effectively hydrogenated resulting in a higher 1,2-PD yield, a lower acetol yield and a lower other by-products yield. It gives a good agreement with the previous discussion that when Ni was added, the 1,2-PD selectivity was higher, the yield of acetol and the other un-desired by-products were lower (see Figure 9). Therefore, the promoting effect of

Ni on methanol steam reforming is the main reason why the addition of Ni can improve the 1,2-PD selectivity. It is also important to note that when Ni/ZnO/Al$_2$O$_3$-OA catalyst was used without Cu being present, essentially no 1,2-PD was formed, and glycerol conversion was only 4.8%. This suggests that Cu is the primarily active component in the catalyst for glycerol dehydration. Yfanti et al. [30] have also reported that when ZnO/Al$_2$O$_3$ catalyst was used for glycerol hydrogenolysis, no glycerol conversion was obtained indicating Cu is necessary for dehydration. Thus, it can be deduced that metallic Ni is inactive for glycerol dehydration and the small amount of glycerol conversion is possibly due to the acidic sites provided by alumina. It can also be known that ZnO is also inactive for glycerol dehydration from the results in Table 5 and the reported data [30]. However, ZnO was reported to play an important role on the catalytic activity as it can favor the formation of small active Cu sites by the Cu-ZnO interactions to improve the activity of Cu catalyst [6]. It is interestingly noted that when Ni/ZnO/Al$_2$O$_3$-OA was used, the methanol conversion was 23.3% which was higher than that using 5%Ni/Cu/ZnO/Al$_2$O$_3$-OA catalyst. It was reported that methanol steam reforming reaction did not occur when ZnO/Al$_2$O$_3$ was used [30]. Therefore, Ni can improve the catalyst activity for methanol steam reforming to produce more hydrogen for acetol hydrogenation, which is likely one of the main reasons that the addition of Ni to Cu/ZnO/Al$_2$O$_3$-OA catalyst can improve the selectivity of 1,2-PD in this process.

$$Conversion_{Glycerol} = 100\% - \frac{n_{Glycerol,in} - n_{Glycerol,f}}{n_{Glycerol,in}} \times 100\%,$$

$$Yield_i = \frac{n_i}{n_{Glycerol,\,in}} \times 100\%,$$

$$Selectivity_i = \frac{Yield_i}{Conversion_{Glycerol}} \times 100\%.$$

Table 5. Promoting effect of Ni on Cu/ZnO/Al$_2$O$_3$-OA activity for glycerol hydrogenolysis and methanol steam reforming [1].

Catalysts	Glycerol Conversion	1,2-PD Selectivity	1,2-PD Yield	Methanol Conversion	Mole Balance
Cu/ZnO/Al$_2$O$_3$-OA [2]	100.0	77.0	77.0	17.7	96.5
Ni/Cu/ZnO/Al$_2$O$_3$-OA [3]	97.4	86.0	83.8	21.6	100.4
Ni/ZnO/Al$_2$O$_3$-OA [4]	4.8	0.0	0.0	23.3	94.6

[1] Conditions: 220 °C, 1.5 MPa N$_2$, 100 g feedstock mixture, 20 wt% glycerol, 32.2 wt% of water and 47.8 wt% methanol (water/methanol molar ratio =1.2), 3 g catalyst, 500 RPM, 24 h. No sample was taken over the reaction time; [2] Cu/Zn/Al molar ratio = 35/35/30; [3] Ni/Cu/Zn/Al molar ratio = 5/32.5/32.5/30; [4] Ni/Zn/Al molar ratio= 10/60/30.

It has been discussed that the major undesired by-products are formed by the condensation reactions between acetol and alcohols; therefore, rapid hydrogenation of acetol to 1,2-PD plays a key role in order to obtain a high 1,2-PD selectivity. To study the promoting effect of Ni on acetol hydrogenation reaction, experiments using an aqueous acetol solution as the feedstock with a constant hydrogen pressure supplied into the reaction system over the reaction time was carried out using both Cu/ZnO/Al$_2$O$_3$-OA catalyst and Ni/Cu/ZnO/Al$_2$O$_3$-OA catalyst. Figure 11 and Table 6 illustrate the effect of Ni on the acetol hydrogenation reaction. A pseudo-first-order reaction kinetics was applied and the rate constant for acetol hydrogenation reaction was calculated using the following equation, where k' is the pseudo-first-order rate at a constant hydrogen pressure:

$$-\frac{d[acetol]}{dt} = k[acetol]P_{H_2} => \ln[acetol] = -k't + \ln[acetol_{t\,=\,0}].$$

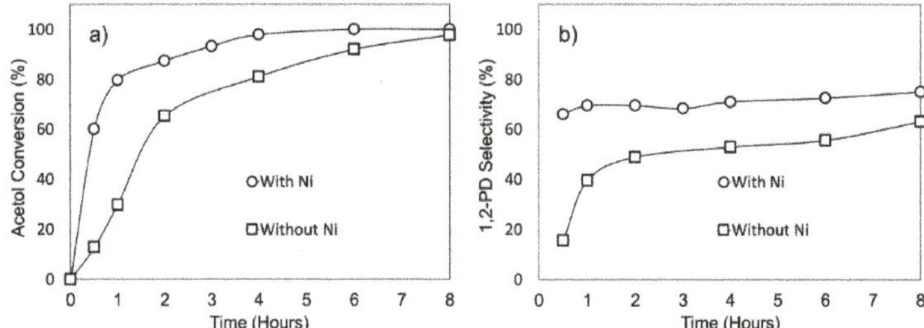

Figure 11. Promoting effect of Ni on Cu/ZnO/Al$_2$O$_3$-OA for acetol hydrogenation: (**a**) acetol conversion; (**b**) 1,2-PD selectivity. Conditions: 200 °C, 500 RPM, 100 g feedstock mixture, 20 wt% aqueous acetol, 1 g catalyst, H$_2$ pressure 2.8 PMa. With Ni: Ni/Cu/Zn/Al (molar) = 5/32.5/32.5/30, without Ni: Cu/Zn/Al = 35/35/30.

Table 6. Promoting effect of Ni on Cu/ZnO/Al$_2$O$_3$-OA for acetol hydrogenation [1].

Catalysts	Acetol Conversion	1,2-PD Selectivity	1,2-PD Yield	Others Yield	Rate Constant (s^{-1})
Cu/ZnO/Al$_2$O$_3$-OA [2]	97.7	63.1	61.6	36.1	1.231×10^{-4}
5% Ni/Cu/ZnO/Al$_2$O$_3$-OA [3]	100.0	75.1	75.1	24.9	2.721×10^{-4}

[1] Conditions: 200 °C, 500 RPM, 100 g feedstock mixture, 20 wt% aqueous acetol, 1 g catalyst, H$_2$ pressure 2.8 MPa, 8 h reaction time; [2] Cu/Zn/Al molar ratio = 35/35/30; [3] Ni/Cu/Zn/Al molar ratio = 5/32.5/32.5/30.

Using the catalyst with Ni added, the acetol hydrogenation reaction is significantly faster than that without Ni added (Figure 11a). In fact, the pseudo-first-order rate constant using the Ni catalyst is more than two times larger (Table 6). As the acetol hydrogenation reaction rate is increased, the acetol concentration in the reaction mixture is decreased resulting in slower side reactions caused by acetol to form other by-products. Therefore, the selectivity of 1,2-PD using the Ni/Cu/ZnO/Al$_2$O$_3$-OA catalyst is higher over the reaction time as illustrated in Figure 11b. To investigate the nature of the by-product formation, the chromatogram of the final product sample (8 h) was compared with the samples obtained for the hydrogenolysis process with methanol steam reforming and the hydrogenolysis process with insufficient external hydrogen added (i.e., 1.4 MPa) as depicted in Figure 12. It can be seen that the retention times of the by-products from the acetol hydrogenation are the same compared with the other two reactions. Hence the by-product formation is due to side reactions of acetol when insufficient hydrogen is provided. Therefore, the selectivity of 1,2-PD is believed to be strongly dependent on the rate of acetol hydrogenation. It is interesting to point out that no ethylene glycol is formed during acetol hydrogenation suggesting that ethylene glycol is formed due to C-C cleavage of glycerol rather than 1,2-PD or acetol. This result is in agreement with a recent mechanistic study of glycerol hydrogenolysis [31]. Since Ni improves the catalytic activity for acetol hydrogenation and also improves the methanol steam reforming to produce more hydrogen for in situ hydrogenation, the addition of Ni improves the selectivity to 1,2-PD in glycerol hydrogenolysis.

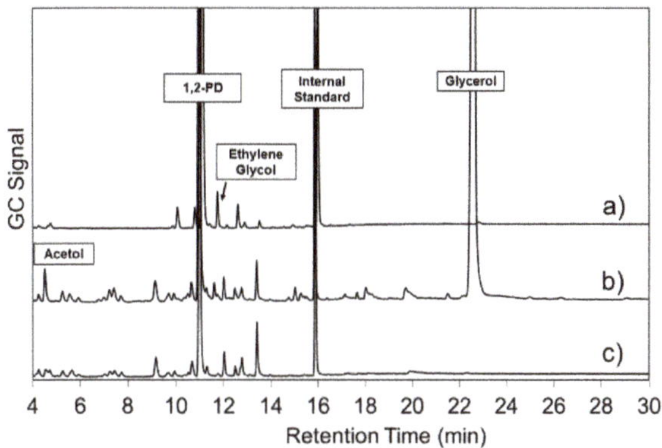

Figure 12. Chromatograms of the final sample of: (**a**) glycerol hydrogenolysis with methanol steam reforming, (**b**) glycerol hydrogenolysis with insufficient molecular hydrogen added, (**c**) acetol hydrogenation. All traces have been displaced for clarity.

3. Materials and Methods

3.1. Materials and Methods for Catalyst Preparation

Ethanol was purchased from Fisher Scientific Canada (Toronto, ON, Canada) (high-performance liquid chromatography grade). The other chemicals were purchased from Sigma Aldrich Co. Canada (Oakville, ON, Canada) and all the gases were purchased from Praxair Canada Inc. (Mississauga, ON, Canada). The procedures to prepare $Cu/ZnO/Al_2O_3$ catalysts via two different precipitation methods, i.e., oxalate gel-co-precipitation and Na_2CO_3 co-precipitation have been described in our previous paper [9] referred to as $Cu/ZnO/Al_2O_3$-OA and $Cu/ZnO/Al_2O_3$-Na respectively. To prepare the $Cu/ZnO/Al_2O_3$-Na catalyst, an aqueous mixture of metal nitrates solution was prepared with the designated metal molar ratio under vigorous stirring, the total metal concentration was 0.5 M. An aqueous solution of 0.5 M sodium carbonate was added drop-wise into the metal nitrate solution until the pH of the solution was equal to 9.0. The slurry was then filtered and washed by de-ionized water until the pH of the filtrate water became 7.0. Then the filtered cake was dried at 110 °C overnight and calcined under a stationary air environment at 450 °C for 4 h. $Ni/Cu/ZnO/Al_2O_3$ catalysts were also prepared via oxalate gel-co-precipitation method referred to as $Ni/Cu/ZnO/Al_2O_3$-OA. To prepare $Cu/ZnO/Al_2O_3$-OA and $Ni/Cu/ZnO/Al_2O_3$-OA catalysts, 20% excess ethanol solution of 0.5 M oxalic acid (anhydrous, ≥97.0%) was quickly injected into an ethanol solution mixture of $Ni(NO_3)_2.6H_2O$ (≥98.5), $Cu(NO_3)_2.2.5H_2O$ (≥98.0%), $Zn(NO_3)_2.6H_2O$ (≥98.0%) and $Al(NO_3)_3.9H_2O$ (≥98.0%) with designated metal molar ratio under vigorous stirring, the total metal concentration was 0.5 M. The precipitation mixture was then aged at room temperature under stirring for 2 h and filtered. The filtered cake was then dried in air at 110 °C for 24 h. The dried particles were powdered and screened via a sieve with 250 μm opening and then calcined in stationary air at 150 °C for 1 h, 200 °C for 1 h, 250 °C for 1 h, 300 °C for 1 h and 360 °C for 4 h [46].

3.2. Materials and Methods for Catalyst Characterization

NH_3 temperature program desorption (TPD), N_2O reactive frontal chromatography (RFC) and temperature program reduction (TPR) experiments were carried out using an Altamira AMI-200 instrument (Pittsburgh, PA, USA). For the NH_3 TPD experiments, approximately 120 mg of the catalyst was loaded into a U shaped quartz reactor for each test. The catalyst was firstly reduced under a flow of 5% H_2 balanced with argon at a volumetric flow rate of 30 mL/min at 300 °C for 2 h. After reduction, the catalyst was cooled down to 25 °C. Then, 20 pulses of 5% NH_3 balanced with argon were injected into the U tube to saturate all the acidic sites of the catalyst, then the catalyst was heated to 1000 °C at a heating rate of 10 °C/min. The NH_3 desorption profile was determined by analyzing the effluent gas through a thermo conductivity detector (TCD). To carry out the TPR experiments, the catalyst amount was calculated to meet the characteristic p value below 30 K (see Supplementary Material A2 for calculation of characteristic p value), then the catalyst was firstly heated to 200 °C and kept at 200 °C for 60 min under a flow rate of 30 mL/min argon stream to remove all the moisture and other species physically absorbed on the catalyst surface. Then, the catalyst was heated under a 30 mL/min flow stream of 5% H_2 balanced with argon at a heating rate of 5 °C/min until 800 °C. For the N_2O RFC experiments, a TPR step was firstly carried out to 300 °C at a heating rate of 5 °C/min to reduce all Cu^{2+} to Cu^0; then the catalyst was cooled down to 60 °C. The surface Cu^0 was then oxidized to Cu^{1+} under a flow of 1% N_2O balanced with argon gas mixture at a flow rate of 30 mL/min. Then, another TPR program was carried out to 300 °C at a heating rate of 5 °C/min to reduce the entire surface Cu^{1+} to Cu^0. The H_2 uptake was then calculated to estimate the copper surface area. Transmission electron microscopy (TEM) was carried out on a FEI Titan 80–300 TEM (Hillsboro, OR, USA) equipped with an aberration corrector for the imaging lens (CEOS). The XRD experiments were carried out on a Bruker D8 Focus model (Madison, WI, USA), with Cu kα radiation and a wavelength of 1.54 Å; the 2θ angle was 15°–55° with a ramp of 0.02° per minute. The TGA profiles were obtained on a TA Instrument, SDT Q600 (New Castle, DE, USA). Approximately 10–15 mg catalyst sample was used for each test, the experiments were conducted under a continuous air flow (40 mL/min), with a temperature range of 30–600 °C at a heating rate of 5 °C/min. The metal content of the catalysts was determined using a Teledyne Leeman Labs high dispersion ICP (Hudson, NH, USA). The DRIFT spectra were collected on a Varian 660 infrared spectrometer (Mulgrave, VIC, Australia) equipped with a DRIFT reaction cell and a liquid nitrogen cooled HgCdTe (MCT) detector. The catalyst was pre-reduced in situ at 300 °C for 1 h in 10% H_2/He, followed by a purge with He at the same temperature to remove the surface-adsorbed H_2. After cooling down to the ambient temperature, CO was admitted at a desired pressure using a vacuum/adsorption system. The spectra were collected after steady state was reached, and the spectral resolution is 4 cm^{-1} and the number of scans is 64.

3.3. Materials and Methods for Catalysts Activity Test

The catalytic reactions were carried out in a 300 mL Parr Instrument 4560 Series mini bench top reactor constructed of hastelloy (Moline, IL, USA). For the glycerol hydrogenolysis reaction with in situ hydrogen produced from methanol steam reforming, 100 g of feedstock mixture containing 20 wt% of glycerol, 32.2 wt% of water and 47.8 wt% of methanol (water/methanol molar ratio = 1.2) were placed into the reactor. Unless specified, 3 g of catalyst (3 wt% with respect to the weight of total reactant mixture) was pre-reduced by a hydrogen stream (ultra-high purity) at 300 °C for 3 h in a quartz tubular reactor enclosed by a furnace. Then the reduced catalysts were transferred into the reaction mixture rapidly. The reactor was firstly flushed with nitrogen three times and pressurized to a desired pressure at ambient temperature before being heated to the desired temperature. Liquid samples were taken during the reaction time via a sampling valve. For the acetol hydrogenation reaction, 100 g of aqueous acetol solution containing 20 wt% of acetol was placed into the reactor, 1 g of catalyst was reduced and added into the reaction mixture, gaseous hydrogen was used to pressurize the reaction and the hydrogen pressure was kept constant over the reaction time. An Agilent 6890N gas chromatograph integrated with a DB-WAX megabore capillary column (Wilminton, DE, USA, 30 m × 0.53 mm I.D. × 10 μm film thickness) and a flame ionization detector (FID) was used to analyze all the liquid samples. Approximately 120 mg of product sample was added into 1 mL of 1,4-butanediol n-butanol solution mixture with 1,4-butanediol concentration of 5 g/L.

4. Conclusions

In this study, the promoting effect of Ni on a $Cu/ZnO/Al_2O_3$-OA catalyst was investigated in a glycerol hydrogenolysis process to produce 1,2-PD using in situ hydrogen produced via methanol steam reforming. The utilization of methanol in the crude glycerol from a biodiesel production process as the hydrogen source will avoid the additional cost for transportation and storage of molecular hydrogen, and the safety risks related to the usage of high pressure hydrogen. The catalyst prepared via oxalate gel-coprecipitation method has a smaller particle size and higher acidity resulting in a higher glycerol conversion and 1,2-PD selectivity. The addition of Ni onto a $Cu/ZnO/Al_2O_3$-OA catalyst resulted in a lower glycerol conversion due to the loss of Cu surface area and strong acidic sites. However, the 1,2-PD selectivity was improved as Ni was added due to several reasons: first of all the addition of Ni can improve the catalytic activity on the methanol steam reforming reaction to produce more hydrogen for acetol hydrogenation reducing the undesired by-products formation by some acetol condensation reactions; secondly, the loss of Cu surface area and strong acidic sites caused by Ni addition can suppress the glycerol dehydration reaction resulting in a slower rate of acetol yield; the other reason is that Ni improves the catalytic activity for acetol hydrogenation to 1,2-PD.

Supplementary Materials: The following are available online at http://www.mdpi.com/2073-4344/9/5/412/s1, A1: Calculation of methanol content in crude glycerol, A2: Calculation of P value, A3: Verification for absence of mass transfer limitations, Table S1: Space Time Yields for glycerol hydrogenolysis with in situ H_2 from methanol steam reforming over $Cu/ZnO/Al_2O_3$-Na catalyst, Tables S2–S4: Space Time Yields for glycerol hydrogenolysis with in situ H_2 from methanol steam reforming over $Cu/ZnO/Al_2O_3$-OA catalysts, Tables S5–S7: Space Time Yields for glycerol hydrogenolysis with in situ H_2 from methanol steam reforming over $Ni/Cu/ZnO/Al_2O_3$-OA catalysts, Table S8: Space Time Yields for acetol hydrogenation over $Cu/ZnO/Al_2O_3$-OA catalyst, Table S9: Space Time Yields for acetol hydrogenation over $Ni/Cu/ZnO/Al_2O_3$-OA catalyst.

Author Contributions: Conceptualization, Y.L. and F.T.T.N.; methodology, Y.L.; validation, G.L.R. and F.T.T.N.; formal analysis, Y.L. and X.G.; writing—original draft preparation, Y.L.; writing—review and editing, G.L.R. and F.T.T.N.; supervision, G.L.R. and F.T.T.N.; project administration, G.L.R. and F.T.T.N.; funding acquisition, G.L.R. and F.T.T.N.

Acknowledgments: The financial support from Natural Science and Engineer Research Council of Canada (NSERC) Discovery Grant Program is gratefully acknowledged. X. Guo acknowledges the financial support from Shanghai Municipal Education Commission (International Visiting Scholar Program, No. 2014-56) for his research stay at University of Waterloo. We also acknowledge Shandong Dingyu Bio-energy Co. Ltd. for providing the industrial data.

Conflicts of Interest: The authors declare no conflict of interest.

References

1. Ma, F.; Hanna, M.A. Biodiesel production: A review. *Bioresour. Technol.* **1999**, *70*, 1–15. [CrossRef]
2. Pagliaro, M.; Rossi, M. *The Future of Glycerol*, 2nd ed.; The Royal Society of Chemistry: Cambridge, UK, 2010; pp. 1–170.
3. Corma, A.; Iborra, S.; Velty, A. Chemical routes for the transformation of biomass into chemicals. *Chem. Rev.* **2007**, *107*, 2411–2502. [CrossRef]
4. Chaminand, J.; Djakovitch, L.; Gallezot, P.; Marion, P.; Pinel, C.; Rosier, C. Glycerol hydrogenolysis on heterogeneous catalysts. *Green Chem.* **2004**, *6*, 359–361. [CrossRef]
5. Dasari, M.A.; Kiatsimkul, P.; Sutterlin, W.R.; Suppes, G.J. Low-pressure hydrogenolysis of glycerol to propylene glycol. *Appl. Catal. A* **2005**, *281*, 225–231. [CrossRef]
6. Wang, S.; Liu, H. Selective hydrogenolysis of glycerol to propylene glycol on Cu-ZnO catalysts. *Catal. Lett.* **2007**, *117*, 62–67. [CrossRef]
7. Xia, S.; Yuan, Z.; Wang, L.; Chen, P.; Hou, Z. Hydrogenolysis of glycerol on bimetallic Pd-Cu/solid-base catalysts prepared via layered double hydroxides precursors. *Appl. Catal. A* **2011**, *403*, 173–182. [CrossRef]
8. Xia, S.; Nie, R.; Lu, X.; Wang, L.; Chen, P.; Hou, Z. Hydrogenolysis of glycerol over Cu0.4/Zn5.6−xMgxAl2O8.6 catalysts: The role of basicity and hydrogen spillover. *J. Catal.* **2012**, *296*, 1–11. [CrossRef]
9. Liu, Y.; Pasupulety, N.; Gunda, K.; Rempel, G.L.; Ng, F.T.T. Glycerol hydrogenolysis to 1,2-propanediol by $Cu/ZnO/Al_2O_3$ catalysts. *Top. Catal.* **2014**, *57*, 1454–1462. [CrossRef]
10. Ma, L.; He, D.; Li, Z. Promoting effect of rhenium on catalytic performance of Ru catalysts in hydrogenolysis of glycerol to propanediol. *Catal. Commun.* **2008**, *9*, 2489–2495. [CrossRef]
11. Jiang, T.; Zhou, Y.; Liang, S.; Liu, H.; Han, B. Hydrogenolysis of glycerol catalyzed by Ru-Cu bimetallic catalysts supported on clay with the aid of ionic liquids. *Green Chem.* **2009**, *11*, 1000–1006. [CrossRef]

12. Graetz, J. New approaches to hydrogen storage. *Chem. Soc. Rev.* **2009**, *38*, 73–82. [CrossRef]

13. D'Hondt, E.; Van de Vyver, S.; Sels, B.F.; Jacobs, P.A. Catalytic glycerol conversion into 1,2-propanediol in absence of added hydrogen. *Chem. Commun.* **2008**, 6011–6012. [CrossRef]

14. Vaidya, P.D.; Rodrigues, A.E. Glycerol reforming for hydrogen production: A review. *Chem. Eng. Technol.* **2009**, *32*, 1463–1469. [CrossRef]

15. Wen, G.; Xu, Y.; Ma, H.; Xu, Z.; Tian, Z. Production of hydrogen by aqueous-phase reforming of glycerol. *Int. J. Hydrogen Energy* **2008**, *33*, 6657–6666. [CrossRef]

16. Barbelli, M.L.; Santori, G.F.; Nichio, N.N. Aqueous phase hydrogenolysis of glycerol to bio-propylene glycol over Pt–Sn catalysts. *Bioresour. Technol.* **2012**, *111*, 500–503. [CrossRef]

17. Roy, D.; Subramaniam, B.; Chaudhari, R.V. Aqueous phase hydrogenolysis of glycerol to 1,2-propanediol without external hydrogen addition. *Catal. Today* **2010**, *156*, 31–37. [CrossRef]

18. Pendem, C.; Gupta, P.; Chaudhary, N.; Singh, S.; Kumar, J.; Sasaki, T.; Datta, A.; Bal, R. Aqueous phase reforming of glycerol to 1,2-propanediol over Pt-nanoparticles supported on hydrotalcite in the absence of hydrogen. *Green Chem.* **2012**, *14*, 3107–3113. [CrossRef]

19. Musolino, M.G.; Scarpino, L.A.; Mauriello, F.; Pietropaolo, R. Selective transfer hydrogenolysis of glycerol promoted by palladium catalysts in absence of hydrogen. *Green Chem.* **2009**, *11*, 1511–1513. [CrossRef]

20. Musolino, M.G.; Scarpino, L.A.; Mauriello, F.; Pietropaolo, R. Glycerol hydrogenolysis promoted by supported palladium catalysts. *ChemSusChem* **2011**, *4*, 1143–1150. [CrossRef] [PubMed]

21. Gandarias, I.; Arias, P.L.; Requies, J.; El Doukkali, M.; Güemez, M.B. Liquid-phase glycerol hydrogenolysis to 1,2-propanediol under nitrogen pressure using 2-propanol as hydrogen source. *J. Catal.* **2011**, *282*, 237–247. [CrossRef]

22. Gandarias, I.; Arias, P.L.; Fernández, S.G.; Requies, J.; El Doukkali, M.; Güemez, M.B. Hydrogenolysis through catalytic transfer hydrogenation: Glycerol conversion to 1,2-propanediol. *Catal. Today* **2012**, *195*, 22–31. [CrossRef]

23. Gandarias, I.; Requies, J.; Arias, P.L.; Armbruster, U.; Martin, A. Liquid-phase glycerol hydrogenolysis by formic acid over Ni–Cu/Al$_2$O$_3$ catalysts. *J. Catal.* **2012**, *290*, 79–89. [CrossRef]

24. Gandarias, I.; Fernández, S.G.; El Doukkali, M.; Requies, J.; Arias, P.L. Physicochemical study of glycerol hydrogenolysis over a Ni-Cu/Al$_2$O$_3$ catalyst using formic acid as the hydrogen source. *Top. Catal.* **2013**, *56*, 995–1007. [CrossRef]

25. Liu, Y. Catalytic Glycerol Hydrogenolysis to Produce 1,2-Propanediol with Molecular Hydrogen and in situ Hydrogen Produced from Steam Reforming. Ph.D. Thesis, University of Waterloo, Waterloo, ON, Canada, 2014.

26. Liu, Y.; Ng, F.T.T.; Rempel, G.L. Glycerol hydrogenolysis to 1,2-propanediol with in situ hydrogen produced from methanol steam reforming. In *Advances in Chemistry of Energy & Fuels*, 250th ed.; American Chemical Society National Meeting & Exposition, Boston, MA, USA, 19 August 2015; American Chemical Society: Washington, DC, USA, 2015; ENFL356.

27. Gaurav, A.; Leite, M.L.; Ng, F.T.T.; Rempel, G.L. Transesterification of triglyceride to fatty acid alkyl esters (biodiesel): Comparison of utility requirements and capital costs between reaction separation and catalytic distillation configurations. *Energy Fuels* **2013**, *27*, 6847–6857. [CrossRef]

28. De Sousa Maia, A.C.; e Silva, I.S.; Stragevitch, L. Liquid-liquid equilibrium of methyl esters of fatty acid/methanol/glycerol and fatty acid ethyl esters/ethanol/glycerol: A case study for biodiesel application. *Int. J. Chem. Eng. Appl.* **2013**, *4*, 285–289. [CrossRef]

29. Vasiliadou, E.S.; Yfanti, V.-L.; Lemonidou, A.A. One-pot tandem processing of glycerol stream to 1,2-propanediol with methanol reforming as hydrogen donor reaction. *Appl. Catal. B* **2015**, *163*, 258–266. [CrossRef]

30. Yfanti, V.-L.; Vasiliadou, E.S.; Lemonidou, A.A. Glycerol hydro-deoxygenation aided by in situ H$_2$ generation via methanol aqueous phase reforming over a Cu-ZnO-Al$_2$O$_3$ catalyst. *Catal. Sci. Technol.* **2016**, *6*, 5415–5426. [CrossRef]

31. Yfanti, V.-L.; Lemonidou, A.A. Mechanistic study of liquid phase glycerol hydrodeoxygenation with in-situ generated hydrogen. *J. Catal.* **2018**, *368*, 98–111. [CrossRef]

32. Yfanti, V.-L.; Ipsakis, D.; Lemonidou, A.A. Kinetic study of liquid phase glycerol hydrodeoxygenation under inert conditions over a Cu-based catalyst. *React. Chem. Eng.* **2018**, *3*, 559–571. [CrossRef]

33. Yfanti, V.-L.; Ipsakis, D.; Lemonidou, A.A. Kinetic model of glycerol hydrodeoxygenation under inert conditions over copper catalyst. *Mater. Today Proc.* **2018**, *5*, 27482–27490. [CrossRef]

34. Chiu, C.; Tekeei, A.; Ronco, J.M.; Banks, M.; Suppes, G.J. Reducing byproduct formation during conversion of glycerol to propylene glycol. *Ind. Eng. Chem. Res.* **2008**, *47*, 6878–6884. [CrossRef]
35. Van Ryneveld, E.; Mahomed, A.S.; van Heerden, P.S.; Friedrich, H.B. Direct hydrogenolysis of highly concentrated glycerol solutions over supported Ru, Pd and Pt catalyst systems. *Catal. Lett.* **2011**, *141*, 958–967. [CrossRef]
36. Bienholz, A.; Schwab, F.; Claus, P. Hydrogenolysis of glycerol over a highly active CuO/ZnO catalyst prepared by an oxalate gel method: Influence of solvent and reaction temperature on catalyst deactivation. *Green Chem.* **2010**, *12*, 290–295. [CrossRef]
37. Shen, J.; Song, C. Influence of preparation method on performance of Cu/Zn-based catalysts for low-temperature steam reforming and oxidative steam reforming of methanol for H_2 production for fuel cells. *Catal. Today* **2002**, *77*, 89–98. [CrossRef]
38. Zhang, X.; Wang, L.; Yao, C.; Cao, Y.; Dai, W.; He, H.; Fan, K. A highly efficient $Cu/ZnO/Al_2O_3$ catalyst via gel-coprecipitation of oxalate precursors for low-temperature steam reforming of methanol. *Catal. Lett.* **2005**, *102*, 183–190. [CrossRef]
39. Iwasa, N.; Masuda, S.; Takezawa, N. Steam reforming of methanol over Ni, Co, Pd and Pt supported on ZnO. *React. Kinet. Catal. Lett.* **1995**, *55*, 349–353. [CrossRef]
40. Qi, C.; Amphlett, J.C.; Peppley, B.A. Methanol steam reforming over NiAl and Ni (M) Al layered double hydroxides (M = Au, Rh, Ir) derived catalysts. *Catal. Lett.* **2005**, *104*, 57–62. [CrossRef]
41. Jiang, T.; Huai, Q.; Geng, T.; Ying, W.; Xiao, T.; Cao, F. Catalytic performance of Pd–Ni bimetallic catalyst for glycerol hydrogenolysis. *Biomass Bioenergy* **2015**, *78*, 71–79. [CrossRef]
42. Banu, M.; Sivasanker, S.; Sankaranarayanan, T.M.; Venuvanalingam, P. Hydrogenolysis of sorbitol over Ni and Pt loaded on NaY. *Catal. Commun.* **2011**, *12*, 673–677. [CrossRef]
43. Pompeo, F.; Santori, G.F.; Nichio, N.N. Hydrogen production by glycerol steam reforming with Pt/SiO_2 and Ni/SiO_2 catalysts. *Catal. Today* **2011**, *172*, 183–188. [CrossRef]
44. Damyanova, S.; Spojakina, A.; Jiratova, K. Effect of mixed titania-alumina supports on the phase composition of $NiMo/TiO_2$-Al_2O_3 catalysts. *Appl. Catal. A* **1995**, *125*, 257–269. [CrossRef]
45. Niemantsverdriet, J.W. *Spectroscopy in Catalysis: An Introduction, Third, Completely Revised and Enlarged*, 3rd ed.; Wiley-VCH Verlag GmbH & Co. KGaA: Weinheim, Germany, 2007; pp. 13–21.
46. Sun, Q.; Zhang, Y.; Chen, H.; Deng, J.; Wu, D.; Chen, S. A novel process for the preparation of Cu/ZnO and $Cu/ZnO/Al_2O_3$ ultrafine catalyst: Structure, surface properties, and activity for methanol synthesis from CO_2+H_2. *J. Catal.* **1997**, *167*, 92–105. [CrossRef]
47. Wang, L.; Liu, Q.; Chen, M.; Liu, Y.; Cao, Y.; Fan, K.N. Structural evolution and catalytic properties of nanostructured Cu/ZrO_2 catalysts prepared by oxalate gel-coprecipitation technique. *J. Phys. Chem. C* **2007**, *111*, 16549–16557. [CrossRef]
48. Hadjiivanov, K.; Venkov, T.; Knözinger, H. FTIR spectroscopic study of CO adsorption on Cu/SiO_2: Formation of new types of copper carbonyls. *Catal. Lett.* **2001**, *75*, 55–59. [CrossRef]
49. Li, D.; Sakai, S.; Nakagawa, Y.; Tomishige, K. FTIR study of CO adsorption on Rh/MgO modified with Co, Ni, Fe, or CeO_2 for the catalytic partial oxidation of methane. *Phys. Chem. Chem. Phys.* **2012**, *14*, 9204–9213. [CrossRef]
50. Poncelet, G.; Centeno, M.A.; Molina, R. Characterization of reduced α-alumina-supported nickel catalysts by spectroscopic and chemisorption measurements. *Appl. Catal. A* **2005**, *288*, 232–242. [CrossRef]
51. Garland, C.W.; Lord, R.C.; Troiano, P.F. Infrared spectrum of carbon monoxide chemisorbed on evaporated nickel films. *J. Phys. Chem.* **1965**, *69*, 1195–1203. [CrossRef]
52. Xi, J.; Wang, Z.; Lu, G. Improvement of Cu/Zn-based catalysts by nickel additive in methanol decomposition. *Appl. Catal. A* **2002**, *225*, 77–86. [CrossRef]
53. Cheng, W. Reaction and XRD studies on Cu based methanol decomposition catalysts: Role of constituents and development of high-activity multicomponent catalysts. *Appl. Catal. A* **1995**, *130*, 13–30. [CrossRef]
54. Wu, H.; Chung, S. Kinetics of hydrogen production of methanol reformation using $Cu/ZnO/Al_2O_3$ catalyst. *J. Comb. Chem.* **2007**, *9*, 990–997. [CrossRef]
55. Meher, L.C.; Gopinath, R.; Naik, S.N.; Dalai, A.K. Catalytic hydrogenolysis of glycerol to propylene glycol over mixed oxides derived from a hydrotalcite-type precursor. *Ind. Eng. Chem. Res.* **2009**, *48*, 1840–1846. [CrossRef]

56. Wang, S.; Zhang, Y.; Liu, H. Selective hydrogenolysis of glycerol to propylene glycol on Cu–ZnO composite catalysts: Structural requirements and reaction mechanism. *Chem. Asian J.* **2010**, *5*, 1100–1111. [CrossRef]
57. Jiménez-Morales, I.; Vila, F.; Mariscal, R.; Jiménez-López, A. Hydrogenolysis of glycerol to obtain 1,2-propanediol on Ce-promoted Ni/SBA-15 catalysts. *Appl. Catal. B* **2012**, *117–118*, 253–259. [CrossRef]
58. Yu, W.; Zhao, J.; Ma, H.; Miao, H.; Song, Q.; Xu, J. Aqueous hydrogenolysis of glycerol over Ni–Ce/AC catalyst: Promoting effect of Ce on catalytic performance. *Appl. Catal. A* **2010**, *383*, 73–78. [CrossRef]
59. Vasiliadou, E.S.; Lemonidou, A.A. Investigating the performance and deactivation behaviour of silica-supported copper catalysts in glycerol hydrogenolysis. *Appl. Catal. A* **2011**, *396*, 177–185. [CrossRef]
60. Vasiliadou, E.S.; Eggenhuisen, T.M.; Munnik, P.; de Jongh, P.E.; de Jong, K.P.; Lemonidou, A.A. Synthesis and performance of highly dispersed Cu/SiO_2 catalysts for the hydrogenolysis of glycerol. *Appl. Catal. B* **2014**, *145*, 108–119. [CrossRef]
61. Sato, S.; Akiyama, M.; Takahashi, R.; Hara, T.; Inui, K.; Yokota, M. Vapor-phase reaction of polyols over copper catalysts. *Appl. Catal. A* **2008**, *347*, 186–191. [CrossRef]

 *catalysts*

Article

Influence of Chemical Surface Characteristics of Ammonium-Modified Chilean Zeolite on Oak Catalytic Pyrolysis

Serguei Alejandro-Martín [1,2,*], Adán Montecinos Acaricia [2], Cristian Cerda-Barrera [3] and Hatier Díaz Pérez [3]

1 Departamento de Ingeniería en Maderas, Facultad de Ingeniería, Universidad del Bío-Bío (UBB), Concepción 4030000, Chile
2 Nanomaterials and Catalysts for Sustainable Processes (NanoCatpPS), Universidad del Bío-Bío (UBB), Concepción 4030000, Chile; amontecinos@ubiobio.cl
3 Departamento de Procesos Industriales, Facultad de Ingeniería, Universidad Católica de Temuco, Temuco 4780000, Chile; ccerda@uct.cl (C.C.-B.); hatierdiazp@gmail.com (H.D.P.)
* Correspondence: salejandro@ubiobio.cl; Tel.: +56-41-311-1168

Received: 10 March 2019; Accepted: 27 April 2019; Published: 21 May 2019

Abstract: The influence of chemical surface characteristics of Chilean natural and modified zeolites on Chilean Oak catalytic pyrolysis was investigated in this study. Chilean zeolite samples were characterised by nitrogen absorption at 77 K, X-ray powder diffraction (XRD), and X-ray fluorescence (XRF). The nature and strength of zeolite acid sites were studied by diffuse reflectance infrared Fourier transform (DRIFT), using pyridine as a probe molecule. Experimental pyrolysis was conducted in a quartz cylindrical reactor and bio-oils were obtained by condensation of vapours in a closed container. Chemical species in bio-oil samples were identified by a gas chromatography/mass spectrophotometry (GC/MS) analytical procedure. Results indicate that after the ionic exchange treatment, an increase of the Brønsted acid site density and strength was observed in ammonium-modified zeolites. Brønsted acids sites were associated with an increment of the composition of ketones, aldehydes, and hydrocarbons and to a decrease in the composition of the following families (esters; ethers; and acids) in obtained bio-oil samples. The Brønsted acid sites on ammonium-modified zeolite samples are responsible for the upgraded bio-oil and value-added chemicals, obtained in this research. Bio-oil chemical composition was modified when the pyrolysis-derived compounds were upgraded over a 2NHZ zeolite sample, leading to a lower quantity of oxygenated compounds and a higher composition of value-added chemicals.

Keywords: chilean natural zeolite; Brønsted acid sites; bio-oil upgrade; catalytic pyrolysis

---

1. Introduction

Nowadays, energy is a crucial issue in the global economy; energy drives industrialisation and agricultural activities. As the pillar of growth and development, energy is mainly obtained from fossil fuel, the primary energy source used worldwide. However, there are some issues which are actively debated nowadays: world sustainability; global warming; democratisation of the energy sources; access to cheap and reliable energy; and the physical or economic scarcity of petroleum as the prime energy source [1]. Traditionally, the incremental energy demand has been satisfied using fossil fuels, but in the coming years that energy will be substituted by low carbon and ultimately zero-carbon [2]. Usage of fossil fuel energy release more than 30 billion tons of CO_2 each year, exceeding the level that can be recycled by nature, contributing to global warming [3]. Thus, several countries are launching strategies to develop new alternative energies (Aeolic, solar, hydroelectric, biomass) to decrease current polluting matrices, through scientific and technological development and human resources training programs [4].

Among the alternative energies, biomass contributes to about 12% of the world primary energy supply, and it is recognised as the leading renewable substitute of fossil oil [5]. Biomass could be used as a sustainable source of bioenergy; bio-based products and add-value biochemicals, through thermochemical, biological, chemical conversion, and physical processes [6]. Currently, the thermochemical conversion of biomass is one of the most promising processes to obtain sustainable fuels to solve environmental problems caused by the over-consumption of fossil fuels [7]. Derived products (biochar, bio-oil, and fuel gas) from biomass pyrolysis offers an alternative way to obtain renewable fuel resources [5]. Nevertheless, upgrading processes are necessary as a consequence of bio-oil low heating value, high viscosity, water content, instability, and corrosiveness [8]. Among those processes, zeolite cracking becomes a remarkable way to reduce the oxygen content of bio-oil samples, through decarbonylation, decarboxylation, dehydration, oligomerization, isomerization, and dehydrogenation reactions [9]. Catalytic cracking of biomass pyrolysis vapours and bio-oil promotes oxygen remotion as CO, CO_2, and H_2O, increasing the possibilities of bio-oil usage as a fuel source [10]. Synthetic zeolites (A, Y, H-Y, ZSM-5, and H-ZSM-5) are preferred catalyst in upgrading processes these days. They are claimed as responsible for the increase of desirable chemical compounds (phenols, furans, aromatics, and hydrocarbons) after bio-oil upgrading processes [11,12]. However, synthetic zeolites high cost (e.g., ZSM-5 -> \$120/Kg, Y -> \$180/Kg, MCM-41 -> \$1190/Kg, β-zeolite -> \$1065/Kg) constitutes a non-minor disadvantage [13].

In contrast, natural zeolites are cheap and plentiful, showing prices in the range of \$20 to \$100 per metric ton (in the US). Worldwide, natural clinoptilolite showed an average value of \$242 per metric ton and even 40% lower for natural zeolites from China [1]. In addition, United States resources of natural zeolites have been estimated by 120 million tons of clinoptilolite, chabazite, erionite, mordenite, and phillipsite in near-surface deposits [14]. Although some zeolites are found in nature, they have limited application in catalytic pyrolysis due to their lower performance compared to synthetic ones. Nevertheless, the activity of natural zeolites can be significantly improved through appropriate modifications [15]. Thus, researchers have conducted some investigations incorporating natural zeolites as the catalyst in biomass catalytic pyrolysis studies [16–18]. An increase of bio-oil yield was linked to the usage of natural zeolite (clinoptilolite) as a catalyst in *Euphorbia rigida* pyrolysis [19]. Ammonium chloride modified zeolite (clinoptilolite) was used previously in peanut shells pyrolysis. Ionic exchange procedure enhances Brønsted acid sites (bridging hydroxyls connected to framework aluminium atoms) on the zeolites' internal structure, leading to deoxygenation of the obtained bio-oil samples [9]. Metal oxides nanoparticles (NiO, ZnO, Cu_2O, CaO, and MgO) were incorporated into natural clinoptilolite elsewhere, to study catalytic activity of zeolites samples in the pyrolysis of hardwood lignin. Incorporated nanoparticles on a zeolite framework act as Brønsted and Lewis acid sites, leading to an increase of phenol yield in bio-oil samples [20,21].

The studies above have been conducted to enhance natural zeolites catalytic activity looking to increase bio-oil yield, identify bio-oil derivatives that contribute to bio-oil stability or to obtain bio-based raw materials for chemical industries [22,23]. Nevertheless, the catalytic activity of Chilean natural and modified zeolite on Chilean Oak pyrolysis and their influence on the chemical composition has not been sufficiently investigated yet. Consequently, this research is focused on the analysis of the influence of chemical surface characteristics of Chilean natural zeolite on the composition of bio-oil obtained from the pyrolysis of Chilean Native Oak.

2. Results and Discussion

2.1. Biomass and Natural Zeolite Characterization

Lignocellulosic biomass is a complex composite material where the main constituents are cellulose, hemicellulose, lignin, water, extractives, and inorganic (ash) compounds. Linkage of components above provides structural strength and flexibility to lignocellulosic species [24]. As the percentage of main components varies from one species to other, Chilean oak was characterised according to detailed procedures (see Materials and Methods). The results of proximate (dry basis), ultimate, and elemental analyses are summarised in Table 1.

Table 1. Chilean Oak proximate (dry basis), ultimate, and elemental analyses.

VM [wt%]	FC [wt%]	Ash [wt%]	GCV [a] [MJ·kg^{-1}]	GCV [b] [MJ·kg^{-1}]	Cellulose [%]	Hemicellulose [%] [c]	Lignin [%]	Extractives [%]	C	H	O	N
85.74	12.62	1.64	20.72	24.93	35.38	35.55	27.10	1.97	47.30	6.36	46.34	-

Volatile Matter (VM), Fixed Carbon (FC) Gross calorific value (GCV). [a] Dried at 378 K, 12 h. [b] Bio-char from pyrolysis at 623 K, 30 min. [c] by difference.

Taking into consideration that the main structural chemical components (carbohydrate polymers and oligomers) constitute about 65 to 75%, lignin 18 to 35%, and organic extractives and inorganic minerals usually 4 to 10%, the summarised results of Table 1 are in accordance with reported articles elsewhere [25].

Thermogravimetric analysis of biomass was conducted, in a nitrogen atmosphere, to study the thermal behavior of Chilean Oak. TG/DTG/DTA (thermogravimetric/derivative thermogravimetric/differential thermogravimetric analysis) profiles show an active pyrolysis (hemicellulose and cellulose degradation) zone around 476 to 668 K, and passive pyrolysis (lignin degradation) zone at T > 668 K. Maximum degradation of hemicellulose occurs at 576 K and cellulose maximum degradation rate (0.01 mg·s^{-1}) was achieved at 628 K, removing 59% of the initial mass at this temperature. Thermogravimetric profiles of Chilean Oak were reported in a previous article [18]. Considering the above, pyrolysis experiments were conducted at 723 K to assure the complete thermal transformation of the biomass.

Natural zeolite was characterized by nitrogen absorption at 77 K, X-ray powder diffraction (XRD) and X-ray fluorescence (XRF), as mentioned in the Materials and Methods section. Table 2 summarises physical–chemical characterization results of natural and modified zeolite samples.

Table 2. Physico-chemical characterization of natural and modified Chilean zeolite samples.

Sample	S [a] [m^2·g^{-1}]	SiO$_2$ [b] [% w/w]	Al$_2$O$_3$ [b] [% w/w]	Na$_2$O [b] [% w/w]	CaO [b] [% w/w]	K$_2$O [b] [% w/w]	MgO [b] [% w/w]	TiO$_2$ [b] [% w/w]	Fe$_2$O$_3$ [b] [% w/w]	MnO [b] [% w/w]	P$_2$O$_5$ [b] [% w/w]	CuO [b] [% w/w]	Si/Al
NZ	168.17	71.61	15.18	2.0	3.43	2.03	0.74	0.61	3.99	0.06	0.12	0.03	4.72
NHZ	228.21	74.39	15.42	0.83	2.16	1.99	0.61	0.57	3.57	0.04	0.08	-	4.82
2NHZ	219.14	75.27	15.59	0.46	1.67	1.92	0.59	0.60	3.53	0.05	0.08	-	4.83

[a] Surface area in [m^2·g^{-1}]; [b] by XRF (X-ray fluorescence) [% w/w].

Surface areas were obtained by fitting the Langmuir adsorption model to nitrogen adsorption registered data. Zeolite samples were outgassed at 623 K for 12 h before all adsorption measurements. Nitrogen adsorption isotherms exposed a combination of type I and IV characteristic isotherms, according to the International Union of Pure and Applied Chemistry (IUPAC) classification [26]. A rapid filling of micropores was registered at low relative pressure (P/P$_0$) values and an increase of adsorbed nitrogen at (P/P$_0 \approx 0.9$), possibly due to mesopores. Registered data are in agreement with other results reported elsewhere [27,28].

As Table 2 shows, an increase of surface area was registered for ammonium-exchanged samples, and it could be associated with the decrease of compensation cations (Ca, Na, K, and Mg) composition in the zeolite framework. Lower compensation cations compositions improved nitrogen diffusion on the zeolite pore structure [29]. XRD patterns of natural and ammonium-modified zeolite samples indicated a high degree of crystallinity. Characteristic peaks of clinoptilolite, mordenite, and quartz structures were observed in X-ray patterns, as reported previously [29]. Comparison of registered patterns corroborates that applied ion-exchange procedure did not bring any significant changes in the zeolite framework, as XRD patterns are almost identical. Mordenite main channels (perpendicular 12 MR pores, 6.7 × 7.0 Å; parallel 8 MR pores, 2.9 × 5.7 Å; and side pockets, 2.9 Å diameter) are responsible for diffusivity and accessibility of target compounds to surface acid sites. Characterization results were similar to other Chilean natural zeolites reported previously [30].

2.2. Characterization of Acid Sites by Pyridine-DRIFT

Figure 1 shows collected diffuse reflectance infrared Fourier transform (DRIFT) spectra of pyridine-saturated natural and modified zeolites after a thermal desorption procedure to evaluate Brønsted and Lewis acid site strength. Pyridine-saturated samples were outgassed (at vacuum) before infra-red experiments, increasing the temperature to 573 K from room temperature (RT). Each spectrum was recorded at 293 K after the cooling of zeolite samples. DRIFT technique allowed the identification of Brønsted (Py-B) and Lewis (Py-L) acid sites in zeolite samples, following the interaction of the pyridine molecule with Brønsted acid sites (near 1540 cm^{-1}) and Lewis acid sites (near 1450 cm^{-1}) [31–34]. The registered bands at 1456 cm^{-1} were associated with adsorbed pyridine at the Lewis acid sites. Likewise, registered bands at 1539 cm^{-1} were linked to Brønsted acid sites, whereas the band at 1488 cm^{-1} could be attributed to both acid sites [35,36].

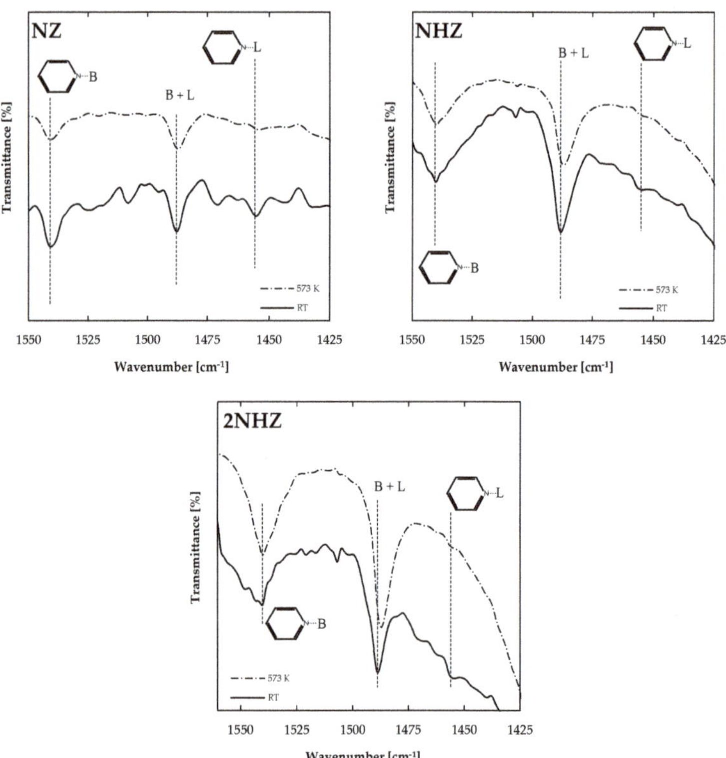

Figure 1. Diffuse reflectance infrared Fourier transform (DRIFT) spectrum of adsorbed pyridine on natural and modified zeolite.

To quantify acid sites, the ratio of Brønsted (B) and Lewis (L) acid sites (AS) was calculated using the following Equation (1):

$$\frac{C_{BAS}}{C_{LAS}} = \left(\frac{1.73}{1.23}\right)\left(\frac{I_{BAS}}{I_{LAs}}\right) \tag{1}$$

where I_{BAS} and I_{LAS} represent the intensity of absorption bands at 1540 and 1456 cm^{-1}, and 1.73 and 1.23 are the extinction coefficients, as reported previously [37,38]. The obtained C_{BAS}/C_{LAS} ratios are summarised in Table 3.

Table 3. Acid sites ratio of natural and modified zeolite samples.

Temperature [K]	C_{BAS}/C_{LAS} Ratio		
	NZ	**NHZ**	**2NHZ**
293	0.97	1.71	2.74
573	1.34	1.36	1.42

The registered data in Table 3 indicates that higher BAS/LAS ratio was observed in ammonium-modified zeolite samples, corroborating that ionic exchange incorporates new Brønsted acid sites at the zeolite framework. Moreover, as the heating temperature increases from 298 K to 573 K, weak acid sites disappear, and only strong acid sites still retain the pyridine probe molecule. Thus, registered peaks after the 573 K outgassing procedure, reveal a higher density of stronger Brønsted acid sites on the 2NHZ sample, in comparison to the NZ sample. As demonstrated here, the modification process can modify the acidity of natural zeolites to increase Brønsted acid sites' density.

2.3. Influence of Chemical Surface Characteristics of Zeolite Samples in Bio-Oil Composition

A semi-quantitative gas chromatography/mass spectrophotometry (GC/MS) analysis of fractionated bio-oil samples was conducted (See Materials and Methods) to identify primary chemical compounds of representative families, to evaluate the influence of chemical surface characteristics of natural and modified zeolite samples on the chemical composition of obtained bio-oil samples. Results are represented in Figure 2.

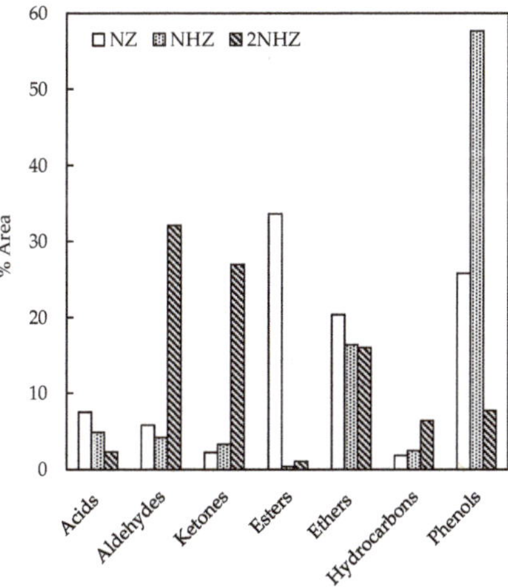

Figure 2. Primary chemical composition of bio-oil samples. Experiment Conditions: Pyrolysis Temp: 723 K; Cat./Oak ratio: 4; Flow 120 ml·min^{-1}; Ramp rate 10 K·min^{-1}.

Figure 2 shows a decrease in the composition of the following families (esters, ethers, and acids) as a result of the catalytic activity of modified (NHZ and 2NHZ) zeolite samples. Esters, ethers, and other oxygenated compounds are unwanted species, as they decrease the bio-oil heating value. Furthermore, a lower acid composition is highly desirable in bio-oil samples, taking into consideration that lower pH promotes aging reactions during the storage of samples.

Compositional fluctuations in stored bio-oil samples are due to potential aging reactions (e.g., alcohols reacting with organic acids forming esters and water, aldehydes reacting with each other to form polyacetal oligomers and polymers, aldehydes or ketones reacting with water to form hydrates, alcohols reacting with aldehydes forming hemiacetals, aldehydes and alcohols reacting to form acetals, and phenol/aldehyde reactions), as suggested by other studies [18,39]. Therefore, a lower content of acids, esters, ethers, and phenols will contribute to higher bio-oil stability during storage. Bio-Oil stability becomes a crucial matter when considering it as a potential substitute for fossil fuel.

On the other hand, an increase in the composition of aldehydes, ketones, and hydrocarbons was registered for obtained bio-oil samples, using 2NHZ as catalysts. The higher content of aldehydes, ketones, and hydrocarbons could be achieved by the direct contribution of Brønsted acid sites from zeolite samples, as reported elsewhere [40]. Furthermore, the lower composition of framework compensation cations on the 2NHZ zeolite sample might improve chemical species diffusion into the zeolite framework. Then, generated species from biomass pyrolysis can reach Brønsted acid sites without difficulty, resulting in new species after an absorbate-site interaction. The contribution of a lower composition of compensation cations to a higher diffusion into Chilean natural zeolites pores was reported elsewhere [41]. However, a higher composition of phenols was observed in obtained bio-oil, using NHZ sample, which is probably due to the lower Brønsted acid site density of this sample and a channel blockade as a result of applied chemical treatment. As mentioned before, aldehydes react with phenols. Thus, a lower content of aldehydes and a higher content of phenols in obtained bio-oil using NHZ sample is a logical outcome.

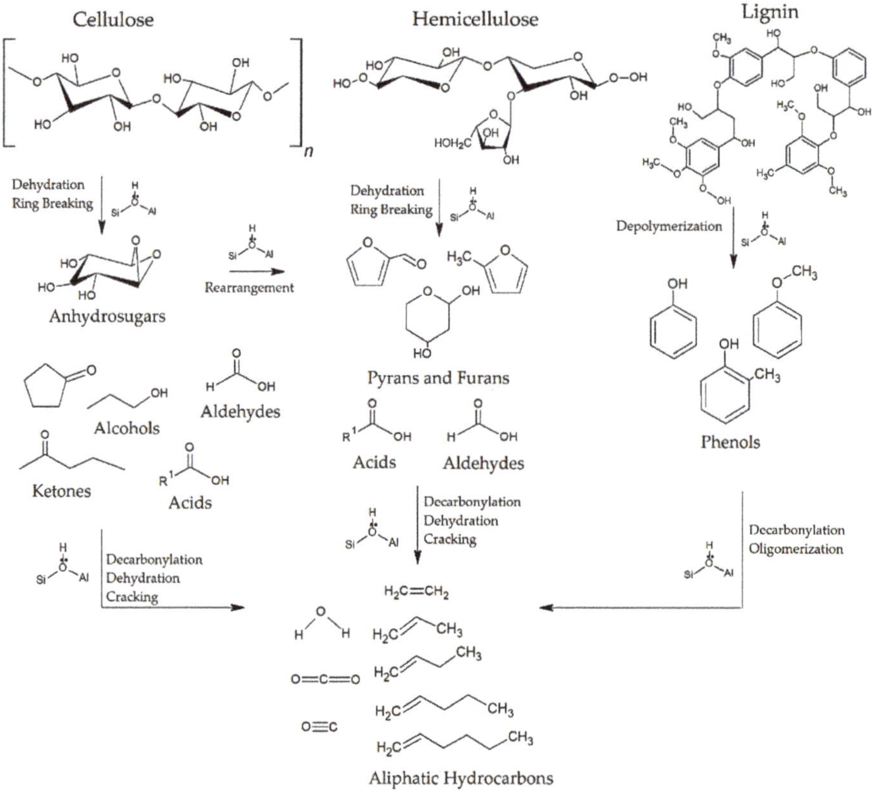

Scheme 1. Simplified reaction pathways of catalytic biomass pyrolysis. Adapted from [40,42–44].

Table 4. Main components of bio-oil obtained from catalytic Oak pyrolysis at 723 K.

*Fam	Compounds	% Area			Compounds	% Area		
		NZ	NHZ	2NHZ		NZ	NHZ	2NHZ
Ald	Butanedial			0.21	2,5-Dimethoxy-4-methylbenzaldehyde	0.88	0.76	
	Syringaldehyde	0.26			2-Furancarboxaldehyde, 5-methyl-			0.23
	Furfural	0.56	0.40	31.03	4-Hydroxy-2-methylbenzaldehyde			0.42
	Furfural, 5-methyl-	0.41	2.17		2,5-Dimethoxy-4-ethylbenzaldehyde	3.62	0.88	
Ket	Acetosyringone	0.68			2-Cyclopenten-1-one, 3,4-dimethyl-			0.72
	1,2-Cyclopentanedione			0.66	2-Cyclopenten-1-one, 3,5,5-trimethyl-			4.81
	Guaiacylacetone	0.55	0.46		2H-Pyran-2-one, 3-acetyl-4-hydroxy-6-methyl-			7.52
	2-Cyclopentene-1,4-dione			0.55	Ethanone, 1-(4-hydroxy-3-methoxyphenyl)-	0.15	0.16	
	2(5H)-Furanone, 5-methyl-			0.27	3,4-Dihydro-6-methyl-2H-pyran-2-one			0.51
	2-Cyclohexen-1-one, 2-methyl-			0.23	3-Cyclopenten-1-one, 2,2,5,5-tetramethyl-			0.34
	1,3-Cyclopentanedione, 2-ethyl-			0.94	3H-Pyrazol-3-one, 2,4-dihydro-2,4-dimethyl-			0.28
	2-Cyclopenten-1-one, 2-hydroxy-		1.12		3H-Pyrazol-3-one, 2,4-dihydro-2,5-dimethyl-			1.75
	2-Acetyl-5-methylfuran			7.10	3-methyl-1, 2-cyclopentanedione	0.88	1.58	
Est	1-Propionylethyl acetate			0.87	Propanoic acid, ethenyl ester	0.27	0.17	
	Butyric acid, neopentyl ester			0.17	Tricosanoic acid, methyl ester	33.33	0.18	
Eth	Benzene, ethoxy-			0.74	Benzene, 2-ethoxy-1,3-dimethoxy-			13.72
	1,2,4-Trimethoxybenzene	6.43	5.37		1,2,3-Trimethoxy-5-methylbenzene	13.94	11.06	
	Benzene, 1-ethoxy-4-methyl-			1.14	Ether, 3-butenyl propyl			0.17
HC	Hexacosane	0.17	0.22	0.56	Nonacosane		0.13	0.45
	4-Decene, 3-methyl-, (E)-			0.19	Octacosane	0.15		0.51
	Docosane	0.23	0.39	0.40	Pentacosane	0.29	0.64	0.72
	Heneicosane		0.23	0.12	Tetracosane	0.45	0.38	0.92
	Hentriacontane			0.33	Triacontane			0.39
	Heptacosane	0.25	0.38	0.52	Tricosane	0.33		0.69
Phe	Phenol, 2-methoxy-	1.47	4.05	5.29	Phenol, 2-methoxy-3-(2-propenyl)-		0.57	
	2-Isopropoxyphenol			0.42	Phenol, 2-methoxy-4-methyl-		3.32	1.13
	2-Methoxy-4-vinylphenol	0.80	1.04		Phenol, 2-methyl-		0.78	
	2-Methoxy-5-methylphenol			0.51	Phenol, 3,4-dimethoxy-	0.36	1.09	
	4-Ethylguaiacol	0.38	6.43		Phenol, 3,4-dimethyl-		0.76	0.08
	Desaspidinol	1.17	1.11		p-Methylguaiacol	1.54	5.65	
	Eugenol		0.54		p-Propylguaiacol		0.85	
	Homopyrocatechol	0.29	1.25		Pyrocatechol	0.42	2.22	
	Isoeugenol	7.76	7.23		Pyrocatechol, 3-methoxy-	0.61	2.19	
	Methoxyeugenol	1.40	0.91		Pyrocatechol, 3-methyl	0.13	1.14	
	Phenol	0.14	0.38		Syringol	9.24	14.97	

*Fam: Families; Ald: Aldehydes; Ket: Ketones; Est: Esters; Eth: Ethers; HC: Hydrocarbons; Phe: Phenols.

Chemical species within the zeolite framework are subjected to interactions with Brønsted acid sites. As a result of those interactions, Cellulose is decomposed (by a depolymerisation process through transglycosylation) into aldehydes, anhydrosugars (mainly levoglucosan), ketones, and alcohols. Hemicellulose is transformed into pyrans and furans, and Lignin is converted into phenols [45]. Thus, the interaction of adsorbed chemical species with Brønsted acid sites of zeolite samples, lead to the generation of hydrocarbons, carbon monoxide, carbon dioxide, water, and coke formation through several ring-opening reactions [45]. Another route goes from anhydrosugars, obtained from the cellulose upon pyrolysis, to furans by a rearrangement. Then, furans initiate a sequence of reactions that lead to hydrocarbons as well. A general idea of the abovementioned potential reaction pathways is illustrated in Scheme 1.

Based on experimental results obtained here, Chilean natural and modified zeolites lead to intermediates, such as aldehydes, ketones, ethers, phenols, and aliphatics. Thus, Chilean zeolites could be considered as an alternative catalyst for the obtention of sustainable value-added chemical compounds. On the other hand, synthetic zeolites usually lead to an increase of aromatics compounds in the bio-oil composition, due to their crystalline structure [46,47]. Moreover, the mesoporous and acidic zeolites lead to the production of shorter chain hydrocarbon due to their high cracking ability. Furthermore, microporous and less acidic zeolites favor the production of long chain hydrocarbons as the cracking process occurred mainly on the zeolite external surface [48]. Finally, reported hydrocarbons

(HC) fractions in Table 4, using 2NHZ, confirmed the abovementioned affirmations since C15-C30 hydrocarbons were registered in acquired chromatographic data.

The GC/MS procedure identified a large number of chemical compounds, after interpretation of registered data, using the automatic library search from National Institute of Standards and Technology (NIST) 2008. As a consequence of natural and modified zeolite catalytic activity, the composition of obtained bio-oils varies from one sample to the other, as Table 4 shows. Chemicals summarised here, are mainly oxygenated species with different functional groups including hydroxyl, phenolic hydroxyl, carboxyl, carbonyl, methoxy, ethoxy, oxygen-containing heterocyclic and unsaturated double bonds. Similar compounds were reported elsewhere [49–51].

The highest composition of furfural in obtained bio-oil samples was obtained when 2NHZ zeolite sample was used. The furfural composition was increased by several times, compared with those obtained when the parent natural zeolite sample was used. As reported in Table 3, the 2NHZ zeolite sample posses a higher density of Brønsted, among all samples analyzed in this study. Consequently, the increase of furfural composition can be associated with the incorporation of Brønsted acid sites in the 2NHZ sample. Brønsted acid sites were linked to the improvement of the degradation of polysaccharides to intermediates for furfural. Then, modified natural zeolites can be considered as a potential catalyst to obtain furfural via catalytic pyrolysis of biomass. As a value-added chemical, furfural is a relevant by-product of bio-oil, and it is widely used in medicines, resins, food additives, and fuel additives manufacture [52].

As mentioned before, bio-oil quality can be improved through deoxygenation reactions using Brønsted acid sites of zeolites, to obtain lower oxygenated compounds in bio-oil. Thus, a qualitative analysis was conducted to confirm the catalytic activity of zeolite samples. Table 5 shows a qualitative compositional analysis of obtained bio-oil samples, using natural and modified zeolite samples on the Oak catalytic slow pyrolysis investigated in this study.

Table 5. Qualitative analysis of bio-oil obtained from catalytic Oak pyrolysis.

Compounds	Percentage [%]		
	NZ	NHZ	2NHZ
Oxygenated	66.0	63.8	46.8
Non-oxygenated	17.0	14.5	14.4
Unknown	17.0	21.7	38.7

As Table 5 shows, a lower percentage of oxygenated compounds was registered in the bio-oil samples from catalytic assays using the 2NHZ sample. The lower percentage of oxygenated compounds can be associated with surface reactions promoted by Brønsted active acid sites in the modified zeolite framework [9,53]. Thus, the surface Brønsted acid sites possess a significant role in the catalytic pyrolysis, leading the transformation of original species from Oak pyrolysis to valuable by-products (hydrocarbons, aldehydes, and ketones).

Oxygenated species have several unsaturated bonds, which are responsible for several reactions in stored bio-oils. Moreover, a lower content of oxygenated compounds is required to avoid aging issues during bio-oils storage, considering the primary reactions of bio-oil aging (esterification, transesterification, homopolymerization, hydration, hemiacetal formation, acetalization, and phenol/aldehyde reactions) [18].

3. Materials and Methods

Biomass samples from a Chilean native oak tree (donated by Miraflores Angol Ltda., Angol, Chile) were subjected to size reduction (d_p < 2 mm), oven-dried at 313 K for 48 h and stored in a desiccator until further use. Samples were characterized by standard methods to obtain: humidity (Una Norma Española-European Norm (UNE-EN) 14774), ash content (UNE-EN 14775), calorific value (UNE-EN 14918), elemental analysis (UNE-EN 15104) and proximate analysis (American Society of

Testing Materials (ASTM) D 3172-73(84)) [54,55]. Thermogravimetric analysis was conducted in a thermobalance (Shimadzu DTG-60H, CROMTEK, Santiago, Chile).

Chilean natural zeolite (14% clinoptilolite, 74% mordenite, 12% quartz) was obtained from Minera FORMAS, Santiago, Chile. Zeolite samples were ground and sieved to 0.3 to 0.425 mm, then rinsed with ultrapure water (from an EASY pure RF II system), oven-dried at 398 K for 24 h, and stored in dry conditions. First, ammonium-modified zeolites were obtained from natural zeolite by an ion-exchange procedure with ammonium sulfate (0.1 mol·dm^{-3}), using a solution/solid ratio of 10:1 in a temperature-controlled water bath at 363 K for 3 h. Finally, modified samples were rinsed with ultra-pure water for 4 h, replacing the water at 2 and 3 h. The sample obtained here was named as NHZ. A second ammonium-modified zeolite sample (2NHZ) was obtained from NHZ, applying the same aforementioned ion-exchange procedure. More details can be found in a previous article [18]. Lastly, zeolite samples were oven-dried at 378 K for 24 h and stored in a desiccator until further use.

To characterize zeolite samples, the following quantitative methods were conducted: nitrogen absorption at 77 K, XRD, and XRF. The surface area was calculated from adsorption isotherms data, obtained with a Micromeritics Gemini 3175 (UDEC, Concepción, Chile).

To evaluate mineralogical content and structural framework of zeolite samples, XRD analysis was performed using a Bruker AXS Model D4 ENDEAVOR diffractometer (UDEC, Concepción, Chile), equipped with a copper X-ray tube and a Ni filter. Powdered natural and modified zeolite samples were mounted on quartz plates and stepped scanned over the angular range of 3 to 70° (2θ), a step size of 0.02 and a time/step of 0.2 s. X-ray generator was fixed at 40 kV and 20 mA. In the same way, X-ray fluorescence allowed the determination of the bulk chemical composition of natural and modified zeolite samples, using a RIGAKU Model 3072 spectrometer (UDEC, Concepción, Chile. Characterization procedures were reported elsewhere [29,56,57].

A procedure was developed to confirm the nature and strength of zeolite acid sites using pyridine as a probe molecule. In contrast to NH3-TPD, pyridine adsorption discriminates between Lewis and Brønsted sites. The pyridine adsorption was followed by DRIFT using a Nicolet (Alta Tecnología, Chile) Avatar 370 MCT with a smart collector accessory, a mid/near-infrared source, and a mercury cadmium telluride (MCT-A) photon detector at 77 K (liquid N_2), according to the procedure reported before [18].

A quartz cylindrical reactor was used in the experimental pyrolysis system, as shown in Figure 3. Biomass samples (9 g) were heated (ramp rate 10 K·min^{-1}) from room temperature to 623 K and kept isothermal for 30 min under a nitrogen flow rate of 60 cm^3·min^{-1}. Zeolite samples were out-gassed at 623 K for 2 h before pyrolysis experiments. Bio-oils were obtained after condensation of pyrolysis vapors in a closed flask at 253 K. Pyrolysis product fractions (bio-oil and biochar) were quantified gravimetrically, and the gaseous fraction was calculated by difference.

As known, a large number of chemical species were commonly identified in bio-oil samples. Thus, a fractionation method was conducted here to separate some of the wood-extractive derived compounds in seven fractions, considering reported articles before [25,58,59]. Bio-oil samples were diluted (ratio 1:1) with ultrapure water and centrifuged at 3300 rpm for 5 min in an LW SCIENTIFIC (Santiago, Chile) ULTRA 8V apparatus. After remotion of the water-soluble supernatant, the water-insoluble phase of the bio-oil samples was placed in an solid-phase extraction (SPE) cartridge (UCT-CUSIL 500 mg/10 cm^3). Then, the SPE column was eluted (using a manifold vacuum system) with 10 cm^3 of each solvent (Lichrosolv grade) according to the following sequence: hexane, cyclohexane, diethyl ether, dichloromethane, ethyl acetate, acetone, and acetonitrile. Eluted samples from SPE cartridge were investigated by GC/MS to identify chemical species, using a Shimadzu (Cromtek, Chile) QP2010-plus apparatus. GC/MS was configured following the procedure reported elsewhere [18]. Chromatographic data were processed using GCMSsolution (v2.53) (Shimadtzu, Santiago, Chile) and mass spectra laboratory databases (NIST08 and NIST08s).

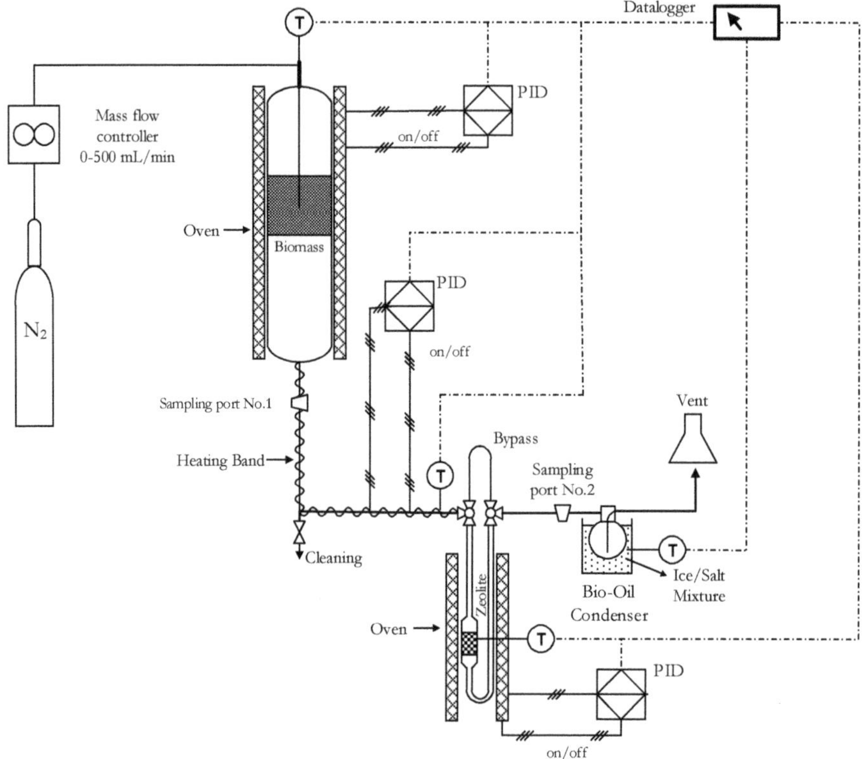

Figure 3. Experimental setup.

4. Conclusions

The catalytic pyrolysis of Chilean Oak over natural and modified zeolites was investigated in this study. As demonstrated here, the acidity of modified zeolites changed the reaction chemistry and products distribution. The Brønsted acid sites on ammonium-modified zeolite samples are responsible for the upgraded bio-oil and value-added chemicals, obtained in this research. Bio-oil chemical composition was modified when the pyrolysis-derived compounds were upgraded over 2NHZ zeolite sample, leading to a lower quantity of oxygenated compounds and a higher composition of value-added chemicals (e.g., furfural), among other compounds. Consequently, modified Chilean natural zeolites can be considered as a potential catalyst to obtain value-added chemicals via catalytic pyrolysis of biomass.

Author Contributions: S.A.-M., A.M.A., and H.D.P. conceived and designed the experiments, A.M.A. and H.D.P. performed the experiments, S.A-M., A.M.A., and C.C.-B. analyzed the data, S.A.-M. and A.M.A. wrote the paper.

Funding: This research was funded by the Chilean National Fund for Scientific and Technological Development (Fondecyt Grant No. 11140781, Fondequip Grant No. 170077), the Nanomaterials and Catalysts for Sustainable Processes (NanoCatpPS) research group from University of Bio-Bio, Chile and the University of Bio-Bio, Concepción, Chile.

Acknowledgments: The authors wish to acknowledge to Daniel J. Nowakowski and Scott W. Banks from EBRI, Aston University in Birmingham U.K., for this valuable collaboration on the DRIFT experiments.

Conflicts of Interest: The authors declare no conflict of interest. The founding sponsors had no role in study design, collection, analyses, and interpretation of data, manuscript writing, the decision to publish the results.

References

1. Borsatto, F.; Inglezakis, V.J. Natural zeolite markets and strategic considerations. In *Handbook of Natural Zeolites*; Bentham Science Publishers: Bacau, Romania, 2012; pp. 11–27. ISBN 9781608054466.
2. Fankhauser, S.; Jotzo, F. Economic growth and development with low-carbon energy. *Wiley Interdiscip. Rev. Clim. Chang.* **2017**, *9*, e495. [CrossRef]
3. Zeng, S.; Liu, Y.; Liu, C.; Nan, X. A review of renewable energy investment in the BRICS countries: History, models, problems and solutions. *Renew. Sustain. Energy Rev.* **2017**, *74*, 860–872. [CrossRef]
4. Rahman, F.A.; Aziz, M.M.A.; Saidur, R.; Bakar, W.A.W.A.; Hainin, M.R.; Putrajaya, R.; Hassan, N.A. Pollution to solution: Capture and sequestration of carbon dioxide (CO_2) and its utilization as a renewable energy source for a sustainable future. *Renew. Sustain. Energy Rev.* **2017**, *71*, 112–126. [CrossRef]
5. Demirbaş, A.; Arin, G. An overview of biomass pyrolysis. *Energy Sources* **2002**, *24*, 471–482. [CrossRef]
6. Papari, S.; Hawboldt, K. A review on condensing system for biomass pyrolysis process. *Fuel Process. Technol.* **2018**, *180*, 1–13. [CrossRef]
7. Hosoya, T.; Kawamoto, H.; Saka, S. Solid/liquid- and vapor-phase interactions between cellulose- and lignin-derived pyrolysis products. *J. Anal. Appl. Pyrolysis* **2009**, *85*, 237–246. [CrossRef]
8. Zanuttini, M.S.; Lago, C.D.; Querini, C.A.; Peralta, M.A. Deoxygenation of m-cresol on Pt/γ-Al2O3 catalysts. *Catal. Today* **2013**, *213*, 9–17. [CrossRef]
9. Gurevich Messina, L.I.; Bonelli, P.R.; Cukierman, A.L. In-situ catalytic pyrolysis of peanut shells using modified natural zeolite. *Fuel Process. Technol.* **2017**, *159*, 160–167. [CrossRef]
10. Rezaei, P.S.; Shafaghat, H.; Daud, W.M.A.W. Production of green aromatics and olefins by catalytic cracking of oxygenate compounds derived from biomass pyrolysis: A review. *Appl. Catal. A Gen.* **2014**, *469*, 490–511. [CrossRef]
11. Mohammed, I.Y.; Kazi, F.K.; Yusup, S.; Alaba, P.A.; Sani, Y.M.; Abakr, Y.A. Catalytic intermediate pyrolysis of Napier grass in a fixed bed reactor with ZSM-5, HZSM-5 and zinc-exchanged zeolite-a as the catalyst. *Energies* **2016**, *9*, 246. [CrossRef]
12. Imran, A.; Bramer, E.A.; Seshan, K.; Brem, G. Catalytic flash pyrolysis of biomass using different types of zeolite and online vapor fractionation. *Energies* **2016**, *9*, 187. [CrossRef]
13. ACS Material Advanced Chemicals Supplier. Available online: https://www.acsmaterial.com/materials/molecular-sieves.html (accessed on 23 October 2018).
14. Inglezakis, V.J.; Zorpas, A.A. Physical and Chemical Properties. In *Handbook of Natural Zeolites*; EBooks, B., Ed.; Bentham Science Publishers: Bacau, Romania, 2012; pp. 70–102. ISBN 9781608052615.
15. Serrano, D.P.; Melero, J.A.; Morales, G.; Iglesias, J.; Pizarro, P. Progress in the design of zeolite catalysts for biomass conversion into biofuels and bio-based chemicals. *Catal. Rev.* **2018**, *60*, 1–70. [CrossRef]
16. Pütün, E.; Uzun, B.B.; Pütün, A.E. Fixed-bed catalytic pyrolysis of cotton-seed cake: Effects of pyrolysis temperature, natural zeolite content and sweeping gas flow rate. *Bioresour. Technol.* **2006**, *97*, 701–710. [CrossRef] [PubMed]
17. Aho, A.; Kumar, N.; Eränen, K.; Salmi, T.; Hupa, M.; Murzin, D.Y. Catalytic pyrolysis of woody biomass in a fluidized bed reactor: Influence of the zeolite structure. *Fuel* **2008**, *87*, 2493–2501. [CrossRef]
18. Alejandro-Martín, S.; Cerda-Barrera, C.; Montecinos, A.; Martín, S.A.; Cerda-Barrera, C.; Montecinos, A. Catalytic Pyrolysis of Chilean Oak: Influence of Brønsted Acid Sites of Chilean Natural Zeolite. *Catalysts* **2017**, *7*, 356. [CrossRef]
19. Ateş, F.; Pütün, A.E.; Pütün, E. Fixed bed pyrolysis of Euphorbia rigida with different catalysts. *Energy Convers. Manag.* **2005**, *46*, 421–432. [CrossRef]
20. Rajić, N.; Logar, N.Z.; Rečnik, A.; El-Roz, M.; Thibault-Starzyk, F.; Sprenger, P.; Hannevold, L.; Andersen, A.; Stöcker, M. Hardwood lignin pyrolysis in the presence of nano-oxide particles embedded onto natural clinoptilolite. *Microporous Mesoporous Mater.* **2013**, *176*, 162–167. [CrossRef]
21. Milovanovic, J.; Stensrød, R.; Myhrvold, E.; Tschentscher, R.; Stöcker, M.; Lazarevic, S.; Rajic, N. Modification of natural clinoptilolite and ZSM-5 with different oxides and studying of the obtained products in lignin pyrolysis. *J. Serbian Chem. Soc.* **2015**, *80*, 717–729. [CrossRef]
22. Corma Canos, A.; Iborra, S.; Velty, A. Chemical routes for the transformation of biomass into chemicals. *Chem. Rev.* **2007**, *107*, 2411–2502. [CrossRef]

23. French, R.; Czernik, S. Catalytic pyrolysis of biomass for biofuels production. *Fuel Process. Technol.* **2010**, *91*, 25–32. [CrossRef]

24. de Wild, P.; Reith, H.; Heeres, E. Biomass pyrolysis for chemicals. *Biofuels* **2011**, *2*, 185–208. [CrossRef]

25. Mohan, D.; Pittman, C.U.; Steele, P.H. Pyrolysis of Wood/Biomass for Bio-oil: A Critical Review. *Energy Fuels* **2006**, *20*, 848–889. [CrossRef]

26. Gregg, S.J.; Sing, K.S.W. *Adsortion, Surface Area and Porosity*; Academic Press: London, UK, 1982; ISBN 12-300950-2.

27. Englert, A.H.; Rubio, J. Characterization and environmental application of a Chilean natural zeolite. *Int. J. Miner. Process.* **2005**, *75*, 21–29. [CrossRef]

28. Christidis, G.E.; Moraetis, D.; Keheyan, E.; Akhalbedashvili, L.; Kekelidze, N.; Gevorkyan, R.; Yeritsyan, H.; Sargsyan, H. Chemical and thermal modification of natural HEU-type zeolitic materials from Armenia, Georgia and Greece. *Appl. Clay Sci.* **2003**, *24*, 79–91. [CrossRef]

29. Valdés, H.; Alejandro, S.; Zaror, C.A. Natural zeolite reactivity towards ozone: The role of compensating cations. *J. Hazard. Mater.* **2012**, *227–228*, 34–40. [CrossRef]

30. Alejandro, S.; Valdés, H.; Manero, M.-H.H.; Zaror, C.A.A. BTX abatement using Chilean natural zeolite: The role of Bronsted acid sites. *Water Sci. Technol.* **2012**, *66*, 1759–1769. [CrossRef] [PubMed]

31. Leliveld, B.R.G.; Kerkhoffs, M.; Broersma, F.A.; van Dillen, J.A.J.; Geus, J.W.; Koningsberger, D.C. Acidic properties of synthetic saponites studied by pyridine IR and TPD-TG of n-propylamine. *J. Chem. Soc. Trans.* **1998**, *94*, 315–321. [CrossRef]

32. Konan, K.L.; Peyratout, C.; Smith, A.; Bonnet, J.-P.; Magnoux, P.; Ayrault, P. Surface modifications of illite in concentrated lime solutions investigated by pyridine adsorption. *J. Colloid Interface Sci.* **2012**, *382*, 17–21. [CrossRef]

33. Simon-Masseron, A.; Marques, J.P.; Lopes, J.M.; Ribeiro, F.R.; Gener, I.; Guisnet, M. Influence of the Si/Al ratio and crystal size on the acidity and activity of HBEA zeolites. *Appl. Catal. A Gen.* **2007**, *316*, 75–82. [CrossRef]

34. Osman, A.I.; Abu-Dahrieh, J.K.; Rooney, D.W.; Halawy, S.A.; Mohamed, M.A.; Abdelkader, A. Effect of precursor on the performance of alumina for the dehydration of methanol to dimethyl ether. *Appl. Catal. B Environ.* **2012**, *127*, 307–315. [CrossRef]

35. Ajaikumar, S.; Pandurangan, A. Efficient synthesis of quinoxaline derivatives over $ZrO_2/MxOy$ (M = Al, Ga, In and La) mixed metal oxides supported on MCM-41 mesoporous molecular sieves. *Appl. Catal. A Gen.* **2009**, *357*, 184–192. [CrossRef]

36. Karthik, M.; Magesh, C.J.; Perumal, P.T.; Palanichamy, M.; Arabindoo, B.; Murugesan, V. Zeolite-catalyzed ecofriendly synthesis of vibrindole A and bis(indolyl)methanes. *Appl. Catal. A Gen.* **2005**, *286*, 137–141. [CrossRef]

37. Kostyniuk, A.; Key, D.; Mdleleni, M. Effect of Fe-Mo promoters on HZSM-5 zeolite catalyst for 1-hexene aromatization. *J. Saudi Chem. Soc.* **2018**. [CrossRef]

38. Tamura, M.; Shimizu, K.; Satsuma, A. Comprehensive IR study on acid/base properties of metal oxides. *Appl. Catal. A Gen.* **2012**, *433–434*, 135–145. [CrossRef]

39. Adam, J.; Blazsó, M.; Mészáros, E.; Stocker, M.; Nilsen, M.H.; Bouzga, A.; Hustad, J.E.; Gronli, M.; Oye, G. Pyrolysis of biomass in the presence of Al-MCM-41 type catalysts. *Fuel* **2005**, *84*, 1494–1502. [CrossRef]

40. Rahman, M.M.; Liu, R.; Cai, J. Catalytic fast pyrolysis of biomass over zeolites for high quality bio-oil – A review. *Fuel Process. Technol.* **2018**, *180*, 32–46. [CrossRef]

41. Alejandro, S.; Valdés, H.; Manéro, M.-H.M.H.; Zaror, C.A. Oxidative regeneration of toluene-saturated natural zeolite by gaseous ozone: The influence of zeolite chemical surface characteristics. *J. Hazard. Mater.* **2014**, *274*, 212–220. [CrossRef]

42. Wang, K.; Kim, K.H.; Brown, R.C. Catalytic pyrolysis of individual components of lignocellulosic biomass. *Green Chem.* **2014**, *16*. [CrossRef]

43. Liu, S.-N.; Cao, J.-P.; Zhao, X.-Y.; Wang, J.-X.; Ren, X.-Y.; Yuan, Z.-S.; Guo, Z.-X.; Shen, W.-Z.; Bai, J.; Wei, X.-Y. Effect of zeolite structure on light aromatics formation during upgrading of cellulose fast pyrolysis vapor. *J. Energy Inst.* **2018**. [CrossRef]

44. Zheng, A.; Zhao, Z.; Chang, S.; Huang, Z.; Wu, H.; Wang, X.; He, F.; Li, H. Effect of crystal size of ZSM-5 on the aromatic yield and selectivity from catalytic fast pyrolysis of biomass. *J. Mol. Catal. A Chem.* **2014**, *383–384*, 23–30. [CrossRef]

45. Veses, A.; Aznar, M.; Martínez, I.; Martínez, J.D.; López, J.M.; Navarro, M.V.; Callén, M.S.; Murillo, R.; García, T. Catalytic pyrolysis of wood biomass in an auger reactor using calcium-based catalysts. *Bioresour. Technol.* **2014**, *162*, 250–258. [CrossRef]

46. Che, Q.; Yang, M.; Wang, X.; Chen, X.; Chen, W.; Yang, Q.; Yang, H.; Chen, H. Aromatics production with metal oxides and ZSM-5 as catalysts in catalytic pyrolysis of wood sawdust. *Fuel Process. Technol.* **2019**, *188*, 146–152. [CrossRef]

47. Che, Q.; Yang, M.; Wang, X.; Yang, Q.; Rose Williams, L.; Yang, H.; Zou, J.; Zeng, K.; Zhu, Y.; Chen, Y.; Chen, H. Influence of physicochemical properties of metal modified ZSM-5 catalyst on benzene, toluene and xylene production from biomass catalytic pyrolysis. *Bioresour. Technol.* **2019**, *278*, 248–254. [CrossRef]

48. López, A.; de Marco, I.; Caballero, B.M.; Laresgoiti, M.F.; Adrados, A.; Torres, A. Pyrolysis of municipal plastic wastes II: Influence of raw material composition under catalytic conditions. *Waste Manag.* **2011**, *31*, 1973–1983. [CrossRef] [PubMed]

49. Chen, D.; Zhou, J.; Zhang, Q.; Zhu, X. Evaluation methods and research progresses in bio-oil storage stability. *Renew. Sustain. Energy Rev.* **2014**, *40*, 69–79. [CrossRef]

50. Holladay, J.E.; White, J.F.; Bozell, J.J.; Johnson, D. Top Value-Added Chemicals from Biomass Volume II - Results of Screening for Potential Candidates from Biorefinery Lignin. *Pacific Northwest Natl. Lab.* **2007**, *II*, 87. [CrossRef]

51. Imam, T.; Capareda, S. Characterization of bio-oil, syn-gas and bio-char from switchgrass pyrolysis at various temperatures. *J. Anal. Appl. Pyrolysis* **2012**, *93*, 170–177. [CrossRef]

52. Zhang, H.; Liu, X.; Lu, M.; Hu, X.; Lu, L.; Tian, X.; Ji, J. Role of Brønsted acid in selective production of furfural in biomass pyrolysis. *Bioresour. Technol.* **2014**, *169*, 800–803. [CrossRef]

53. Mihalcik, D.J.; Mullen, C.A.; Boateng, A.A. Screening acidic zeolites for catalytic fast pyrolysis of biomass and its components. *J. Anal. Appl. Pyrolysis* **2011**, *92*, 224–232. [CrossRef]

54. Erol, M.; Haykiri-Acma, H.; Küçükbayrak, S. Calorific value estimation of biomass from their proximate analyses data. *Renew. Energy* **2010**, *35*, 170–173. [CrossRef]

55. Parikh, J.; Channiwala, S.A.; Ghosal, G.K. A correlation for calculating HHV from proximate analysis of solid fuels. *Fuel* **2005**, *84*, 487–494. [CrossRef]

56. Alejandro, S.; Valdés, H.; Zaror, C.A. Natural Zeolite Reactivity Towards Ozone: The Role of Acid Surface Sites. *J. Adv. Oxid. Technol.* **2011**, *14*, 182–189. [CrossRef]

57. Valdés, H.; Farfán, V.J.; Manoli, J.A.; Zaror, C.A. Catalytic ozone aqueous decomposition promoted by natural zeolite and volcanic sand. *J. Hazard. Mater.* **2008**, *165*, 915–922. [CrossRef] [PubMed]

58. Garcia-Perez, M.; Chaala, A.; Pakdel, H.; Kretschmer, D.; Roy, C. Characterization of bio-oils in chemical families. *Biomass and Bioenergy* **2007**, *31*, 222–242. [CrossRef]

59. Kanaujia, P.K.; Sharma, Y.K.; Garg, M.O.; Tripathi, D.; Singh, R. Review of analytical strategies in the production and upgrading of bio-oils derived from lignocellulosic biomass. *J. Anal. Appl. Pyrolysis* **2014**, *105*, 55–74. [CrossRef]

Communication

Enhancement of Light Olefins Selectivity Over N-Doped Fischer-Tropsch Synthesis Catalyst Supported on Activated Carbon Pretreated with $KMnO_4$

Zhipeng Tian [1,2], Chenguang Wang [1,*], Zhan Si [1], Chengyan Wen [1], Ying Xu [1], Wei Lv [1], Lungang Chen [1], Xinghua Zhang [1] and Longlong Ma [1]

[1] CAS Key Laboratory of Renewable Energy, Guangdong Provincial Key Laboratory of New and Renewable Energy Research and Development, Guangzhou Institute of Energy Conversion, Chinese Academy of Sciences, Guangzhou 510640, China; tianzp@ms.giec.ac.cn (Z.T.); sizhan43@126.com (Z.S.); wency1@ms.giec.ac.cn (C.W.); xuying@ms.giec.ac.cn (Y.X.); lvwei@ms.giec.ac.cn (W.L.); chenlg1981@ms.giec.ac.cn (L.C.); zhangxh@ms.giec.ac.cn (X.Z.); mall@ms.giec.ac.cn (L.M.)

[2] University of Chinese Academy of Sciences, Beijing 100049, China

* Correspondence: wangcg@ms.giec.ac.cn; Tel.: +86-20-37029721

Received: 11 May 2019; Accepted: 29 May 2019; Published: 31 May 2019

Abstract: Ammonium iron citrate was used as an iron precursor in order to prepare N-doped catalysts supported on $KMnO_4$ pretreated activated carbon (10MnK-AC). Iron nitride was synthesized in company with the formation of α-Fe_2O_3 on 10MnK-AC. The characterizations of the catalysts show that nitrogen atoms were doped into iron lattice rather than the networks of the carbon support. The performance of Fischer-Tropsch synthesis to light olefins (FTO) suggest an improvement in O/P ratio (olefins to paraffins molar ratio of C_2–C_4) over the iron catalysts supported on 10MnK-AC. The further promotion of light olefins selectivity (up to 44.7%) was obtained over FeN-10MnK-AC catalyst owing to the collaborative contribution of the electron donor effect of nitrogen and the suppression effect on the second hydrogenation over 10MnK-AC support.

Keywords: nitrogen-doping; iron nitrides; light olefins; CO hydrogenation; $KMnO_4$ pretreatment

---

1. Introduction

Fischer-Tropsch synthesis to light olefins (FTO) from syngas derived from coal, natural gas and biomass is a typical approach to obtain the building blocks for chemical industry [1]. Especially, the production of fungible fuels from CO-rich syngas derived from biomass is a promising way for the sustainable development of the world energy economy. Light olefins obtained from the FTO reaction can further be oligomerized to gasoline and diesel fuels, which is more controllable than a one-step FTS reaction [2]. High isomerization to iso-paraffins during the olefin oligomerization contributes to the high-octane rate of gasoline [3]. Iron-based supported catalysts have been widely studied because of their low cost, better selectivity towards light alkenes and proper water-gas shift (WGS) reaction activity for CO-rich syngas [4–6]. FTO catalytic performance is highly related to the catalyst properties such as active phases generated during the reduction and reaction processes [7–9], promoters [10,11] and the interactions with support for supported iron catalysts. The theory that iron carbide is the active phase in Fischer-Tropsch synthesis is widely accepted [12]. However, catalyst sintering and carbon deposition on active phase lead to a relatively poor catalyst stability for iron-based catalysts at such severe reaction conditions [13,14]. In order to extend the lifetime of catalyst, a novel iron catalyst supported on $KMnO_4$ pretreated activated carbon (AC) has been prepared in our previous work [15].The better stability of iron particles and higher light olefins selectivity is obtained due to the promotion of potassium and manganese retained on the AC surface. The existence of a manganese promoter also inhibits the dissociation of hydrogen which makes the hydrogenation of the C^* intermediate less active. The competitive adsorption of hydrogen on manganese oxides retards the further hydrogenation of unsaturated hydrocarbons and enhances light olefins selectivity. Meanwhile, as the structure promoter, the bonding force between iron species and MnO_x improves dispersity and stability of active sites. Alkali promoters such as Na, K, etc. can facilitate the formation of iron carbides, and it reduces the production of methane. Recently, the introduction of nitrogen atoms into iron-based supported catalysts for Fischer-Tropsch synthesis have been applied to increase the selectivity of lower olefins [16–20]. It is found that doping nitrogen atoms into the matrix of carbon materials can change the electronic environment and increase FTO performances. N atoms acted as the electron donor suppresses second hydrogenation and improves light olefins selectivity. In contrast, Bao's group [21] has investigated the enhancement of iron nitrides supported on carbon nanotube (CNT) walls catalysts. They synthesized a series of FeN/CNT catalysts by heating conventional Fe/CNT catalysts at 450 °C in ammonia stream and found that iron nitride (Fe_2N) exhibits higher activity and C_2–C_4 alkenes selectivity than iron carbides, which also verifies the proposition suggested by Anderson et al [22]. Moreover, the effects of Mn and K added to FeN/CNT catalyst that prohibit second hydrogenation and facilitate CO adsorption are discussed [23].

Following these previous studies, herein we have prepared an N-doped iron catalyst supported on $KMnO_4$ pretreated AC and compared its FTO performance with an N-doped catalyst supported on AC and other catalysts in our previous studies [15]. Unlike the heating treatment in ammonia stream [21], we have introduced N atoms into the Fe lattice by using ammonium iron citrate as a precursor. A series of catalyst characterizations were conducted and ascertained the position of nitrogen atoms in catalysts. The synergistic effect between N-doped active phase and MnK-AC support was discussed as well by associating characterization with FTO reaction results.

2. Results and Discussion

As is shown in Figure 1, the morphology of the catalysts with different supports (AC and $KMnO_4$-treated AC) and iron precursors (iron nitrate and ammonium iron citrate) were determined by Powder X-ray diffraction (XRD) analysis. Obviously, all these catalysts showed typical peaks at 26.5° and 44.6°, which can be attributed to graphitic carbon. Further, the intensity of the broad peak around 25° can be attributed to the amorphous carbon in AC. The relative ratio of amorphous carbon and graphitic carbon can be determined by I_D/I_G ratio in Raman spectra (Table 1). The I_D/I_G ratio was determined by the peak intensity of D band (1350 cm^{-1}) and G band (1575 cm^{-1}) in Raman spectra. The D band means the disorder in carbon materials and the G band means the graphitization degree of carbon materials. The I_D/I_G ratios of Fe-10MnK-AC and FeN-10MnK-AC are lower than those of Fe-AC and FeN-AC catalysts. It implies that $KMnO_4$ pretreatment on the AC support increases its graphitization degree. On the other hand, no obvious changes in I_D/I_G ratio are observed by comparing the results of Fe- catalysts and N-doped FeN- catalysts. That is to say, nitrogen doping has no effect on the properties of carbon support. This result is opposite to the common sense that nitrogen introduction in carbon material decreases its graphitization degree and improves the defectiveness [24,25]. Probably, N atoms attached to the metal oxides rather than carbon support. Table 1 also offers the metal element contents of Fe, Mn and K analyzed by inductive coupled plasma-atomic emission spectroscopy (ICP-AES), which are almost the same with the desired loadings within the preparation error. The pore properties are provided in Table 1 as well. From the specific surface area and pore size, no difference between undoped and N-doped catalysts. However, there is a large decrease of specific area and a growth of pore size compared to the untreated catalysts to $KMnO_4$-pretreated catalysts. This is due to the high content of manganese and the relative low content of porous AC on $KMnO_4$-pretreated catalysts (Fe-10MnK-AC and FeN-10MnK-AC). This results in the decline in adsorption ability of reactants, which further reduces the reactivity.

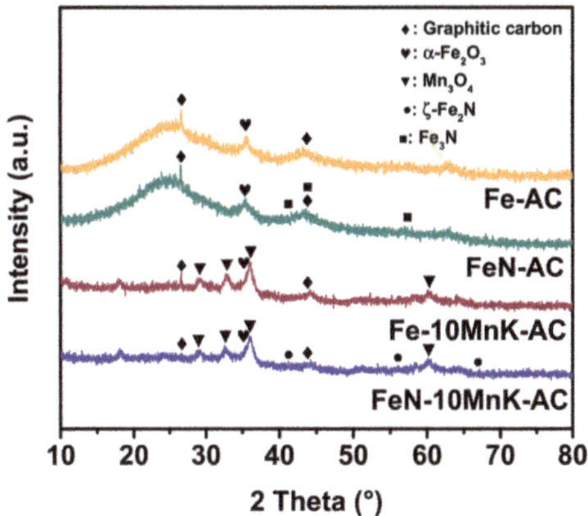

Figure 1. XRD patterns of catalysts: Fe-AC, FeN-AC, Fe-10MnK-AC and FeN-10MnK-AC. Graphitic carbon (♦, JCPDS No. 75-2078), α-Fe$_2$O$_3$ (♥, JCPDS No. 73-0603), Mn$_3$O$_4$ (▼, JCPDS No. 75-1560), ε-Fe$_3$N (■, JCPDS No. 83-0876) and ζ-Fe$_2$N (●, JCPDS No. 50-0958) are marked in the patterns.

Table 1. Elemental contents and properties of supported catalysts.

Catalysts	Element Contents (wt %)[1]			N atom. (%)[2]	I_D/I_G[3]	Iron Average Particle Size (nm)	Surface Area (m²/g)	Pore Size (nm)
	Fe	Mn	K					
Fe-AC	10.1	-	-	0.4	1.42	3.2 ± 1.0	575.3	2.58
FeN-AC	9.7	-	-	1.6	1.45	5.4 ± 1.4	549.3	2.56
Fe-10MnK-AC	10.1	29.3	5.3	0.4	1.34	2.0 ± 0.9	190.6	3.34
FeN-10MnK-AC	9.4	26.7	4.6	2.1	1.35	4.2 ± 1.5	187.1	3.37

[1] Determined by ICP-AES. [2] Calculated by X-ray photoelectron spectroscopy (XPS) spectra. [3] Calculated by the peak intensity of D band (1350 cm^{-1}) and G band (1575 cm^{-1}) in Raman spectra.

As reported by our previous work [15], the major peak at 35.6° can be assigned to the (110) crystal plane of α-Fe$_2$O$_3$ (JCPDS No. 73-0603). And three characteristic peaks at 28.9°, 32.4° and 36.1° can be ascribed to Mn$_3$O$_4$ (JCPDS No. 75-1560). This implies that different supports and precursors have no significant effect on the crystal structure of iron oxides on catalysts. Interestingly, two tiny peaks emerged at 40.8° and 57.0° in the XRD spectrum of FeN-AC can be attributed to iron nitride ε-Fe$_3$N (JCPDS No. 83-0876). Compared with the result of Fe-AC, it is believed that part of iron oxide transformed into iron nitride during calcination by using ammonium iron citrate as precursor. In this procedure, ammonium iron citrate was decomposed and the volatilized NH$_3$ replaced oxygen atoms in iron oxide by nitridation process. Then the formation of iron nitride is preceded through the incorporation of nitrogen into the iron lattice [22]. Specially, based on the XRD spectrum analysis result, another type of iron nitride (ζ-Fe$_2$N, JCPDS No. 50-0958) is acquired on FeN-10MnK-AC catalyst. The difference of iron nitride crystal phase is probably ascribed to the existence of manganese oxides on the surface of carbon support. The presence of Mn facilitates the formation of ζ-Fe$_2$N phase according to previous studies [23]. Only a small fraction of iron nitrides was obtained on FeN-AC and FeN-10MnK-AC catalysts according to the lower intensity of the peaks attributed to ε-Fe$_3$N and ζ-Fe$_2$N. In order to give a detailed analysis of the nitrogen atoms and the interaction with iron, XPS analysis was conducted. As is illustrated in the XPS survey spectra (Figure 2a), both FeN-AC and FeN-10MnK-AC show a relatively weak N1s photoelectron peak at 400.0 eV, which are revealed more specifically in high resolution spectra of the N1s region (Figure 2b). The nitrogen content of each catalyst is calculated and determined by the relative peak area of the XPS survey spectra, which are 0.4% (Fe-AC), 1.6% (FeN-AC), 0.4% (Fe-10MnK-AC) and 2.1% (FeN-10MnK-AC), respectively. However, the N content is too low so that the N1s peaks are hard to deconvolute. On the basis of the

Fe2p spectra shift to higher binding energies (Figure 2c), it can be inferred that the interaction between Fe and N is formed on the catalysts when using ammonium iron citrate as precursor. It means that in the crystal lattice of iron oxides, N atoms partially replace O atoms and get into the Fe lattice to form iron nitrides [26,27]. In brief, nitrogen atoms are not doped into carbon networks but iron lattice. Additionally, from the results of transmission electron microscopy (TEM) images and corresponding energy dispersive spectroscopy (EDS) mapping of FeN-10MnK-AC (Figure 3), we can easily find out that Fe, Mn, K and N dispersed on the carbon substrate homogeneously. The particle size distribution of FeN-10MnK-AC catalyst is presented as well. The average particle size is 4.2 ±1.5 nm, which is larger than that of Fe-10MnK-AC catalyst (2.0 ± 0.9 nm). The average particle size of Fe-AC and FeN-AC are 3.2 ± 1.0 nm and 5.4 ± 1.4 nm separately. It shows a similar trend with those catalysts supported on $KMnO_4$ pretreated AC. Using ammonium iron citrate as iron precursor leads to the increase of iron particle size. The effective contact area decreases and leads to the reduction of catalyst activity, which can be observed in the FTO performances (Table 2).

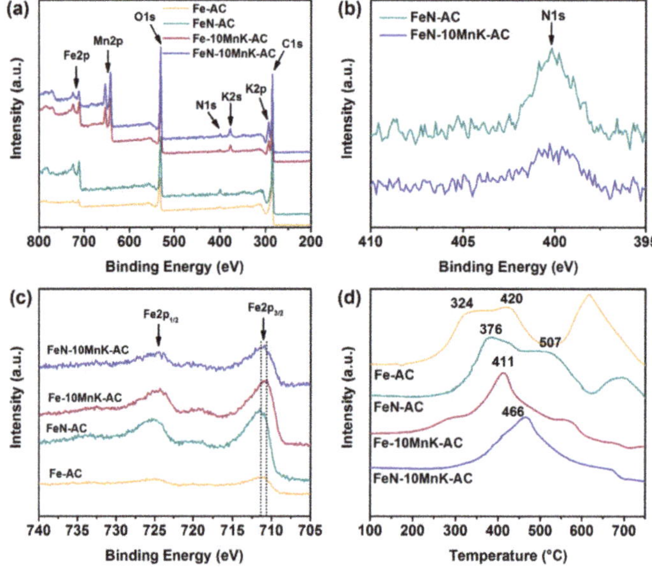

Figure 2. (**a**) XPS survey spectra of the Fe-AC, FeN-AC, Fe-10MnK-AC and FeN-10MnK-AC catalysts. The sharp photoelectron peaks at 712.1 eV (Fe2p), 642.2 eV (Mn2p), 531.1 eV (O1s), 378.0 eV (K2s), 294.1 eV (K2p), 285.0eV (C1s) and a weak photoelectron peak at 400.0 eV (N1s) are marked in the spectra. (**b**) High resolution XPS spectra of the N1s region for FeN-AC and FeN-10MnK-AC catalysts. (**c**) Fe2p XPS spectra and (**d**) hydrogen temperature programmed reduction (H_2-TPR) profiles of Fe-AC, FeN-AC, Fe-10MnK-AC and FeN-10MnK-AC catalysts.

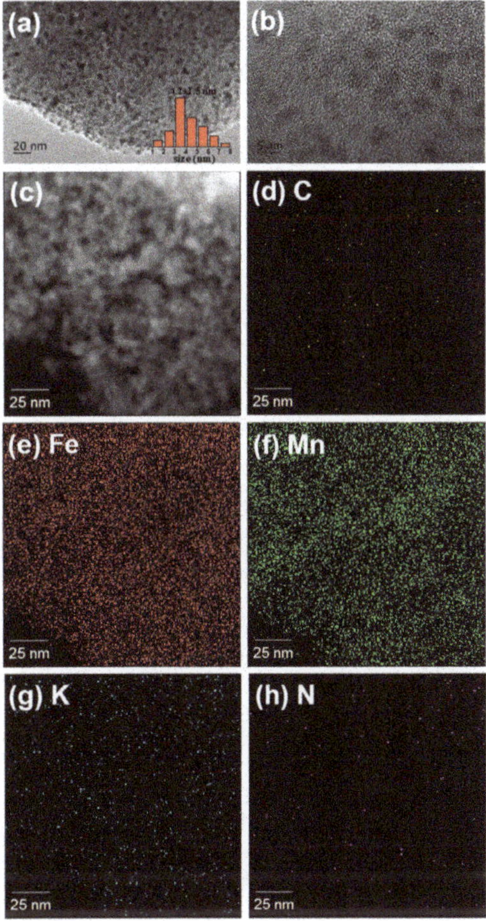

Figure 3. (**a**) TEM and (**b**) high resolution TEM (HRTEM) images of FeN-10MnK-AC catalysts. (**c**) High-angle annular dark field scanning transmission electron microscopy (HAADF-STEM) image and corresponding elemental mapping of FeN-10MnK-AC catalysts: (**d**) C; (**e**) Fe; (**f**) Mn; (**g**) K; (**h**) N.

Table 2. FTO performances of iron catalysts and N-doped iron catalysts.

Catalysts	Fe-AC	FeN-AC	Fe-10MnK-AC	FeN-10MnK-AC
GHSV (h^{-1})	15,000	15,000	15,000	15,000
CO conversion (%)	64.8	59.7	51.0	38.5
H_2 conversion (%)	60.6	57.4	29.5	17.1
H_2/CO usage	0.97	1.00	0.55	0.46
CO_2 selectivity (% C)	61.7	46.0	60.7	47.2
FTY [1] ($mol_{CO} \cdot g_{Fe}^{-1} \cdot h^{-1}$)	1.77	1.66	1.48	1.12
Product selectivity (wt %)				
CH_4	28.9	27.5	21.1	22.5
$C_2^=-C_4^=$	31.2	30.8	30.1	44.7
$C_2^0-C_4^0$	23.4	23.7	11.9	8.9
C_{5+}	16.5	18.1	36.9	23.9
O/P [2]	1.2	1.1	2.5	5.0

[1] FTY is the Fe time yield which is calculated as the mole number of CO molecules converted per gram of iron per hour. [2] O/P is the ratio of olefins to paraffins calculated by molar fraction of C_2-C_4 hydrocarbons.

The reduction performances of these prepared catalysts were analyzed by a temperature programmed reduction measurement in hydrogen (H_2-TPR). This is highly related to the intension of metal-support and metal-metal interactions. As to the comparison of Fe-AC and FeN-AC catalysts, N-doped iron catalysts show a stronger interaction that retards the reduction procedure and promotes the reduction peaks (from 324 °C to 376 °C and 420 °C to 507 °C), which are attributed to the $Fe_2O_3 \rightarrow Fe_3O_4$ and $Fe_3O_4 \rightarrow FeO$ transformation stages. The mixed spinel oxides (Fe, Mn)O_x with strong interaction leads to broad reduction peaks of Fe-10MnK-AC and FeN-10MnK-AC catalysts, which impedes the reduction of iron oxide to some extent. Similarly, the H_2-TPR profiles of Fe-10MnK-AC and FeN-10MnK-AC catalysts show an increase in reduction peaks (from 411 °C to 466 °C). The improvement between traditional iron catalysts and N-doped catalysts means that the existence of slight amount of N in the lattice of iron oxide or Fe-Mn mixed oxides inhibits the reducibility of supported catalysts. On the other hand, a considerable difference in catalyst reducibility also proves the interaction between doped nitrogen and metal promoters (Mn and K).

The intrinsic effects of the $KMnO_4$ pretreatment on carbon support, nitrogen doping, and their collaboration could be investigated by associating FTO performances and catalysts properties. Herein, the catalytic performances of the iron catalysts as well as N-doped iron catalysts that loaded on AC and $KMnO_4$ pretreated AC have been analyzed and compared. As presented in Table 2, the absorption and activation degree of CO and H_2 on the active sites of the catalysts were presented as CO conversion, H_2 conversion and the effective utilization of these two kinds of feed gas, which is denoted as the H_2/CO usage ratio. CO conversions of the iron catalysts supported on $KMnO_4$ pretreated AC (Fe-10MnK-AC and FeN-10MnK-AC) decreased compared with iron catalysts supported on AC, which is probably caused by the metal-support interaction between iron and manganese that retards the CO dissociation. H_2-TPR results show that the iron oxides supported on 10MnK-AC are difficult to reduce, which also verifies the less active catalytic surface. In addition, H_2/CO usage dramatically reduced from 1 to about 0.5. When H_2/CO usage ratio decreases, the chemisorption and dissociation of H_2 is relatively lower than that of CO. And the hydrogenation activity on the active sites is reduced. Considering the catalyst properties, it implies that manganese existed on pretreated AC hinders the dissociation rate of hydrogen. It effectively suppresses second hydrogenation of alkenes and the formation of CH_4 [28]. As shown in the hydrocarbon distributions of these catalysts, selectivity of methane decreased from 28.9% (Fe-AC) and 27.5% (FeN-AC) to 21.1% (Fe-10MnK-AC) and 22.5% (FeN-10MnK-AC). Besides, the O/P ratio also exhibits a promotion owing to the addition of manganese and potassium in the pretreatment process of support with $KMnO_4$ solution. These catalysts supported on 10MnK-AC suppressed H_2 adsorption and dissociation on active sites [29]. From the hydrogenation distribution, it can be concluded that these catalysts (Fe-10MnK-AC and FeN-10MnK-AC) have higher C_{5+} selectivities. This is regarded as a decrease of hydrogenation activity which leads to the higher amount of unsaturated intermediates, and this causes the increase of carbon chain growth ability.

According to the XRD and XPS results, a slight amount of nitrogen was introduced through the decomposition of ammonium iron citrate during calcination procedure. Unlike previous studies [17–19], nitrogen atoms were originated from the iron precursor with NH_4^+ group rather than exogenous NH_3 source, which leads to a more affiliative contact with iron. This is associated with the XRD results that iron nitrides were formed along with the formation of iron oxide. No evident changes of I_D/I_G ratio on Raman spectra between iron catalysts and catalysts doping with nitrogen is observed. It implies that the introduction of nitrogen atoms has no effects on the properties of carbon support and nitrogen atoms existed in the crystal lattice of iron oxide. Compared with Fe-AC, FeN-AC have nearly no shift on hydrocarbon distribution and the ratio of olefins to paraffins (O/P). Only a decrease of CO_2 selectivity from 61.7% to 46.0% was observed. Fe-10MnK-AC and FeN-10MnK-AC exhibits a similar trend on CO_2 selectivity as well. It signifies that nitrogen doping weakens WGS reaction activity, which favors the effective yield of desired hydrocarbons. Furthermore, an enormous shift to C_2–C_4 hydrocarbons selectivity is detected by comparing Fe-10MnK-AC and FeN-10MnK-AC catalysts. This is especially the case insofar as up to 44.7% of light olefins selectivity is obtained over FeN-10MnK-AC, which is higher than those over Fe-AC (31.2%), FeN-AC (30.8%) and Fe-10MnK-AC (30.1%) catalysts. The generation of C_2–C_4 alkanes is further retarded owing to the combined effect of promoter Mn and K as well as the doping agent N. The O/P ratio is improved at the same time. Nitrogen doping in Fe-Mn mixed oxides further enhances the catalyst surface basicity since it plays a role of electron donor and weakens H_2 dissociation during reaction, which is also the underlying reason why a lower H_2/CO usage ratio is achieved on FeN-10MnK-AC than Fe-10MnK-AC [30]. More explicitly, the introduced nitrogen atoms modifies the electron environment around the active sites. Electron pairs of nitrogen atoms occupy the electron hole pairs of adsorbed H^+ ions. This retards the formation of C–H bond and ulteriorly limits the second hydrogenation of unsaturated hydrocarbons adsorbed on active sites. As is described in catalyst characterization, ζ-Fe_2N with an orthorhombic structure is formed on FeN-10MnK-AC with the help of manganese while ε-Fe_3N is formed on FeN-AC. During the reaction, the dissociative CO diffuses into the lattice of iron nitrides and the iron carbonitrides phase is generated. It is regarded as a more stable active phase under FTO reaction conditions [21]. From the comparison of FTO performances, a synergistic effect of ζ-Fe_2N and the promoters Mn and K serve as a vital role in impeding the excessive hydrogenation of intermediate C* attached to FTO active sites. The competitive adsorption to hydrogen of manganese and the electron donation effect of nitrogen increase the FTO performances on FeN-10MnK-AC catalyst.

3. Materials and Methods

Catalysts used in this work were prepared by a facile method with some modifications reported by our group [15]. Briefly, the powdered wooden activated carbon (200 mesh, ~1400m^2/g, pore volume: 0.67 mL/g, purity: ≥98%, purchased form Sunson Activated Carbon Technology CO., Ltd., Shanghai, China) was firstly pretreated by boiling water to remove ash. And then, washed-out AC was redispersed in 10 wt % HNO_3 aqueous solution and refluxed at 80 °C for 5 h in order to remove residual metal ions and improve oxygen-containing groups on the surface of AC to enhance iron particle dispersity during impregnation and calcination process. The as-treated AC was filtered and washed until pH = 7 and dried overnight.

The dried AC was treated by $KMnO_4$ aqueous solution. Namely, 3 g AC was dispersed in 200 mL 0.1 mol/L $KMnO_4$ aqueous solution and placed in a water bath at 70 °C for 30 min with magnetic stirring. After washing with deionized water for several times, the obtained AC was dried overnight at 120 °C and denoted as 10MnK-AC.

Fe-AC and Fe-10MnK-AC catalysts were prepared by incipient wetness impregnation method. Pretreated AC and 10MnK-AC were separately impregnated with $Fe(NO_3)_3 \cdot 9H_2O$ solution (nominal Fe loading: 10 wt %). The suspensions were subject to magnetic stirring at room temperature for 24 h in order to evaporate the solvent and then dried at 120 °C for 12 h. Finally, samples were calcined under static N_2 at 300 °C for 3 h (heating rate: 2 °C/min) and Fe-AC and Fe-10MnK-AC catalysts were received. In contrast, FeN-AC and FeN-10MnK-AC were prepared with the same method except that the iron precursor was replaced with ammonium ferric (III) citrate, which is also regarded as the source of nitrogen that comes from the decomposition of NH_4^+ [31–33].

The morphologies of the catalysts were observed by a TEM (JEOL-2100F, Tokyo, Japan) at 200 kV accelerating voltage and the elemental mappings of the catalysts were obtained by EDS attached to TEM. N_2 adsorption-desorption analysis was carried out at 77 K to obtain pore properties such as specific surface area, total pore volume and pore size distribution on Micromeritics ASAP2460, Norcross, GA, USA. Powder X-ray diffraction (XRD) patterns by X-ray diffraction radiation (X'pert Pro MPD with Cu Kα (λ = 0.154 nm) radiation, 2θ=5-80°, Philips, Almelo, Netherland) was used to measure the phase and structure of the catalysts we prepared. XPS spectra were conducted using a Thermo ESCALAB 250XI, USA. The metal contents were determined by inductive coupled plasma-atomic emission spectroscopy (ICP-AES, OPTIMA 8000DV, PerkinElmer, Waltham, MA, USA). Raman spectra were measured on a Raman microscope (LabRAM HR800-LS55, HORIBA, Paris, France) using 532 nm laser excitation at ambient temperature. H_2-TPR was conducted in a U-shape quartz tube reactor with a thermal conductivity detector (TCD) with a H_2 flow rate of 50 ml/min by Quantachrome CPB-1, Boynton Beach, FL, USA.

The Fischer-Tropsch synthesis performance of the catalysts we prepared were conducted in a fixed bed reactor system. The reactor's inner diameter is 5 mm and the length is 250 mm. For every reaction experiment, 0.2 g catalyst was firstly diluted with 0.8 g quartz sand (60–80 mesh) homogeneously and sealed in the center of the reactor with quartz wool. Then, catalyst was reduced on site by syngas (a $H_2/CO/N_2$ ratio of 47.5/47.5/5 by volume and N_2 was used as the internal standard) at 350 °C and ambient pressure for 12 h and FT synthesis reaction was conducted under the same syngas at 320 °C, 2 MPa. And the gas hourly space velocity (GHSV) was set to 15,000 h^{-1} by regulating the gas flow rate. An online gas chromatography (GC7890B, Agilent, Santa Clara, CA, USA) equipped with an flame ionization detector (FID) and Thermal conductivity detector (TCD) was applied to analyze gaseous hydrocarbons and other gases (CO, H_2, CO_2, N_2, etc.) in the tail gas. The liquid phase was collected by a cold trap. The oil phase was separated from water phase and analyzed by an offline gas chromatography (GC2010, Shimadzu, Kyoto, Japan). Mostly the oil phase contains long chain alkanes, alkenes and small amount of alcohols. Catalyst activity was determined by CO conversion and hydrocarbons selectivity (by means of mass fraction) in all products was calculated on a carbon basis without CO_2. CO conversion was calculated as

$$X_{CO} = \frac{N_{CO,\,in} - N_{CO,\,out}}{N_{co,\,in}} \times 100\% \tag{1}$$

where $N_{CO,\,in}$ and $N_{CO,\,out}$ represent the molar flow rates of CO at the inlet and outlet of the reactor. Hydrocarbon selectivity (S_{CnHm}) was calculated on a carbon basis without CO_2:

$$S_{C_nH_m} = \frac{M_{C_nH_m}}{\sum_1^n M_{C_nH_m}} \times 100\% \tag{2}$$

where M_{CnHm} represent the weight of specific hydrocarbons with the carbon number n in the products. To ensure the validity of data, carbon balance should be better than 95%.

4. Conclusions

In conclusion, we have synthesized the N-doped FTO catalysts using ammonium iron citrate as precursor. XRD, XPS and Raman tests showed that the nitrogen atoms do not diffuse into carbon support but the iron lattice in order to form iron nitrides during catalyst calcination. The interaction between N and Fe was proven to exist from another perspective by the peak shift in XPS Fe2p spectra. The synergistic effect of N-doped active phases and the $KMnO_4$ pretreated carbon support was discussed by comparing their catalytic properties and performances with the undoped catalyst and untreated catalyst. A higher light olefins selectivity (44.7%) than Fe-AC, FeN-AC and Fe-10MnK-AC catalysts were achieved over FeN-10MnK-AC catalyst. FTO performances of Fe-AC and FeN-AC justifies that nitrogen doping alone does not improve the O/P ratio and C_2–C_4 olefins selectivity. The combined effect of inhibiting the second hydrogenation of alkenes probably associated with the competitive adsorption of hydrogen and electron donation effect of nitrogen. A comprehensive understanding of the effect on FTO catalyst performances of iron nitride needs to be further investigated.

Author Contributions: Z.T. and C.W. performed the experiments. Z.T., C.W. L.M. proposed and designed the experiments and corresponding characterizations. Y.X., W.L., L.C. and X.Z. contributed to analysis and discuss the experiment data. Z.T. and Z.S. contributed to write and polish this manuscript.

Funding: This work was supported by NSFC (National Natural Science Foundation of China Project (51476175 and 51776206) and CAS Pioneer Hundred Talents Program.

Conflicts of Interest: The authors declare no conflict of interest.

References

1. Torres Galvis, H.M.; de Jong, K.P. Catalysts for Production of Lower Olefins from Synthesis Gas: A Review. *ACS Catal.* **2013**, *3*, 2130–2149. [CrossRef]
2. Quann, R.J.; Green, L.A.; Tabak, S.A.; Krambeck, F.J. Chemistry of olefin oligomerization over ZSM-5 catalyst. *Ind. Eng. Chem. Res.* **1988**, *27*, 565–570. [CrossRef]
3. Zhang, Q.; Wang, T.; Weng, Y.; Zhang, H.; Vitidsant, T.; Li, Y.; Zhang, Q.; Xiao, R.; Wang, C.; Ma, L. Direct conversion of simulated propene-rich bio-syngas to liquid iso-hydrocarbons via FT-oligomerization integrated catalytic process. *Energy Convers. Manag.* **2018**, *171*, 211–221. [CrossRef]
4. Davis, B.H. Fischer− Tropsch synthesis: Comparison of performances of iron and cobalt catalysts. *Ind. Eng. Chem. Res.* **2007**, *46*, 8938–8945. [CrossRef]
5. Chernavskii, P.A.; Kazak, V.O.; Pankina, G.V.; Perfiliev, Y.D.; Li, T.; Virginie, M.; Khodakov, A.Y. Influence of copper and potassium on the structure and carbidisation of supported iron catalysts for Fischer–Tropsch synthesis. *Catal. Sci. Technol.* **2017**, *7*, 2325–2334. [CrossRef]
6. Xie, J.; Torres Galvis, H.M.; Koeken, A.C.; Kirilin, A.; Dugulan, A.I.; Ruitenbeek, M.; de Jong, K.P. Size and Promoter Effects on Stability of Carbon-Nanofiber-Supported Iron-Based Fischer–Tropsch Catalysts. *ACS Catal.* **2016**, *6*, 4017–4024. [CrossRef] [PubMed]
7. De Smit, E.; Cinquini, F.; Beale, A.M.; Safonova, O.V.; van Beek, W.; Sautet, P.; Weckhuysen, B.M. Stability and reactivity of ε− χ− θ iron carbide catalyst phases in fischer− tropsch synthesis: Controlling μc. *J. Am. Chem. Soc.* **2010**, *132*, 14928–14941. [CrossRef]
8. Dictor, R.A.; Bell, A.T. Fischer-Tropsch synthesis over reduced and unreduced iron oxide catalysts. *J. Catal.* **1986**, *97*, 121–136. [CrossRef]
9. Niemantsverdriet, J.; Van der Kraan, A.; Van Dijk, W.; Van der Baan, H. Behavior of metallic iron catalysts during Fischer-Tropsch synthesis studied with Mössbauer spectroscopy, x-ray diffraction, carbon content determination, and reaction kinetic measurements. *J. Phys. Chem.* **1980**, *84*, 3363–3370. [CrossRef]
10. Zhang, Q.; Kang, J.; Wang, Y. Development of novel catalysts for Fischer–Tropsch synthesis: tuning the product selectivity. *ChemCatChem* **2010**, *2*, 1030–1058. [CrossRef]
11. Cheng, Y.; Lin, J.; Wu, T.; Wang, H.; Xie, S.; Pei, Y.; Yan, S.; Qiao, M.; Zong, B. Mg and K dual-decorated Fe-on-reduced graphene oxide for selective catalyzing CO hydrogenation to light olefins with mitigated CO_2 emission and enhanced activity. *Appl. Catal. B Environ.* **2017**, *204*, 475–485. [CrossRef]
12. Huo, C.-F.; Li, Y.-W.; Wang, J.; Jiao, H. Insight into CH4 formation in iron-catalyzed Fischer− Tropsch synthesis. *J. Am. Chem. Soc.* **2009**, *131*, 14713–14721. [CrossRef] [PubMed]
13. De Smit, E.; Weckhuysen, B.M. The renaissance of iron-based Fischer–Tropsch synthesis: on the multifaceted catalyst deactivation behaviour. *Chem. Soc. Rev.* **2008**, *37*, 2758–2781. [CrossRef]
14. Bartholomew, C.H. Mechanisms of catalyst deactivation. *Appl. Catal. A Gen.* **2001**, *212*, 17–60. [CrossRef]
15. Tian, Z.; Wang, C.; Si, Z.; Ma, L.; Chen, L.; Liu, Q.; Zhang, Q.; Huang, H. Fischer-Tropsch synthesis to light olefins over iron-based catalysts supported on KMnO4 modified activated carbon by a facile method. *Appl. Catal. A Gen.* **2017**, *541*, 50–59. [CrossRef]
16. Park, H.; Youn, D.H.; Kim, J.Y.; Kim, W.Y.; Choi, Y.H.; Lee, Y.H.; Choi, S.H.; Lee, J.S. Selective Formation of Hägg Iron Carbide with g-C3N4 as a Sacrificial Support for Highly Active Fischer–Tropsch Synthesis. *ChemCatChem* **2015**, *7*, 3488–3494. [CrossRef]
17. Gu, B.; He, S.; Zhou, W.; Kang, J.; Cheng, K.; Zhang, Q.; Wang, Y. Polyaniline-supported iron catalyst for selective synthesis of lower olefins from syngas. *J. Energy Chem.* **2017**, *26*, 608–615. [CrossRef]
18. Chen, X.; Deng, D.; Pan, X.; Hu, Y.; Bao, X. N-doped graphene as an electron donor of iron catalysts for CO hydrogenation to light olefins. *Chem. Commun.* **2015**, *51*, 217–220. [CrossRef]
19. Schulte, H.J.; Graf, B.; Xia, W.; Muhler, M. Nitrogen- and Oxygen-Functionalized Multiwalled Carbon Nanotubes Used as Support in Iron-Catalyzed, High-Temperature Fischer-Tropsch Synthesis. *ChemCatChem* **2012**, *4*, 350–355. [CrossRef]
20. Xiong, H.; Moyo, M.; Motchelaho, M.A.; Tetana, Z.N.; Dube, S.M.; Jewell, L.L.; Coville, N.J. Fischer–Tropsch synthesis: iron catalysts supported on N-doped carbon spheres prepared by chemical vapor deposition and hydrothermal approaches. *J. Catal.* **2014**, *311*, 80–87. [CrossRef]
21. Yang, Z.; Guo, S.; Pan, X.; Wang, J.; Bao, X. FeN nanoparticles confined in carbon nanotubes for CO hydrogenation. *Energy Environ. Sci.* **2011**, *4*, 4500. [CrossRef]

22. Anderson, R.B.; Shultz, J.; Seligman, B.; Hall, W.K.; Storch, H. Studies of the Fischer—Tropsch Synthesis. VII. Nitrides of Iron as Catalysts1. *J. Am. Chem. Soc.* **1950**, *72*, 3502–3508. [CrossRef]

23. Yang, Z.; Pan, X.; Wang, J.; Bao, X. FeN particles confined inside CNT for light olefin synthesis from syngas: Effects of Mn and K additives. *Catal. Today* **2012**, *186*, 121–127. [CrossRef]

24. Li, N.; Wang, Z.; Zhao, K.; Shi, Z.; Gu, Z.; Xu, S. Large scale synthesis of N-doped multi-layered graphene sheets by simple arc-discharge method. *Carbon* **2010**, *48*, 255–259. [CrossRef]

25. Ismagilov, Z.R.; Shalagina, A.E.; Podyacheva, O.Y.; Ischenko, A.V.; Kibis, L.S.; Boronin, A.I.; Chesalov, Y.A.; Kochubey, D.I.; Romanenko, A.I.; Anikeeva, O.B. Structure and electrical conductivity of nitrogen-doped carbon nanofibers. *Carbon* **2009**, *47*, 1922–1929. [CrossRef]

26. Dorjgotov, A.; Ok, J.; Jeon, Y.; Yoon, S.-H.; Shul, Y.G. Activity and active sites of nitrogen-doped carbon nanotubes for oxygen reduction reaction. *J. Appl. Electrochem.* **2013**, *43*, 387–397. [CrossRef]

27. Prieto, P.; Marco, J.; Sanz, J. Synthesis and characterization of iron nitrides. An XRD, Mössbauer, RBS and XPS characterization. *Surf. Interface Anal.* **2008**, *40*, 781–785. [CrossRef]

28. Al-Sayari, S.A. Catalytic conversion of syngas to olefins over Mn–Fe catalysts. *Ceram. Int.* **2014**, *40*, 723–728. [CrossRef]

29. Miller, D.G.; Moskovits, M. A study of the effects of potassium addition to supported iron catalysts in the Fischer-Tropsch reaction. *J. Phys. Chem.* **1988**, *92*, 6081–6085. [CrossRef]

30. Yeh, E.; Schwartz, L.; Butt, J. Silica-supported iron nitride in Fischer-Tropsch reactions: II. Comparison of the promotion effects of potassium and nitrogen on activity and selectivity. *J. Catal.* **1985**, *91*, 241–253. [CrossRef]

31. Torres Galvis, H.M.; Bitter, J.H.; Davidian, T.; Ruitenbeek, M.; Dugulan, A.I.; de Jong, K.P. Iron particle size effects for direct production of lower olefins from synthesis gas. *J. Am. Chem. Soc.* **2012**, *134*, 16207–16215. [CrossRef] [PubMed]

32. Galvis, H.M.T.; Bitter, J.H.; Khare, C.B.; Ruitenbeek, M.; Dugulan, A.I.; de Jong, K.P. Supported iron nanoparticles as catalysts for sustainable production of lower olefins. *Science* **2012**, *335*, 835–838. [CrossRef] [PubMed]

33. Galvis, H.M.T.; Koeken, A.C.; Bitter, J.H.; Davidian, T.; Ruitenbeek, M.; Dugulan, A.I.; de Jong, K.P. Effects of sodium and sulfur on catalytic performance of supported iron catalysts for the Fischer–Tropsch synthesis of lower olefins. *J. Catal.* **2013**, *303*, 22–30. [CrossRef]

 *catalysts*

Article

Application of Microwave in Hydrogen Production from Methane Dry Reforming: Comparison Between the Conventional and Microwave-Assisted Catalytic Reforming on Improving the Energy Efficiency

Seyyedmajid Sharifvaghefi [1,†], Babak Shirani [1,†], Mladen Eic [1] and Ying Zheng [2,*]

[1] Department of Chemical Engineering, University of New Brunswick, Fredericton, NB E3B 5A3, Canada

[2] Department of Chemical and Biochemical Engineering, Thompson Engineering Building, Western University, London, ON N6A 5B9, Canada

* Correspondence: ying.zheng@uwo.ca; Tel.: +1(519)661-82138

† These authors share equal first authorship.

Received: 22 June 2019; Accepted: 18 July 2019; Published: 20 July 2019

Abstract: The microwave-assisted dry reforming of methane over Ni and Ni–MgO catalysts supported on activated carbon (AC) was studied with respect to reducing reaction energy consumption. In order to optimize the reforming reaction using the microwave setup, an inclusive study was performed on the effect of operating parameters, including the type of catalysts' active metal and their concentration in the AC support, feed flow rate, and reaction temperature on the reaction conversion and H_2/CO selectivity. The methane dry reforming was also carried out using conventional heating and the results were compared to those of microwave heating. The catalysts' activity was increased under microwave heating and as a result, the feed conversion and hydrogen selectivity were enhanced in comparison to the conventional heating method. In addition, to improve the reactants' conversion and products' selectivity, the thermal analysis also clarified the crucial importance of microwave heating in enhancing the energy efficiency of the reaction compared to the conventional heating.

Keywords: dry reforming methane (DRM); methane; carbon dioxide; microwave; conversion; catalyst; selectivity; thermal integration

---

1. Introduction

Applying methane dry reforming process for hydrogen production has received significant attention during the last decades as a result of increasing demand for clean and renewable energy [1–5]. In addition, the main reactants of the reforming reaction, methane and carbon dioxide, have tremendous effects on global warming, which necessitate controlling their concentration by converting them to a clean and sustainable source of energy, i.e., hydrogen. In addition to hydrogen, carbon monoxide is also another valuable product of methane dry reforming reaction that is further reacted to produce ultraclean fuels including gasoline, methanol, and dimethyl ether (DME) with negligible hazardous byproducts, e.g., aromatics [6,7]. Consequently, improving the CO_2 reforming of methane is essential in terms of conversion, selectivity, and energy efficiency.

The carbon dioxide reforming of methane is an endothermic reaction, which requires huge external heat source (Equation (1)). Hence, applying novel techniques to perform the reaction with minimum input energy is necessary to improve the process efficiency and reduce the costs.

$$CH_4 + CO_2 \Leftrightarrow 2CO + 2H_2 \qquad \Delta H_{298} = 247 \text{ kJ/mol} \qquad (1)$$

A new technique that received extensive attention in the last decades is based on using microwave energy potential for reforming of methane and carbon dioxide. In comparison to conventional heating methods, microwave applications for producing the required heat for reaction is more energy efficient and less expensive. Utilizing microwaves also provides more advantages than conventional heating including a rapid process heating, reduced processing time and work space, more accurate and uniform heating, and high quality [8–11]. In contrast, microwave heating has limitations and technical problems such as selection of the type of materials with a broad adsorption bandwidth that strongly absorb microwaves [12]. As a result, an extensive study is required on the application of microwave setup to optimize and improve the methane-reforming process in terms of conversion and product selectivity.

The dry reforming reaction takes place in the presence of a catalyst, which enhances the conversion rate, syngas selectivity, and increases the reaction rate. The applied catalysts are based on expensive noble metals (Pt, Pd, and Ru) or less expensive metals including Ni, Co, Fe, Cr, and Mo. The active metals are supported on metal oxides (Al_2O_3, MgO, and TiO_2), rare-earth oxides (CeO_2), zeolites, silicate, and mesoporous carbon materials [13–15]. In spite of the important role of catalysts on increasing the reaction rate and enhancing reactants conversion, the syngas production from methane reforming still needs to be performed at extreme conditions of high pressure and temperature [16,17]. Hence, conventional techniques have to be improved by integrating new techniques in catalytic methane reforming. The microwave heating of catalysts provides similar conversion to those of conventional heating at reduced temperatures by selective heating of active sites. In addition, the coke formation rate is smaller than conventional heating methane reforming, confirming the supremacy of microwave heating methane reforming.

In this study, carbon dioxide reforming of methane was studied in a microwave field. In order to improve adsorption efficiency of the microwaves' energy, activated carbon was used as a support for Ni and MgO active metal catalysts. Combining microwave setup and Ni-MgO/AC catalysts improved reforming reaction in terms of selectivity and conversion rates. In order to optimize the selectivity of products and conversion of reactants and the energy efficiency of the system, the effects of operating parameters such as catalyst bed temperature, feed flowrate, and also the concentration of Ni and addition of MgO were investigated in detail. The catalysts were also conventionally heated to compare the conversion, selectivity, and thermal efficiency with those obtained using microwave.

2. Results and Discussion

Textural properties, reducibility, and Ni dispersion and surface area of the prepared catalysts are shown in Table 1. With an increase in nickel content of the catalysts, the surface area and pore volume was reduced. The H_2 uptake of the catalysts, however, was found to to follow a different trend, increasing with nickel content of 15% and then being reduced for the catalysts with higher nickel content. The surface area for NiMgO/AC was found to be the lowest (539 m^2/g), however the H_2 uptake of this catalyst was only lower than the Ni/MgO/AC catalyst.

Table 1. Main characteristics of the synthesized catalysts.

	Surface Area (m^2/g)	Pore Volume (cc/g)	Average Pore Diameter (Å)	H_2 Uptake (moles/g Catalyst)	Ni Dispersion (%) [a]	Ni Surface Area (m^2/g) [a]
AC	1287	0.894	27.78	-	-	-
5Ni/AC	1004	0.733	29.20	2.80E-06	0.66	2.19
10Ni/AC	915	0.687	30.50	3.40E-06	0.40	1.33
15Ni/AC	871	0.654	31.76	4.56E-06	0.36	1.19
20Ni/AC	833	0.600	28.80	4.29E-06	0.25	0.84
25Ni/AC	665	0.473	28.46	4.16E-06	0.20	0.65
NiMgO/AC	639	0.457	28.96	5.95E-06	0.70	2.32
Ni/MgO/AC	712	0.468	26.28	6.15E-06	0.72	2.40

[a] calculated by assuming H_2/Ni atom ratio of 1.

The XRD profile of the 10Ni/AC, NiMgO/AC, and Ni/MgO/AC samples after 1 h of reduction in H_2 flow is shown in Figure 1. The peak at 26.6 can be ascribed to the AC support. The peaks at 44.4 and 76.3 are attributed to the (111) and (220) diffraction planes of Ni phase. The remaning peaks are assigned to the MgO phase. Although both NiO and MgO have the same charactristric peaks, the intensity of the peaks suggests that the NiO phase in the fresh catalyst is mostly reduced to Ni in all of the tested catalysts.

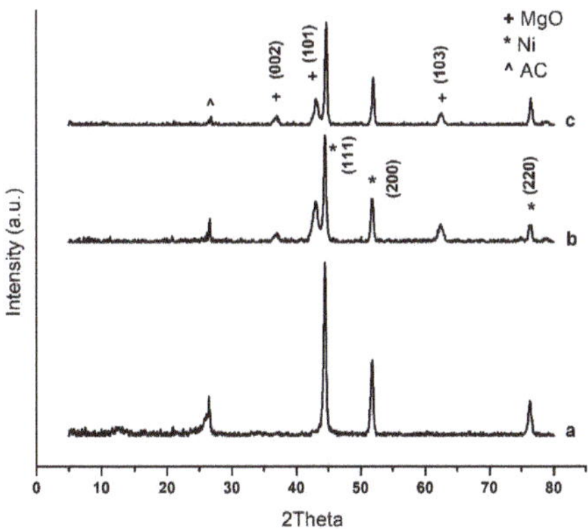

Figure 1. XRD profile of (**a**) 10Ni/AC, (**b**) NiMgO/AC, and (**c**) Ni/MgO/AC.

Temperature programmed reduction (TPR) behaviors of calcined 10Ni/AC, NiMgO/AC, and Ni/MgO/AC catalysts were investigated as shown in Figure 2. H_2 reduction for 10 Ni/AC catalyst shows the typical reduction peaks for NiO at 250 °C and 430 °C, which have been related to the reduction of Ni^{3+} and Ni^{2+} to metallic nickel, respectively [18,19]. For NiMgO/AC and Ni/MgO/AC catalysts, the first reduction peak is also observed at 250 °C, suggesting that part of the NiO phase is not influenced by the MgO. However, the second reduction peak is shifted to 740 °C, which can be assigned to the Ni^{2+} ions on the outermost layer of the MgO [20]. The wider peak for NiMgO/AC at this temperature indicates the reduction of NiO forms located in the subsurface layers of the MgO lattice, which is understandable considering that this catalyst was prepared through co-impregnation of NiO and MgO precursors. The higher dispersion of NiO in NiMgO/AC and Ni/MgO/AC facilitates the formation of smaller crystallites, which are more strongly bonded and, consequently, lowering the reducibility. On 10Ni/AC with lower dispersion, Ni species are not so strongly bonded to the surface, thus presenting a higher reducibility.

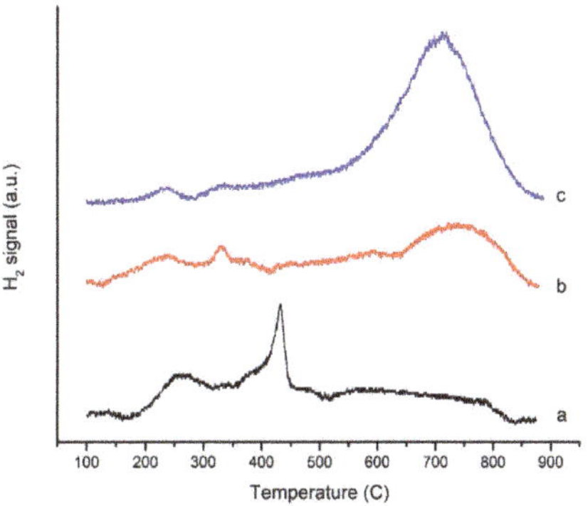

Figure 2. Temperature programmed reduction (TPR) profile of (**a**) 10Ni/AC, (**b**) NiMgO/AC, and (**c**) Ni/MgO/AC.

2.1. Temperature Effect on Conversion and Product Selectivity

The conversions of CH_4 and CO_2 over the synthesized catalysts at different temperatures using microwave as the heating medium, are shown in Figure 3a,b, respectively. It can be seen that the increase in temperature of reaction accelerates the reforming reaction resulting in higher conversion of the reactants at higher temperatures. As evidenced, among the Ni/AC catalysts, 15% nickel loading shows the highest conversion with 76% and 79% for CH_4 and CO_2, respectively. Conversions follow a decreasing trend with higher nickel content. The observed trend can be related to the H_2 uptake and nickel dispersion results by these catalysts, which shows the same trend. The results indicate that with the increase in nickel content up to 15% the number of available sites for the adsorption of the reacting gases was increased. However, with the addition of extra nickel to the catalyst, the nickel particles agglomerate and, therefore, the dispersion of the catalyst was reduced, resulting in lower active sites and lower conversion of the reactants.

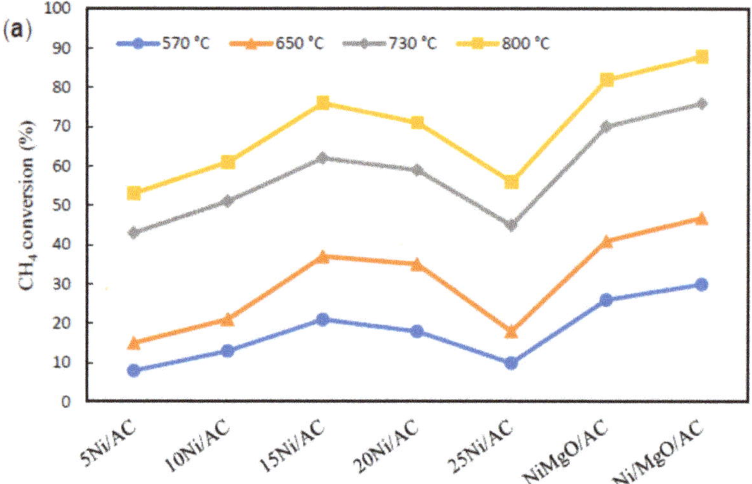

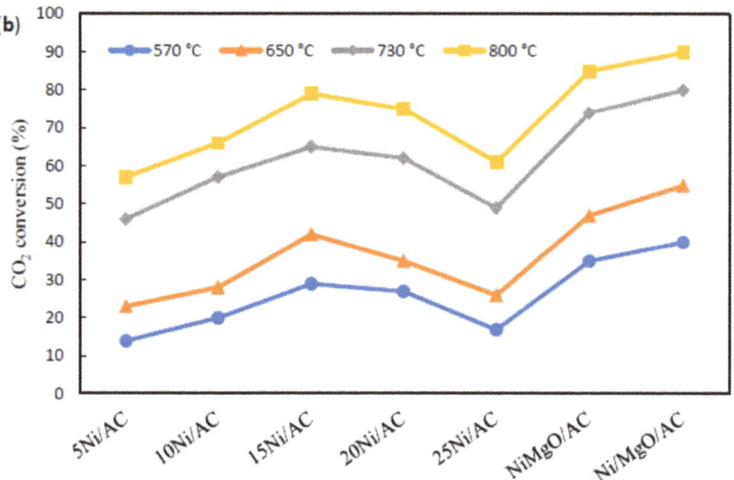

Figure 3. *Cont.*

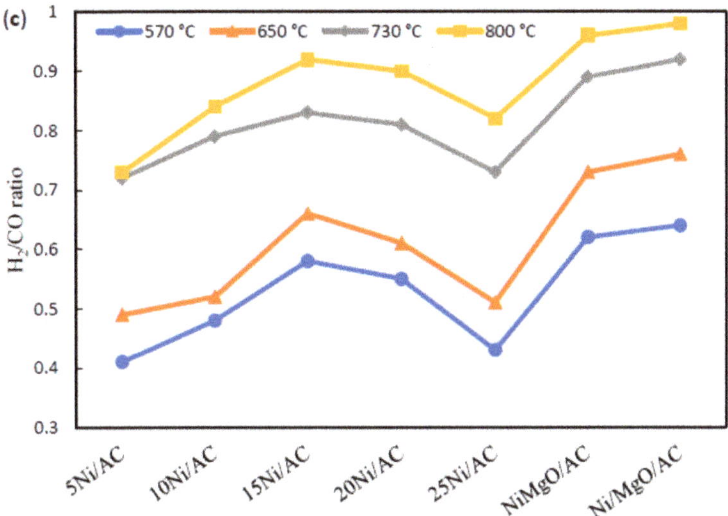

Figure 3. (a) CO_2 and (b) CH_4 conversions and (c) product selectivity as a function of temperature over the prepared catalysts using microwave heating.

The results show that the conversion of CH_4 and CO_2 increases almost linearly with nickel content of the catalysts (up to 15%). Furthermore, the addition of MgO to the catalyst was found to have a positive impact on the activity of the catalysts increasing the CH_4 conversion to 84% and 88% for NiMgO/AC and Ni/MgO/AC catalysts. H_2 uptake and dispersion results for these catalysts suggest that with the addition of MgO, nickel dispersion over the AC support is enhanced and therefore the number of active sites for the reforming reactions is increased. Comparing NiMgO/AC with Ni/MgO/AC, the conversion for the later catalyst was found to be higher. This could be attributed to the fact that with co-impregnation of NiO and MgO, parts of the NiO phase will be located beneath the MgO layers making them inaccessible to the reacting gases. Figure 3c shows the H_2/CO ratio for the tested catalysts at various temperatures. The products ratio follows the same trend as the conversion of the reactants for the catalysts. Moreover, with temperature increase, the ratio was increased for all the catalysts.

2.2. Effect of Heating Mechanism on Methane Reforming

To study the effect of the microwave heating on the dry reforming reaction, the experiments were also carried out using conventional heating over Ni/MgO/AC catalyst. The results are shown in Figure 4. The equilibrium conversions of CO_2 and CH_4 and the equilibrium molar ratio of H_2/CO are also plotted in the Figure 4. Two reactions were considered to perform the equilibrium calculations, i.e., methane reforming by carbon dioxide and reverse water–gas shift reaction (RWGS). It is believed that the reforming of methane starts by decomposition of CH_4 to carbonaceous deposits (Equation (2)). These deposits (at least those which are highly reactive), are gasified in the next step by CO_2 (Equation (3)), giving rise to an in-situ regeneration of the active sites. The reverse water–gas shift reaction also occurs as a side reaction in which some of the H_2 from the CH_4 decomposition reacts with the CO_2 to produce H_2O and CO (Equation (4)). The equilibrium data were obtained from the thermodynamic analysis of the results.

$$CH_4 \rightarrow C + 2H_2 \tag{2}$$

$$C + CO_2 \rightarrow 2CO \tag{3}$$

$$CO_2 + H_2 \Leftrightarrow H_2O + CO \tag{4}$$

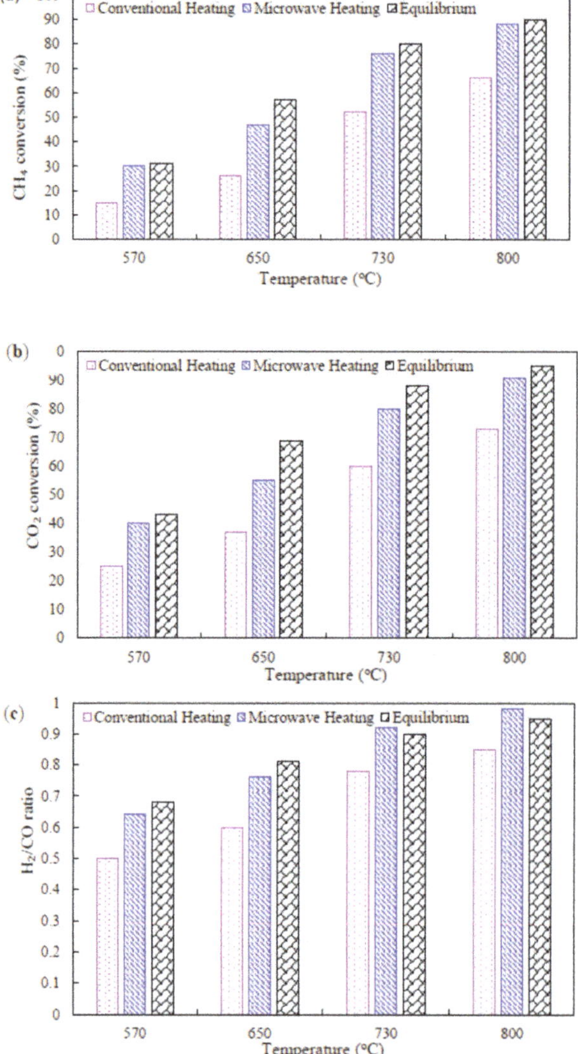

Figure 4. Comparison of microwave and conventional heating and equilibrium conditions for (a) CO_2 and (b) CH_4 conversions and (c) product selectivity as a function of temperature over Ni/MgO/AC catalyst and gas hourly space velocity (GHSV) of 33,000 mL/g cat.h.

Compared to the conventional heating, the conversion of methane and carbon dioxide were both higher using microwave heating. The difference in conversions can be explained by the differences between the heating mechanisms. In conventional heating, heat flux is responsible for energy transfer while microwave energy is directly transferred to the catalyst, without any heat flux. This results in higher temperature inside the catalyst compared to the temperature of the surrounding atmosphere near the surface, which can be translated to higher uniformity of heat distribution with respect to conventional heating [21,22]. Microwave heating also gives rise to micro-plasmas (also called hot spots) within the catalyst bed, where the temperature is higher than the measured average temperature in the bulk catalyst bed [23]. Such hot spots with higher temperature are reported to be responsible for enhancements in reaction rate, higher yields, and improved selectivity of heterogeneous catalytic reactions [24–26]. It is interesting to note that the conversions of CO_2 and CH_4 were always higher under microwave conditions with an almost constant difference with conventional heating. In contrast to conventional heating, microwave heating can generate multiple hot spots at the catalyst surface, where local reaction temperatures are higher than the overall reaction temperature.

In general, the results show that the H_2/CO ratios increased when the temperature was raised. With conventional heating, the H_2/CO ratio was lower than the equilibrium values. In contrast, the H_2/CO ratio was higher than the equilibrium value in the higher temperature range under microwave heating. The H_2/CO ratio was always found to be less than 1.0, an indication that the reverse water–gas shift reaction also occurred as a side reaction in addition to the methane reforming reaction. The variation of H_2/CO ratios depends on the degree of competition between these reactions. Based on the thermodynamic data, both the CH_4 reforming and the reverse water–gas shift reactions are endothermic. By increasing the temperature, both equilibrium constants (K_p) increased, but the enhancement of K_P for the CH_4 reforming reaction was to a greater extent than that for the reverse water–gas shift reaction. This showed higher impact of temperature on the first reaction to produce hydrogen and carbon monoxide, giving a ratio closer to unity. The higher H_2/CO ratio for microwave system can be explained by the creation of hot spots, which is similar to that from the thermal temperature effect. Using microwave heating, the reaction could occur at a higher temperature compared to the measured average temperature, so a higher H_2/CO ratio could be reached, which can be even higher than the equilibrium value at the measured temperature.

2.3. Effect of Varying Flow Rate on Methane Reforming

The effect of varying the flow rate on the conversion of CH_4 and CO_2 at 650 °C for Ni/MgO/AC catalyst is shown in Figure 5. CH_4 conversion decreased from 89% to 47% when gas hourly space velocity (GHSV) varies from 6700 mL/g cat.h to 33,000 mL/g cat.h. The results show a linear relation between the conversion of reactants and the flow rate of the inlet gases. The larger the GHSV, the lesser the contact time and the number of adsorbed reactants on the active sites. Unlike the results shown in Figure 4 where conversions of CO_2 and CH_4 were below equilibrium for GHSV of 33,000 mL/g cat.h, the reduction in flow rate leads to higher conversions than the calculated equilibrium. This shift in equilibrium for the methane reaction is also an indication of the presence of higher temperature at reacting sites (hot spots) than the measured average temperature of the bulk catalyst. With the decrease in GHSV, the H_2/CO ratio was increased. This can be attributed to the increase in gas–solid contact time. With higher contact time, the water–gas shift reaction can occur to consume the produced CO and therefore increase the H_2/CO ratio.

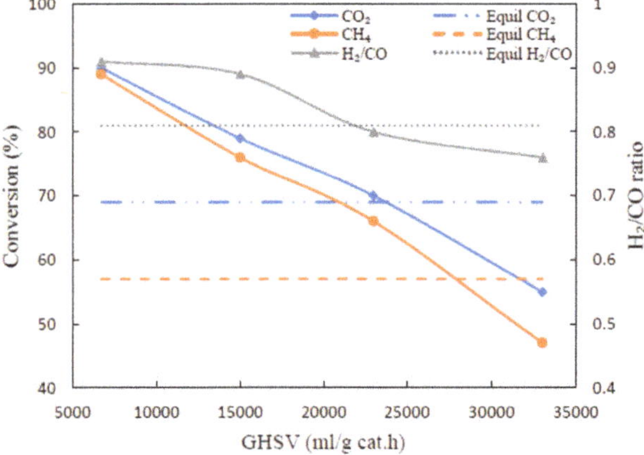

Figure 5. Effect of GHSV on Ni/MgO/AC catalyst performance for dry reforming of methane at 650 °C.

2.4. Reactor Energy Efficiency

Following the analysis presented by Durka et al. [27], an energy balance in macroscopic level is performed over the catalytic bed as follows:

$$\sum_i n_{i_{In}} \int_{T_{amb}}^{T_{in}} C_{p_i}(T)dT + Q_{Input} - Q_{Loss} = \Delta H_r^{T_{amb}} + \sum_i n_{out} \int_{T_{amb}}^{T_{out}} C_{p_i}(T)dT \tag{5}$$

where n_i is the number of moles of reactants and products, T_{amb} is the ambient temperature, T_{in} and T_{out} are the temperature of preheated reactants and products respectively, Q_{Input} is the heat provided to the reactor bed from the energy source, and Q_{Loss} is the heat dissipated to the surrounding. Since summation of the terms in left side of Equation (5) represents the net input energy into the reactor, Durka et al. [27] defined the energy efficiency as the fraction of input heat required to taking place the chemical reaction:

$$\epsilon = \frac{\Delta H_r^{T_{amb}}}{\Delta H_r^{T_{amb}} + \sum_i n_{out} \int_{T_{amb}}^{T_{out}} C_{p_i}(T)dT} \tag{6}$$

where ϵ is the energy efficiency. In the present study, in addition to the reaction between methane and carbon dioxide, reaction 4 is also considered as the side reaction. By considering the carbon dioxide as the limiting reactant, the total heat released from Equations (1) and (4) at ambient temperature is calculated based on the total mole of reacted CO_2. The results of the energy analysis for microwave and conventional heating reactor using 10Ni/AC and Ni/Mgo/AC catalysts are shown in Figure 6.

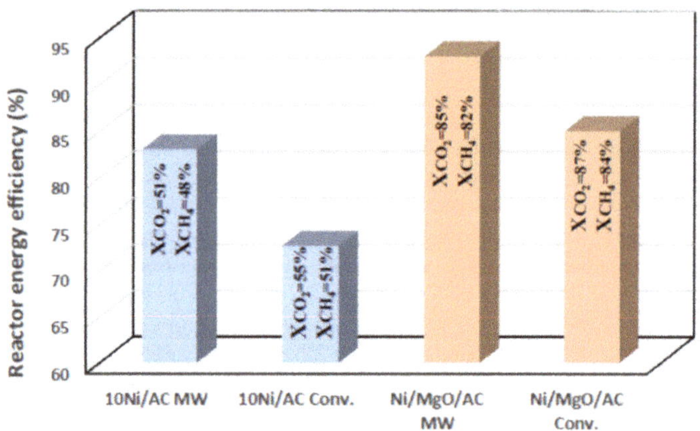

Figure 6. Energy efficiencies of microwave and conventional heating reactors using Ni/AC and Ni/MgO/AC catalysts.

According to Figure 6, similar conversions of carbon dioxide and methane were selected to compare the energy efficiency of the two reactors. As a result, the heat of reaction is constant for both microwave and conventional heating technique and the only factor determining the reactor efficiency is the heat dissipating out of the reactor with the product (Equation (6)). The catalyst bed in the microwave reactor has a lower temperature than the one in the conventional heating reactor at equal conversion. The temperatures of the bed and product gases for microwave and conventional heating reactors applying 10Ni/AC and Ni/MgO/AC catalysts are summarized in Table 2. The results indicate that applying microwave for heating the catalysts results in lower overall reaction temperature at similar conversion. The reason is because of the significant temperature difference between the metal sites and support materials [27]. This temperature difference results in significantly higher temperature metal reaction sites than the support and gas phase, providing similar conversion to conventional reactor. The product temperature leaving the microwave reactor has a significantly lower temperature than those leaving the conventionally heated reactor. Consequently, based on Equation (6), the energy efficiency of the microwave reactor is greater than the conventional reactor. According to Figure 6, applying microwave heating resulted in 10% improvement in energy efficiency for the bed filled with 10Ni/AC catalyst and 8% for the bed of Ni/MgO/AC catalyst.

Table 2. Temperatures of the beds and product gases for microwave and conventional heating reactors.

Catalyst	T_{Bed} (°C)	$T_{outlet\text{-}gas}$ (°C)
10Ni/AC MW	690	216
10Ni/AC Conv.	900	571
Ni/MgO/AC MW	695	212
Ni/MgO/AC Conv.	900	560

Based on the results, it is clear that the microwave performs the reaction at a considerably lower temperature, and it is more energy efficient. However, it should be emphasized that the greater energy efficiency values do not necessarily means that the total energy required by microwave is less than the conventional heating reactors. Precise analysis on magnetron efficiencies and the energy loss in microwave conversion to heat has to be carried out to optimize the consumed energy by the microwave. Another advantage of applying the microwave in providing the reaction energy is the fast stabilization of the bed temperature. In comparison to conventional heating, where a few hours were required for the bed approaching the reaction temperature, the microwave heated the bed in a significantly shorter time and the reaction temperature was reached in a few minutes.

3. Experimental

3.1. Catalyst Preparation and Characterization

The xNi/AC with different Ni metal loading (x showing the Ni loading) were prepared by incipient wet impregnation. To prepare the xNi/AC, an aqueous solution containing the required mass of $Ni(NO_3)_2.6H_2O$ (Sigma-Aldrich) was added to AC support (from Carbon Resources, coal based HCL washed) to make 5, 10, 15, 20, and 25Ni/AC. 10Ni/10MgO/AC (named Ni/MgO/AC herein) and 10Ni-10MgO/AC (named NiMgO/AC herein) catalysts were also prepared using the impregnation technique. The Ni/MgO/AC catalyst was synthesized by step-wise impregnation of first the MgO precursor ($Mg(NO_3)_2.6H_2O$, Sigma-Aldrich) followed by drying and calcination and then the addition of Ni precursor. NiMgO/AC catalyst was prepared by co-impregnation of the MgO and Ni precursors. The catalysts were dried under continuous stirring at 90 °C followed by drying at 120 °C for 24 h. The catalysts were eventually calcined at 700 °C with a ramp rate of 10 °C/min under Argon stream for 5 h. The catalysts were reduced at 600 °C prior to their characterization. The textural properties of the prepared catalysts including BET surface area, pore volume, and pore diameter were measured by analyzing nitrogen adsorption–desorption isotherm at 77 K using Autosorb-1 (Quantachrome Instrument). In addition, the crystalline patterns of the catalysts were determined by X-ray diffraction (XRD) using a Bruker D8 Advance spectrometer. The diffractometer was equipped with a two-circle goniometer housed in a radiation safety enclosure. The X-ray source was a sealed, 2.2 kW Cu X-ray tube (X-ray wavelength 1.54 A), maintained at an operating current of 40 kV and 25 mA. Samples were scanned in the range of 10–80° 2θ. A step size of 0.02° and a step time of 1.0 s were used during the measurements. A peltier-cooled solid-state [Si(Li)] detector (Sol-X) with a useful energy range of 1 to 60 KeV was used as the detector.

The hydrogen temperature programmed reduction (TPR) analysis for fresh catalysts were performed using Autosorb-1. For TPR analysis, about 10 mg catalyst was regenerated in helium flow for 2 h at 120 °C. After removing the moistures, a helium stream that contained 5% hydrogen was introduced to the catalyst, and the system was heated up to 1000 °C with a heating rate of 10 °C/min to reduce the catalysts. The data regarding the temperature were collected using a thermocouple placed in a catalyst bed for accurate measurement. H_2 chemisorption was performed to determine the nickel dispersion and nickel surface area. Further, 10 mg of catalyst was first reduced in-situ at 600 °C. The catalyst was subsequently purged with Ar flow at 600 °C and was then cooled down to room temperature. Hydrogen was introduced again as a mixture of H_2 (2%) and Ar (98%) with total flow rate of 50 mL/min. The catalyst was flushed with argon at 80 °C and the temperature was raised afterwards with only argon flowing. The effluent gas from the catalyst bed was analyzed by mass spectroscopy (RGA 200, SRS) to determine the H_2 concentration. Nickel dispersion and surface area were calculated considering the one-on-one adsorption of hydrogen on nickel atoms and atomic nickel cross-sectional area of 6.49 $\times 10^{-20}$ m^2/Ni-atom by the following equations:

$$Dispersion = S. \, M. \, N/100W \tag{7}$$

$$Surface \, Area = N. \, L. \, C/100W \tag{8}$$

where S is a stoichiometric factor. That means, for example, in the case of Ni and H_2 chemisorption, one molecule of H_2 dissociates to give two atoms to be chemisorbed on Ni atoms. N is the amount of monolayer, experimentally obtained from flow experiments in mol/g cat. M is the molecular weight of the metal (g/mol). W is the weight percentage of metal. L is the Avogadro number and C is the nickel cross-sectional area (m^2/Ni-atom).

3.2. Experimental Setup

Microwave Setup: The schematic diagram of the microwave setup is shown in Figure 7. The dry reforming reaction was carried out in a fixed-bed quartz tube with an inside diameter of 16.9 mm and a length of 9.2 mm. The catalysts (0.9 g) were placed in the waveguide channel and were heated by microwaves reaching the reaction temperature. The quartz wool was used to adjust the location of catalyst bed in the center of waveguide channel. The temperature of the catalyst bed was measured using a thermo-gun (model Milwakee M12) through an open window in front of the catalyst. The thermo-gun was calibrated before the experiments. The system was at atmospheric pressure and the applied microwave power was set to change the temperature of the catalyst. Prior to the catalytic reaction, the catalyst was reduced at 600 °C for 2 h with hydrogen flow of 50 mL/min. After heating the catalysts to the reaction temperature, a mixture of methane-carbon dioxide (1:1) with a total flowrate of 500 mL/min was heated in an electric pre-furnace, which contained a 15 m circular 1/8″ stainless-steel tube, for preheating, and eventually fed to the catalyst bed. About 30 min after feed introduction, when the products' concentration became stable, the gas products were sampled and analyzed using a gas chromatograph (GC, Varian 3400) equipped with a residual gas analyzer (RGA 200). The product sampling was repeated three times and reasonable accuracy (less than 1% difference) in the results was obtained.

Conventional Heating Setup: Similar setup to microwave was used with a difference that in these set of experiments; a quartz tubular U-shaped reactor was used to hold 0.9 g of catalyst. The reactor had a diameter (i.d.) of 7.9 mm and a length of 36.3 mm for holding catalysts. The reactor was heated in a furnace and the bed temperature was measured using a type R thermocouple placed inside the tube in the furnace. Similar operating conditions to that of the catalytic-microwave reaction was applied, and the effects of temperature and metal content were investigated and compared to the results from microwave setup.

Data Evaluation: The methane and carbon dioxide conversions and hydrogen and carbon monoxide selectivities were calculated using the following equations:

$$X_{CH_4} = \frac{n_{T,in} \times y_{CH4,in} - n_{T,out} \times y_{CH4,out}}{n_{T,in} \times y_{CH4,in}} \times 100 \tag{9}$$

$$X_{CO_2} = \frac{n_{T,in} \times y_{CO2,in} - n_{T,out} \times y_{CO2,out}}{n_{T,in} \times y_{CO2,in}} \times 100 \tag{10}$$

$$S_{H_2} = \frac{n_{T,out} \times y_{H2,out}}{n_{T,in} - (n_{T,out} \times y_{CH4,out} + n_{T,out} \times y_{CO2,out})} \times 100 \tag{11}$$

$$S_{CO} = \frac{n_{T,out} \times y_{CO,out}}{n_{T,in} - (n_{T,out} \times y_{CH4,out} + n_{T,out} \times y_{CO2,out})} \times 100 \tag{12}$$

$$R_{H2/CO} = \frac{n_{T,out} \times y_{H2,out}}{n_{T,out} \times y_{CO,out}} \tag{13}$$

where X is the conversion, S is the selectivity, y is the mole fraction, $n_{T,in}$ is the total moles of the input gases, and $n_{T,out}$ is the total moles of output gases.

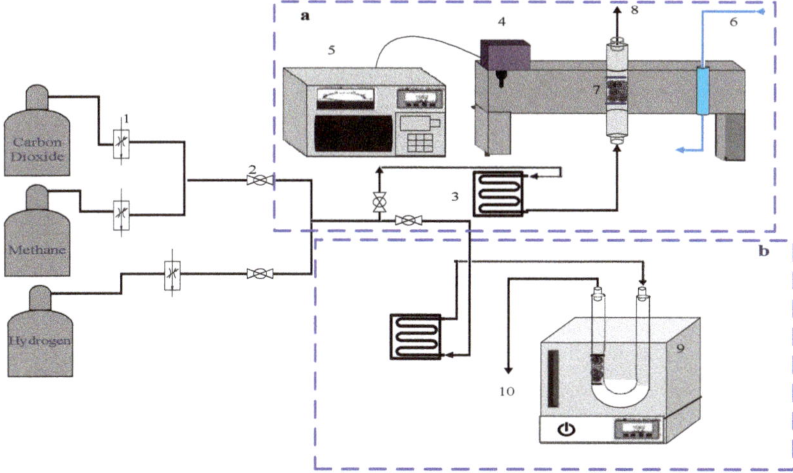

Figure 7. Schematic diagram of (**a**) microwave and (**b**) conventional setups for methane dry reforming. (1) Mass flow controller. (2) Ball valve. (3) Preheater jacket. (4) Microwave magnetron (5) Microwave power supply. (6) Cooling water. (7) Catalyst bed (quartz tube). (8) Products from microwave reactor. (9) Furnace. (10) Product from conventional reactor.

The equilibrium constant is given in terms of mole fractions y or number of moles n of the individual components for both methane reforming (MR) and reverse water gas shift (RWGS) reactions as follows:

$$k_{MR} = \frac{p_{H_2}^2 \cdot p_{CO}^2}{p_{CO_2} \cdot p_{CH_4}} = \frac{y_{H_2}^2 \cdot y_{CO}^2}{y_{CO_2} \cdot y_{CH_4}} \cdot P_{Tot}^2 = \frac{n_{H_2}^2 \cdot n_{CO}^2}{n_{CO_2} \cdot n_{CH_4} \cdot n_{Tot}^2} \cdot P_{Tot}^2 = \exp(-\Delta G_{MR}/RT) \tag{14}$$

$$k_{RWGS} = \frac{y_{H_2O} \cdot y_{CO}}{y_{CO_2} \cdot y_{H_2}} = \frac{n_{H_2O} \cdot n_{CO}}{n_{CO_2} \cdot n_{H_2}} = \exp(-\Delta G_{RWGS}/RT). \tag{15}$$

In order to calculate the equilibrium conversions, the equilibrium constant was calculated by finding the Gibbs free energy. The calculations were performed using MATLAB.

4. Conclusions

Methane reforming by CO_2 was investigated in this study over series of Ni/AC, NiMgO/AC, and Ni/MgO/AC catalysts using microwave as the heating source. Among Ni/AC catalysts with varying nickel loading, 15Ni/AC showed the highest activity. The addition of Mg as the promoter was found to enhance the activity of the catalyst by increasing nickel dispersion. H_2/CO ratio was also increased, indicating that with the addition of Mg, the effect of secondary reaction was less profound. Increasing the reaction temperature was found to improve the conversion of reactants, which was expected as the reforming reaction is endothermic. Increasing the temperature from 650 °C to 730 °C showed the highest increase in conversion and H_2/CO ratio.

Compared to conventional heating, applying microwave heating at the same catalyst bed temperature for both heating systems resulted in significant enhancement of the methane conversion. However, for high flow rate (GHSV of 33,000 mL/g cat.h), the conversions were below the equilibrium value. Decreasing the flow rate, the conversions passed the equilibrium values for the tested temperature (650 °C). The higher conversions were related to the creation of micro-plasmas or hot spots over the active sites increasing the reaction temperature higher than the measured average temperature of catalyst bed.

Reactor energy efficiency based on the outlet gas temperatures was compared between both heating mechanisms for 10Ni/AC and Ni/MgO/AC catalysts. The microwave heating was found to have lower outlet gas temperature and higher efficiency than its conventional counterpart.

Author Contributions: Conceptualization, Y.Z.; data curation, S.S. and B.S.; formal analysis, S.S. and B.S.; methodology, S.S. and B.S.; supervision, M.E. and Y.Z.; writing—original draft, S.S. and B.S.; writing—review and editing, M.E. and Y.Z.

Funding: This research was funded by The Natural Sciences and Engineering Research Council of Canada (NSERC).

Acknowledgments: The authors gratefully acknowledge financial support from the Natural Sciences and Engineering Research Council of Canada (NSERC) Discovery Grant and Tobie Boutot for his technical support.

Conflicts of Interest: The authors declare no conflict of interest.

References

1. Wysocka, I.; Hupka, J.; Rogala, A. Catalytic Activity of Nickel and Ruthenium–Nickel Catalysts Supported on SiO_2, ZrO_2, Al_2O_3, and $MgAl_2O_4$ in a Dry Reforming Process. *Catalysts* **2019**, *9*, 540. [CrossRef]
2. Chun, S.M.; Shin, D.H.; Ma, S.H.; Yang, G.W.; Hong, Y.C. CO_2 Microwave Plasma—Catalytic Reactor for Efficient Reforming of Methane to Syngas. *Catalysts* **2019**, *9*, 292. [CrossRef]
3. Kumar, A. Low Temperature Activation of Carbon Dioxide by Ammonia in Methane Dry Reforming—A Thermodynamic Study. *Catalysts* **2018**, *8*, 481. [CrossRef]
4. Chung, W.C.; Chang, M.B. Dry reforming of methane by combined spark discharge with a ferroelectric. *Energy Convers. Manag.* **2016**, *124*, 305–314. [CrossRef]
5. Hamzehlouia, S.; Jaffer, S.A.; Chaouki, J. Microwave Heating-Assisted Catalytic Dry Reforming of Methane to Syngas. *Sci. Rep.* **2018**, *8*, 8940. [CrossRef] [PubMed]
6. Choudhary, T.V.; Choudhary, V.R. Energy-Efficient Syngas Production through Catalytic Oxy-Methane Reforming Reactions. *Angew. Chem. Int. Ed.* **2008**, *47*, 1828–1847. [CrossRef] [PubMed]
7. Abdullah, A.Z.; Bhatia, S.; Fan, M.-S.; Fan, M. Catalytic Technology for Carbon Dioxide Reforming of Methane to Synthesis Gas. *ChemCatChem* **2009**, *1*, 192–208.
8. Caddick, S.; Fitzmaurice, R.; Li, H.; Nguyen, N.; Fronczek, F.R.; Vicente, M.G.H. Microwave enhanced synthesis. *Tetrahedron* **2009**, *65*, 3315–3323. [CrossRef]
9. Ku, H.; Siores, E.; Taube, A.; Ball, J. Productivity improvement through the use of industrial microwave technologies. *Comput. Ind. Eng.* **2002**, *42*, 281–290. [CrossRef]
10. De la Hoz, A.; Diaz-Ortiz, A.; Moreno, A. Microwaves in organic synthesis. Thermal and non-thermal microwave effects. *Chem. Soc. Rev.* **2005**, *34*, 164–178. [CrossRef]
11. Gude, V.; Patil, P.; Martinez-Guerra, E.; Deng, S.; Nirmalakhandan, N. Microwave energy potential for biodiesel production. *Sustain. Chem. Process.* **2013**, *1*, 5. [CrossRef]
12. Xiong, L.; Yu, M.; Liu, J.; Li, S.; Xue, B. Preparation and evaluation of the microwave absorption properties of template-free graphene foam-supported Ni nanoparticles. *RSC Adv.* **2017**, *7*, 14733–14741. [CrossRef]
13. Yang, Y.C.; Lee, B.J.; Chun, Y.N. Characteristics of methane reforming using gliding arc reactor. *Energy* **2009**, *34*, 172–177. [CrossRef]
14. Heintze, M.; Pietruszka, B. Plasma catalytic conversion of methane into syngas: The combined effect of discharge activation and catalysis. *Catal. Today* **2004**, *89*, 21–25. [CrossRef]

15. Wang, Y.F.; Tsai, C.H.; Chang, W.Y.; Kuo, Y.M. Methane steam reforming for producing hydrogen in an atmospheric-pressure microwave plasma reactor. *Int. J. Hydrogen Energy* **2010**, *35*, 135–140. [CrossRef]
16. Zeng, Y.; Zhu, X.; Mei, D.; Ashford, B.; Tu, X. Plasma-catalytic dry reforming of methane over γ-Al_2O_3 supported metal catalysts. *Catal. Today* **2015**, *256*, 80–87. [CrossRef]
17. Kraus, M.; Egli, W.; Haffner, K.; Eliasson, B.; Kogelschatz, U.; Wokaun, A. Investigation of mechanistic aspects of the catalytic CO_2 reforming of methane in a dielectric-barrier discharge using optical emission spectroscopy and kinetic modeling. *Phys. Chem. Chem. Phys.* **2002**, *4*, 668–675. [CrossRef]
18. Kalai, D.Y.; Stangeland, K.; Jin, Y.; Tucho, W.M.; Yu, Z. Biogas dry reforming for syngas production on La promoted hydrotalcite-derived Ni catalysts. *Int. J. Hydrogen Energy* **2018**, *43*, 19438–19450. [CrossRef]
19. Solsona, B.; Concepción, P.; Nieto, J.M.L.; Dejoz, A.; Cecilia, J.A.; Agouram, S.; Soriano, M.D.; Torres, V.; Jiménez-Jiménez, J.; Castellón, E.R. Nickel oxide supported on porous clay heterostructures as selective catalysts for the oxidative dehydrogenation of ethane. *Catal. Sci. Technol.* **2016**, *6*, 3419–3429. [CrossRef]
20. Parmaliana, A.; Arena, F.; Frusteri, F.; Giordano, N. Temperature-programmed reduction study of NiO?MgO interactions in magnesia-supported Ni catalysts and NiO?MgO physical mixture. *J. Chem. Soc. Faraday Trans.* **1990**, *86*, 2663. [CrossRef]
21. Zlotorzynski, A. The Application of Microwave Radiation to Analytical and Environmental Chemistry. *Crit. Rev. Anal. Chem.* **1995**, *25*, 43–76. [CrossRef]
22. Meredith, R.J. *Engineers' Handbook of Industrial Microwave Heating, Iet*; The Institution of Electrical Engineers: London, UK, 1998.
23. Domínguez, A.; Fernández, Y.; Fidalgo, B.; Pis, J.; Menéndez, J. Biogas to syngas by microwave-assisted dry reforming in the presence of char. *Energy Fuels* **2007**, *21*, 2066–2071. [CrossRef]
24. Stuerga, D.; Gaillard, P. Microwave heating as a new way to induce localized enhancements of reaction rate. Non-isothermal and heterogeneous kinetics. *Tetrahedron* **1996**, *52*, 5505–5510. [CrossRef]
25. Will, H.; Scholz, P.; Ondruschka, B. Heterogeneous Gas-Phase Catalysis Under Microwave Irradiation—A New Multi-Mode Microwave Applicator. *Top. Catal.* **2004**, *29*, 175–182. [CrossRef]
26. Zhang, X.; Hayward, D.O.; Mingos, D.M.P. Effects of Microwave Dielectric Heating on Heterogeneous Catalysis. *Catal. Lett.* **2003**, *88*, 33–38. [CrossRef]
27. Durka, T.; Stefanidis, G.D.; Van Gerven, T.; Stankiewicz, A.I. Microwave-activated methanol steam reforming for hydrogen production. *Int. J. Hydrogen Energy* **2011**, *36*, 12843–12852. [CrossRef]

 *catalysts*

Article

The Effects of Catalyst Support and Temperature on the Hydrotreating of Waste Cooking Oil (WCO) over CoMo Sulfided Catalysts

Hui Wang [1,2,*], Kyle Rogers [2], Haiping Zhang [2,3], Guoliang Li [2], Jianglong Pu [1], Haoxuan Zheng [2], Hongfei Lin [2], Ying Zheng [2,4,*] and Siauw Ng [5]

[1] College of Biological, Chemical Sciences and Engineering, Jiaxing University, 118 Jiahang Road, Jiaxing 314001, China
[2] Department of Chemical Engineering, University of New Brunswick, 15 Dineen Drive, Fredericton, NB E3B 5A3, Canada
[3] Department of Chemical Engineering & Technology, Tianjin University, Tianjin 300072, China
[4] Department of Chemical and Biochemical Engineering, Western University, 1151 Richmond Street, London, ON N6A 3K7, Canada
[5] National Centre for Upgrading Technology, Canmet ENERGY-Devon, 1 Oil Patch Drive, Edmonton, AB T9G 1A8, Canada
* Correspondence: huiwang@zjxu.edu.cn or hui.wang@unb.ca (H.W.); ying.zheng@uwo.ca (Y.Z.)

Received: 25 July 2019; Accepted: 12 August 2019; Published: 15 August 2019

Abstract: Waste cooking oil (WCO) hydrotreating to produce green diesel is good for both the environmental protection and energy recovery problems. The roles of catalyst support and reaction temperature on reactions during WCO hydrotreating process were evaluated over an unsupported and a commercial sulfided cobalt and molybdenum (CoMoS) catalyst supported by a mixture of Al_2O_3, TiO_2, and SiO_2. The presence of catalyst support helped to improve the dispersion and enlarge the surface area of CoMoS, and was found to be a key factor in reducing reaction temperature, in enhancing the hydrodeoxygenation (HDO) and hydrogenation capabilities, and in decreasing polymerization capability. The increase of reaction temperature strongly improved the deoxygenation, hydrogenation, and cracking reaction activities. Compared to the unsupported CoMoS, the supported one exhibited good deoxygenation and hydrogenation capabilities at 340 °C in WCO hydrotreating to produce diesel fraction; however, high temperature operation needs to be carefully controlled because it may cause overcracking and dehydrogenation.

Keywords: catalyst support; CoMo sulfided catalyst; deoxygenation; cracking and polymerization; hydrogenation and dehydrogenation; waste cooking oil

---

1. Introduction

Renewable biofuels can reduce greenhouse gas emissions because of the carbon fixation that occurs during plant growth through the process of photosynthesis [1]. The conversion of waste cooking oil (WCO) into fuel is a good method to solve both the environmental protection and energy recovery problems. Biodiesel—fatty acid methyl ester—is the most common biofuel presently; however, it has some drawbacks: (1) high acidic number and low heat value due to the presence of oxygen and (2) air sensitivity due to the presence of unsaturated hydrocarbons. Green diesel, the product from hydrotreating of triglycerides, is made up of deoxygenated and saturated hydrocarbons, and has similar molecular structures to those of petroleum derived diesel fuels. Therefore, hydrotreating is regarded as the potential route to solve all the problems of biodiesel. Another advantage is that the triglyceride-based biofuels can be compatible with the current infrastructure.

Catalysts are the core for the catalytic reactions, and the reaction routes significantly depend on the catalyst used. MoS_2, a very popular catalyst for petroleum hydrodesulfurization and hydrodenitrogenation, and is also regarded as an effective catalyst for deoxygenation during triglyceride hydrotreating [2–8]. The roles of different promoters on the triglyceride deoxygenation process over MoS_2 have been extensively investigated. The sulfur vacancies formed when MoS_2 was promoted by Ni, which improved the hydrogenation capability and the selectivity of hydrodeoxygenation (HDO) to hydrodecarbonylation/decarboxylation (HDCO) to a higher level. The reason for this improvement is that there is a synergistic effect between MoS_2 and Ni on the unsaturated metal edge sites [9]. Co-promoted MoS_2 exhibited saturated edge sites in hydrogen atmosphere, favoring cracking by C–C hydrogenolysis, and showed a higher selectivity for HDCO over HDO due to the adsorption of carbon atoms in reactant molecules on the sulfur edge adjacent to Co [9,10].

Different catalyst supports [11], even the same support only differing in morphology, orientation, or composition [12–15], had a significant effect on the hydrotreating process. The dispersion of the metal sulfides and the existence of the bond between metal sulfides and support were influenced by the types of catalyst support; therefore, the activity of deoxygenation and the selectivity of the deoxygenation pathways (HDO and HDCO) could be fine-tuned [16–18]. Alumina-supporting sulfided cobalt and molybdenum (CoMoS) did not correlate with the deoxygenation capability, but it enhanced methylation reactions to a large extent [10,19]. Compared to nonacidic support, activated carbon and silica, more acidic CoMoS catalyst supports, such as alumina, enhanced decarbonylation and deesterification reactions of carboxyl groups [16]. Comparing NiMo/SAPO-11 and NiMo/Al$_2$O$_3$, the former had higher selectivity for decarboxylation (41.2%); whereas the latter favored to hydrodeoxygenation (62.5%) [20]. Among SiO_2, Al_2O_3, and TiO_2, NiMo/SiO$_2$ showed lower hydrogenation and higher decarboxylation; whereas NiMo/TiO$_2$ resulted in higher HDO selectivity [17]. The Si/(Si+Al) ratio of catalyst support significantly affected the total acidic sites of catalyst, further changed the deoxygenation reaction rates and pathways [21]. Due to these conflicting results and the lack of studies to discuss non-deoxygenation reactions took place at the same time during WCO hydrotreatment, it is necessary to investigate the roles of catalyst support on triglyceride hydrotreating.

Comparing an unsupported catalyst to a supported catalyst may make it easier to comprehend what role catalyst support plays during WCO deoxygenation process. Besides, operational temperature was found to be the most dominant factor affecting the triglyceride hydrotreating process, which significantly affected product composition [1,22–24]. The previous publications have proven that CoMoS catalysts were efficient deoxygenation catalyst for triglycerides [6–10,12,15,16,19,22]. Moreover, CoMoS catalysts are widely used in petroleum or its fraction hydrotreatment processes; therefore, they are more easily applied in industrial plant. This present work investigated the roles of catalyst support on WCO hydrotreating process by three categories reactions: deoxygenation, hydrogenation/dehydrogenation, and cracking/polymerization by comparing an unsupported CoMoS catalyst and a supported (Al$_2$O$_3$-TiO$_2$-SiO$_2$) CoMoS catalyst at different reaction temperatures. The main objectives were to explore the reaction routes, to identify the effects of support/temperature on product distributions and compositions, and to provide a basis for industrial production using the triglyceride hydrotreating process.

2. Results and Discussion

2.1. Catalyst Characterization

2.1.1. Catalyst Morphology and Basic Properties

The morphology of the unsupported and supported CoMoS catalyst were identified by TEM images (Figure 1). The unsupported CoMoS (Figure 1a) appears to have longer slabs (Figure 1b) than the supported CoMoS.

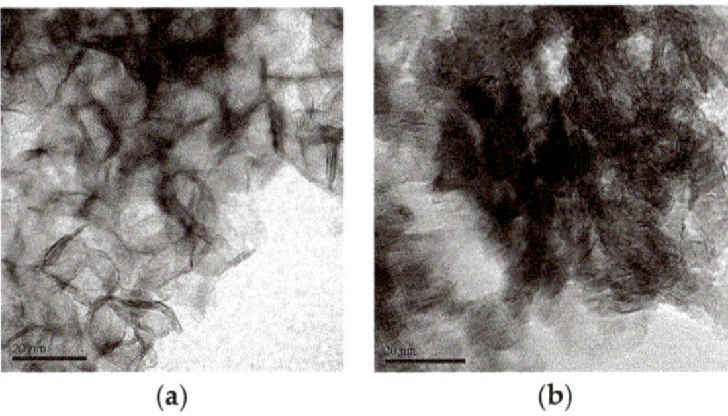

(a) **(b)**

Figure 1. TEM images of unsupported and supported CoMoS catalysts. (**a**) Unsupported catalyst and (**b**) supported catalyst.

The basic properties of the studied CoMoS are shown in Table 1. Regarding the average layer number, the unsupported and supported CoMoS have average layer numbers of 2.9 and 2.3, respectively, and the slab length of the unsupported and supported catalysts was 7.8 nm and 5.8 nm. Using the method described by Calais et al. [25], the ratios of edge metal atoms to edge atoms (including Co and Mo metal atoms and sulfur atoms) on the unsupported and supported CoMoS catalyst were 0.31 and 0.25, respectively. This illustrated that the supported CoMoS had more edge metal sites whereas the unsupported one had more saturated sulfur sites. The atomic ratios of Co to the total metal atoms (Co + Mo) for the unsupported and supported CoMoS were 0.25 and 0.31, respectively. Compare to the unsupported CoMoS, the supported one had a larger surface area but a smaller pore volume. Figure S1 shows that the unsupported catalyst has a bimodal pore size distribution: a sharp peak at 2.5 nm and a broad peak nearby 12 nm; however, only one broad peak near to 6 nm showed in the supported catalyst pore size distribution results.

Table 1. Basic properties of unsupported and supported CoMoS catalysts.

Catalysts	Supported Catalyst	Unsupported Catalyst
Average slab length (nm)	5.84	7.80
Average layer numbers	2.34	2.92
Fraction of edge metal atoms	0.25	0.15
Surface area ($m^2\ g^{-1}$)	150.0	82.2
Pore volume ($cm^3\ g^{-1}$)	0.35	0.53
Co/Mo (wt%/wt%)	3.0–4.0/16–20	10.3/50.4
Co/(Co + Mo) (atom ratio)	0.31	0.25

2.1.2. Active Site Distribution

The temperature-programmed reduction (TPR) spectra shown in Figure 2 are quantitatively summarized in Table S1.

As can be seen in Figure 2, H_2S was not generated over the unsupported catalyst until the temperature was higher than 400 °C. The trials with the supported catalyst generated different results: H_2S was detected right when the consumption of H_2 began. The temperature of the first H_2 consumption peak over the supported catalyst (at 175 to 275 °C) was much lower than that over the unsupported catalyst (at approximately 230 to 430 °C). This may be explained by: (1) the well dispersion of the active sites by the supports where smaller particles were formed; (2) the presence of catalyst support that limits the growth of MoS_2 crystals. These small-sized active phases should show more activity towards the hydrogen reduction and sulfiding since much more contact surfaces are available than the large-sized bulk one (unsupported catalyst). This means that hydrogen could be better dissociatively adsorbed on the active phase for the supported CoMoS catalyst, while much less available hydrogen would participate in the hydrogenation reactions for the unsupported one.

On the other hand, according to the previous study, the nano-sized metals, such as bare nickel nano particles, are easily agglomerated during a reduction process [26]. This might be the reason that the unsupported catalysts could be easily sintered during the reaction which increases the difficulty of hydrogen reduction and decreases the catalytic activity. As shown in Table S1, the amounts of H_2 consumption and H_2S production were also much lower over the supported catalyst than over the unsupported catalyst. This is because for the same amount of catalysts, more portions of active phases are contained in the unsupported catalyst than the supported CoMoS catalyst. However, the supported catalyst displayed a higher activity under low reaction temperature (250–275 °C) due to the lower hydrogen reaction temperature of the supported CoMoS (Figure 2). Furthermore, under hydrotreating a temperature range of 250 to 375 °C, unsaturated sulfur vacancies, formed after H_2S is produced, are the primary active sites that the supported CoMoS have. Comparatively, saturated sulfurs are the main active sites that the unsupported CoMoS have, because there is no H_2S release, which means the consumed hydrogen was adsorbed on saturated sulfur sites. These results that the types of active sites on catalyst surface are in accordance with the fraction of edge metal atoms.

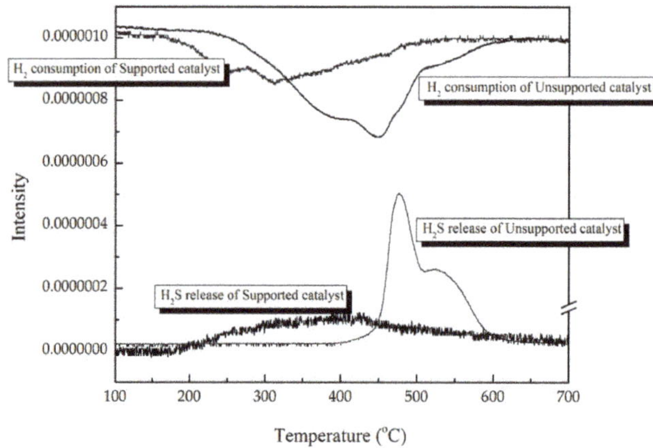

Figure 2. Temperature-programmed reduction (TPR) spectra of the supported and unsupported catalyst.

The unsupported catalyst only contained metal active sites, whereas the supported catalyst was a bifunctional catalyst, having both metal active sites and acidic sites present on the support. There were no strong acids on the catalyst support, while the amounts of weak and medium acids were 2.26 and 0.23 mmol per gram of catalyst. It should be noted that only Lewis acids exist on the support made of alumina and titania. It was reported that the Lewis acids derived from the Mo^{4+} helped the hydrodeoxygenation process owing to the electron acceptors on the Lewis acid sites where the carbonyl group (C=O) could be easily adsorbed [27]. Therefore, the comparatively sufficient reduction of the supported CoMoS catalyst leads to the formation of more Mo^{4+}, which are favorable for the hydrodexoygenation, and on the other hand, the Lewis acids on the support also favors to enhance the hydrodexoygenation in the hydrotreating process. This explains the better catalytic performances of the supported CoMoS than that of the unsupported one.

2.2. Liquid Products

Compositions of Feed and Liquid Products

The WCO feed was filtered before use. The H_2O content, density, and total acid value were 0%, 0.9177 g/cm³, and 0.86 mg KOH/g, respectively. The feed was primarily composed of triglycerides, which were converted into the corresponding fatty acid methyl esters (determined by GC/MS) [28]. C16 saturated fatty acid (palmitic acid) and C18 unsaturated acids were the primary fatty acids in

WCO, accounting for 6.7 ± 0.8% and 89.3 ± 0.8% (72.4 ± 1.1 wt % of oleic acid and 17.5 ± 0.3 wt % of linoleic acid), respectively. Figure 3 shows the product fractions of chemical compounds in the hydrotreated liquid products.

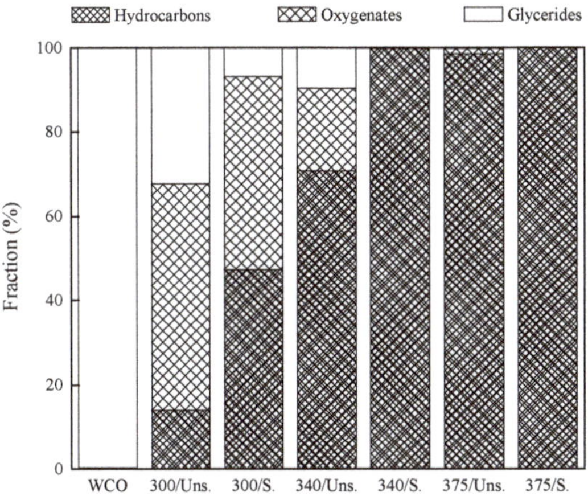

Figure 3. Fraction distribution of liquid products hydrotreated by unsupported and supported CoMoS catalysts. * The number is reaction temperature; Uns and S mean unsupported and supported CoMoS, respectively. e.g., 340-S: the liquid products are upgraded at 340 °C over supported CoMoS. The same abbreviations are used in the remaining Figures.

Over the unsupported catalyst, the hydrocarbon fractions increased when the temperature rose from 300 to 375 °C. Over the supported catalyst, the hydrocarbon fractions first increased and then subsequently were stable at 100% when the temperature rose from 300 to 340 and from 340 to 375 °C, respectively. The hydrocarbon fraction was close to 100% over the unsupported CoMoS when the temperature was at 375 °C, which was higher than 340 °C over the supported CoMoS. The existence of the catalyst support dramatically reduced the threshold value of reaction temperature: the supported catalyst needed ~35 degrees less than the unsupported catalyst to reach a similar deoxygenation extent, which is of great application value in industry. The presence of the catalyst support, Al_2O_3–TiO_2–SiO_2, strongly enhanced the deoxygenation capabilities of CoMoS.

3. Reaction Route Discussion

Considering the significant differences in liquid product compositions that were obtained by doing the reactions with the supported and unsupported catalyst at different temperatures, the reaction routes were primarily examined from three points of interest: deoxygenation (C–O cracking or C–CO cracking), cracking (C–C cracking)/polymerization (C–C chain growth), and hydrogenation (C=C saturation)/dehydrogenation (C–H disassociation). The reaction routes of the hydrotreated WCO over the unsupported CoMoS were investigated in a previous study [14]. The simplified reaction pathways are shown in Scheme 1.

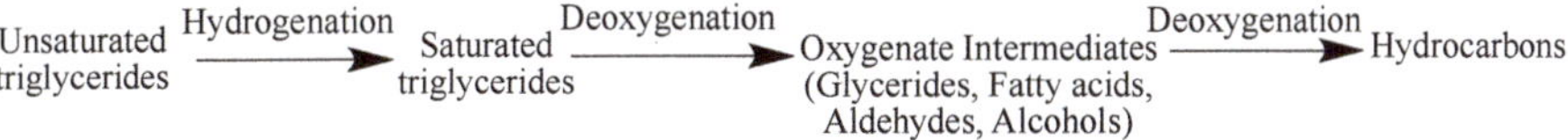

Scheme 1. Simplified reaction pathways during waste cooking oil (WCO) hydrotreating processes.

Unsaturated triglycerides were first subjected to hydrogenation, which transformed unsaturated carbon–carbon double bonds into saturated carbon–carbon single bonds. Oxygen-containing intermediates (including glycerides and fatty acids) and gas hydrocarbon products were then produced through the breaking down of the newly saturated triglycerides. This process occurred very quickly and was irreversible [29]. The by-product of this process, glycerol, was quickly converted into propane or propene and water by hydrotreating. The fatty acids were then deoxygenated (through removal of CO, CO_2, or H_2O) into hydrocarbons under different reaction routes. The produced hydrocarbons could undergo non-deoxygenate reactions. Focus was placed on the deoxygenation, cracking/polymerization, and hydrogenation/dehydrogenation reactions of the supported and unsupported CoMoS catalyst.

3.1. Deoxygenation

3.1.1. Types of Oxygen Containing Compounds

The general goal of studying the WCO hydrotreating was to produce oxygen-free diesel fraction.

As shown in Figure 4, the oxygenate fractions in the products decrease with the increase of reaction temperature when using the unsupported CoMoS as a catalyst. The oxygenates were almost completely removed (less than 0.2% oxygenates were left) in the products over 340 °C by using the supported CoMoS; whereas 1.4% fatty acids existed in the products at 375 °C by using the unsupported CoMoS. This indicates that the presence of the catalyst support caused an enhancement to the deoxygenation reactions. The oxygenate intermediates in the liquid products by the unsupported and supported CoMoS were primarily fatty acids.

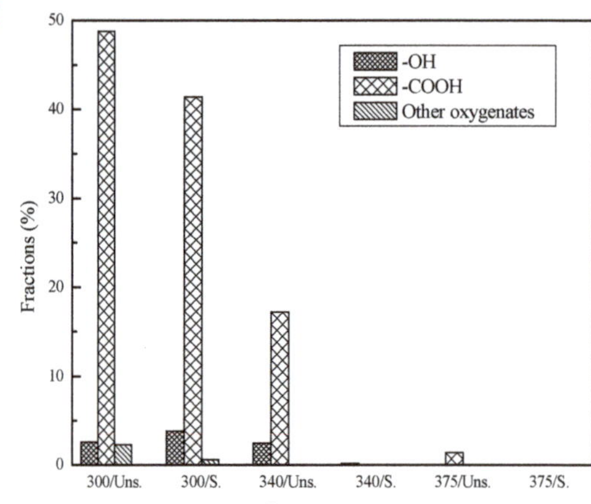

Figure 4. Oxygenates distributions in WCO hydrotreated liquid products.

3.1.2. HDO vs. HDCO

The two reaction routes that fatty acids undergo for deoxygenation are C-O cracking (hydrodeoxygenation, HDO) and C–CO cracking (HDCO, including hydrodecarbonylation (DCO) and hydrodecarboxylation (DCO_2)). Differentiating between DCO and DCO_2 is difficult due to the fact that CO and CO_2 can undergo water–gas shift (CO reacting with water to form H_2 and CO_2). From the point of thermodynamics, the methanation of CO and CO_2 becomes possible below 550 °C and a low temperature is favorable [30,31]. Consequently, under the present reaction conditions (300–375 °C), it is possible that water and methane are produced from the reaction of CO and/or CO_2 to H_2 [32]. Therefore, only two deoxygenation routes were considered, HDO and HDCO [33]. Stearic acid was used as an example to illustrate the deoxygenation routes (see Scheme 2).

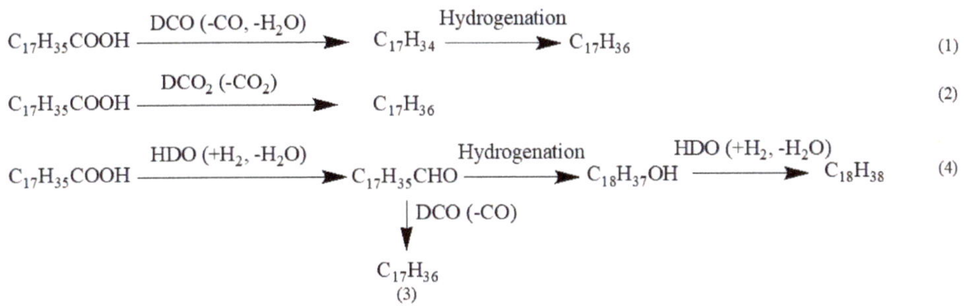

$$C_{17}H_{35}COOH \xrightarrow{\text{DCO (-CO, -H}_2\text{O)}} C_{17}H_{34} \xrightarrow{\text{Hydrogenation}} C_{17}H_{36} \qquad (1)$$

$$C_{17}H_{35}COOH \xrightarrow{\text{DCO}_2\text{ (-CO}_2\text{)}} C_{17}H_{36} \qquad (2)$$

$$C_{17}H_{35}COOH \xrightarrow{\text{HDO (+H}_2\text{, -H}_2\text{O)}} C_{17}H_{35}CHO \xrightarrow{\text{Hydrogenation}} C_{18}H_{37}OH \xrightarrow{\text{HDO (+H}_2\text{, -H}_2\text{O)}} C_{18}H_{38} \qquad (4)$$

$$\downarrow \text{DCO (-CO)}$$

$$C_{17}H_{36}$$
$$(3)$$

Scheme 2. Deoxygenation routes during WCO hydrotreating processes.

Unsaturated C17 hydrocarbon is produced first due to DCO by releasing CO and water [34], and then C17 olefin can be converted to paraffin (reaction (1) in Scheme 2). The decarboxylation of fatty acid does not require hydrogen and results in saturated C17 and a by-product of CO_2 (reaction (2) in Scheme 2) [35,36], or the saturated C17 (reaction (3) in Scheme 2) is formed from the reduction of fatty acid to aldehyde [37]. C18 hydrocarbon and by-product, H_2O, are produced through hydrodeoxygenation (reaction (4) in Scheme 2) [38,39], where the alcohol or aldehyde intermediates can be respectively formed from the reduction of aldehydes or fatty acids. The ratio of the occurrences of the HDO reactions to the HDCO reactions can be measured by relating the concentration of even number hydrocarbons to odd number hydrocarbons in the products. Based on the composition of WCO, which consists of a majority of C18 fatty acids (91%), this method is suitable for obtaining the HDO/HDCO ratios. The results were confirmed by the ratios of C16/C15 because the C16 content in the WCO is 8.0% (7.5% saturated and 0.5% unsaturated C16). The errors were less than 3.0% between C18/C17 and C16/C15 ratios.

The comparison between the C18/C17 ratios of different products is shown in Figure 5.

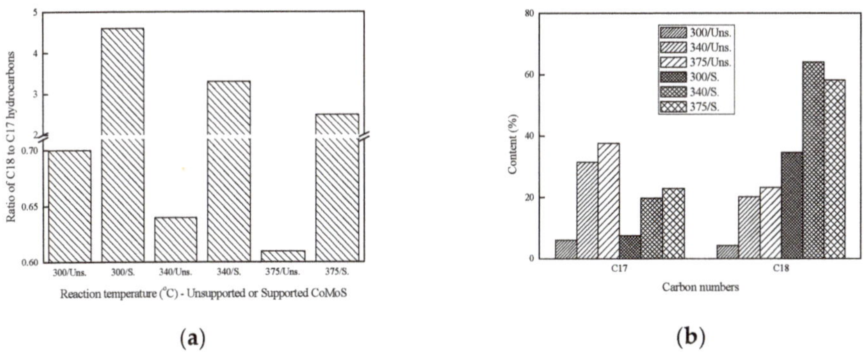

(**a**)   (**b**)

Figure 5. C17 and C18 distributions in WCO hydrotreated liquid products. (**a**) Mass ratio of C18/C17 hydrocarbons. (**b**) C17 and C18 distributions.

It was noticed that the C18/C17 ratios of the products obtained over the supported CoMoS were much higher than those over the unsupported CoMoS as shown Figure 5. This suggests that HDO was the main reaction pathway of the supported CoMoS catalyst (oxygen was removed primarily in the form of H_2O). On the other hand, this result also suggests that HDCO was the main reaction pathway of the unsupported CoMoS (oxygen was removed primarily in the form of CO or CO_2). The presence of the CoMoS catalyst support significantly influenced the triglyceride deoxygenation process, probably because the Lewis acidic sites on the support strengthened the hydrogenolysis capability [16,35], which increased the hydrogen concentration on the catalyst surface, and further enhanced the HDO capability. Lewis acidic sites on the support (Al_2O_3–TiO_2–SiO_2) might also have increased the occurrence of hydrogenation-dehydration and hydrogenation steps (reactions (3), (6), and (7) in Scheme 2) because the fatty acids bonded with Lewis acidic sites on the catalyst support facilitating the removal of the –OH [35], which led to an enhancement of the HDO pathway.

As shown in Figure 2, the supported catalyst contained abundant sulfur vacancies, whereas the unsupported catalyst contained none until the reaction temperature reached 400 °C. Therefore, at the reaction temperature of 300 to 375 °C, the main active sites of the supported catalyst were sulfur vacancy, whereas the unsupported catalyst was dominated by saturated sites. The distinct active sites are responsible for different main reaction pathways. Combining TPR results and liquid product compositions, the active phases on the supported catalyst have a good ability to dissociate H_2 to remove the oxygen by HDO, while the active phases on the unsupported catalyst do not have enough ability to activate the adsorbed hydrogen, resulting in the occurrence of HDCO. In other words, HDO took place on the active sites that were unsaturated and oxygen was removed by the sulfur vacancies residing on the catalyst surface. Conversely, HDCO took place on the sulfur-saturated sites. Ryymin et al observed that reduction reactions (e.g., hydrodeoxygenation and hydrogenation) occurred on unsaturated sites independently and decarbonylation occurred on sulfur-saturated sites when the reactions occurred such as deoxygenation of phenol and methyl heptanoate over a sulphided $NiMo/Al_2O_3$ catalyst [35]. For both the supported and unsupported CoMoS, the influence of temperature on HDCO was the same: HDCO was found slightly more favorable under higher temperature.

3.2. Hydrocracking (C–C Cracking) and Polymerization (C–C Chain Growth)

The cracking and polymerization selectivity of the unsupported and supported CoMoS products at different temperatures were evaluated and the experiment results are shown in Figure 6. WCO was composed of more than 96 wt % of C16 and C18 fatty acids. The C15 to C18 hydrocarbons were therefore the expected corresponding deoxygenated products if no cracking and/or polymerization occurred [33]. Hydrocarbons with carbon numbers lower than 15 and carbon numbers higher than 18 were thus considered to be the products of hydrocracking and polymerization, respectively. In comparison with the feed, the hydrotreated products had a higher $C15^-$ and a lower C15–18 distribution as shown in Figure 6.

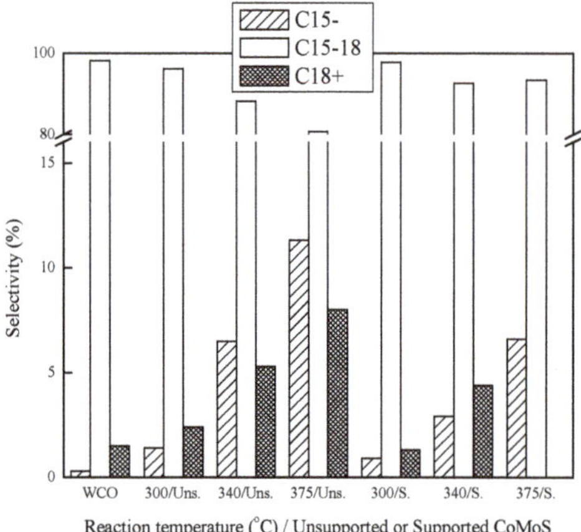

Figure 6. Hydrocracking and polymerization reactions during WCO hydrotreating processes.

All liquid products were primarily composed of C15–18 (over 80%). Reaction temperature was a key factor in the cracking/polymerization of WCO hydrotreating using a CoMoS catalyst. Bezergianni et al. made the same observation [1,2,23,40]. When the reaction temperature increased from 300 to

375 °C over the unsupported CoMoS, the product fraction of hydrocarbons that were lighter than C15 increased from approximately 1.4 wt % to 11.3 wt %, while C15-18 decreased from 96.2 wt % to 80.7 wt %. The same effect also occurred for the supported CoMoS: when the reaction temperature increased from 300 to 375 °C, the C15$^-$ fraction increased from 0.9 wt % to 6.6 wt % and C15-18 decreased from 97.8 wt % to 93 wt %.

Over the unsupported CoMoS, the following changes were observed when the reaction temperature increased from 300 to 375 °C: 15.5 wt % of C15-18 was converted into 9.9 wt % of C15$^-$ and 5.6 wt % of C18$^+$. The degree of polymerization increased with increasing reaction temperature. An interesting observation was that at 375 °C only cracking reactions occurred during WCO hydrotreating over the supported CoMoS—polymerization was not observed. Compared to hydrocarbons with lower carbon number, the higher ones are more reactive during cracking process. Therefore, the disappearance of C18$^+$ hydrocarbons at a high temperature over the supported CoMoS may be explained by the further cracking of the polymerized hydrocarbons into the lighter ones. This is also agreed with the fact that over the supported CoMoS catalyst, the C15-18 hydrocarbons did not undergo a significant decrease from 340 to 375 °C, while the C15$^-$ hydrocarbons increased substantially. On the other hand, a distinct increase in C18$^+$ hydrocarbons was observed from 340 to 375 °C over the unsupported CoMoS catalyst. This further confirms that the supported CoMoS had higher hydrocracking ability and indicates that Lewis acidic sites on the support (Al$_2$O$_3$–TiO$_2$–SiO$_2$) may have an important impact on hydrocracking reaction capabilities.

3.3. Hydrogenation (C=C Saturation) and Dehydrogenation (C–H Disassociation)

The content of compounds containing C=C bonds is important when used as fuels, because olefin content is related to fuel instability. The degree of hydrogenation was revealed by the selectivity of alkanes found in the liquid products, as shown in Figure 7.

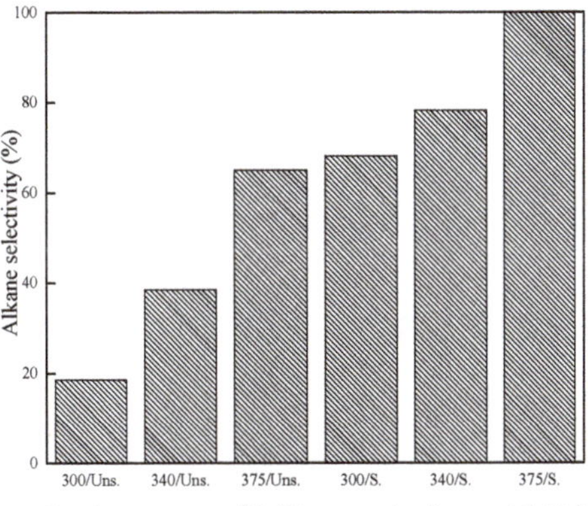

Figure 7. Hydrogenation or dehydrogenation reactions during WCO hydrotreating processes.

According to Figure 7, the alkane selectivity increased with increasing reaction temperature for all trials both over the supported and unsupported catalyst. The majority of oxygenates were removed from the product obtained by upgrading at 375 °C over the unsupported catalyst; however, there was still a certain amount of unsaturated hydrocarbons (around 35 wt %) that were left (shown in Figure 7). When using the supported CoMoS at 300 °C, 340 °C, and 375 °C, more than 68 wt %, 78 wt %, and 100 wt % hydrocarbons were successfully saturated. This suggests that the supported CoMoS caused a higher hydrogenation activity compared to the unsupported CoMoS. The presence of catalyst support may promote the hydrogenation capabilities [35]. The reason for this is probably because acidic sites on the catalyst support could make the C=C bond active, the good dispersion of active phases helping to dissociate hydrogen, and could migrate to the C=C bonds by spill over to form C–C bonds.

Besides, small amount of cyclic hydrocarbons (less than 1.1%) were found in the liquid products upgraded over both supported and unsupported catalysts when reaction temperature were higher at 375 °C.

The stability of the catalyst is an important part of the research to show the efficiency of the best catalyst; limit to the length of this paper, the authors published the reusability of the supported catalyst under the best condition in the journal of "Molecular Catalysis" [8].

4. Experimental

4.1. Catalyst Preparation and Evaluation

The synthesis of unsupported catalyst were the same as in the previous studies (using a hydrothermal method with $Na_2S·9H_2O$, HCl, MoO_3, and $Co(NO_3)_2·6H_2O$) [9,29]. The hydrothermal reaction conditions were at 320 °C for 2 h. The supported CoMoS studied in this work were with a mixture of Al_2O_3, SiO_2, and TiO_2 as the support and purchased from Liaoning Haitai Sci-Tech Development Co., Ltd. (Fushun, Liaoning, China).

Before evaluation, the catalyst was sulfided at 280 °C under 1.5 MPa with a flow rate of 80 mL/min hydrogen and a 0.1 mL min^{-1} dodecane solution containing 2% dimethyl disulfide (DMDS) for 2 h. Hydrotreating of WCO was carried out in an autoclave (Parker Autoclave Engineers Inc., Erie, PA, USA, 1L) under a pressure of 9 MPa hydrogen (99.99%) and at temperatures of 300, 340, and 375 °C. The mass ratio of catalyst active component to feed was 1:200 and the reaction time was 8 h. A blank testing without catalyst was carried out at 375 °C and the thermal reaction effect was not considered because of the low glyceride/oxygen conversion [29].

4.2. Catalyst Characterization

An Autosorb-1 (ANTON PAAR USA INC., Ashland, VA, USA) was used to test the physical properties of the investigated CoMoS. All detailed testing information is the same as the previous publication [9,29].

An Autosorb-1 combined with a RGA-200 (Stanford Research Systems, Sunnyvale, CA, USA) was used to detect NH_3-TPD and TPR, which reveal the amount and strength of active sites on the catalysts, respsectively.

TEM (JEOL 2011, JEOL Ltd., Otemachi, Chiyoda, Tokyo, Japan) was used to determine the morphology of CoMoS, and EDX (Genesis 4000 spectrometer) was used to estimate the elements and their approximate composition in the unsupported CoMoS. The elemental compositions of the supported CoMoS were determined by A JEOL JXA-733 Electron Microprobe analyzer (JEOL Ltd., Otemachi, Chiyoda, Tokyo, Japan). Image analysis software was used to measure the slab length and layer numbers of CoMoS. The average slab length and layer numbers were calculated from 100^+ slabs from different particles.

The sulfur content of the unsupported and supported catalyst was detected using an elemental analyzer (CHN-932, LECO Empowering Results, St. Joseph, MI, USA) and an Antek Nitrogen/Sulfur Analyzer (NS-9000, Antek Instruments, Inc., Houston, TX, USA), respectively.

4.3. Product Analysis

The qualitative chemical compositions of liquid products were analyzed by a gas chromatograph (Shimadzu GC-17A, Shimadzu Scientific Instruments, Kyoto, Japan) coupled to a mass spectrometer (Shimadzu MS-QP5000) (GC/MS, Shimadzu Scientific Instruments, Kyoto, Japan).

The quantitative testing of hydrocarbons in liquid samples was determined by external standard method using a Varian 450 gas chromatograph (Agilent Technologies, Inc., Santa Clara, CA, USA).

The quantitative testing of oxygenates was determined as follows. The total oxygen content in the feed and in the liquid products was calculated by the difference of carbon and hydrogen contents that were examined by a CHN-932 elemental analyzer (LECO Empowering Results, St. Joseph, MI, USA) according to ASTM D 5291. The free fatty acids were analyzed by both gas chromatography (Shimadzu GC-17A, Shimadzu Scientific Instruments, Kyoto, Japan) and titration (ZD-2A Automatic potentiometric titrator, Rex, Shanghai, China) according to ASTM D 664. The contents of alcohols and aldehydes were calculated from GC/MS results by the peak area and response factor of 1.3 and 1.4, respectively [41]. The oxygen balance was used to calculate the content of unreacted glycerides.

4.4. Calculations

Hydrogenation activity is reflected by alkane selectivity (Equation (1)).

$$Alkane\ selectivity = \frac{Mass\ of\ Alkanes}{Mass\ of\ (Alkanes + Alkenes)} \tag{1}$$

All detailed testing information is the same as the previous publication [9,29,42,43].

5. Conclusions

This research investigated the roles of the catalyst support and reaction temperature on WCO hydrotreating process by evaluating the deoxygenation, cracking/polymerization, and hydrogenation/dehydrogenation performances of an unsupported CoMoS catalyst and an Al_2O_3-TiO_2-SiO_2 supported CoMoS catalyst.

In the deoxygenation process, the hydrocarbon fraction was close to 100 wt % in the products that were upgraded at 375 °C over the unsupported CoMoS and 340 °C over the supported CoMoS, respectively. HDO was the main reaction pathway of the supported CoMoS catalyst and HDCO was the main reaction pathway of the unsupported CoMoS. The increase in reaction temperature did not affect the main reaction pathways, yet it slightly enhanced HDCO capability.

In the hydrocracking and polymerization process, the unsupported catalyst was correlated with higher levels of polymerization when compared to the supported catalyst because of lack of the acidic support. The increase in reaction temperature enhanced the cracking and polymerization capability for the unsupported catalyst, but only cracking capability for the supported catalyst.

In the hydrogenation and dehydrogenation process, the dehydrogenation reaction occurred when the reaction temperature was above 375 °C. Compared to the unsupported CoMoS, the supported CoMoS exhibited higher hydrogenation capability at 300–375 °C.

During WCO hydrotreating process, the presence of the CoMoS catalyst support enhanced HDO, reduced the polymerization degree, and decreased the reaction temperature. The increase in reaction temperature promoted the deoxygenation, hydrogenation, and cracking reaction capabilities. The supported CoMoS exhibited good deoxygenation and hydrogenation capabilities under 340 °C in WCO hydrotreating to produce diesel fraction; however, high temperature operation needs to be carefully controlled because it may cause overcracking and dehydrogenation.

Supplementary Materials: The following are available online at http://www.mdpi.com/2073-4344/9/8/689/s1, Figure S1: BJH pore size distribution of the supported and unsupported catalyst, Table S1: The amount of active sites on the supported and unsupported catalyst.

Author Contributions: Conceptualization: H.W., H.Z. (Haiping Zhang), H.L., S.N. and Y.Z.; Data curation: H.W., H.Z. (Haiping Zhang), G.L. and K.R.; Formal analysis: H.W., H.Z. (Haiping Zhang), H.L., G.L. and K.R.; Funding acquisition: Y.Z. and H.W.; Investigation: H.W., H.Z. (Haiping Zhang), G.L., K.R. and H.Z. (Haoxuan Zheng); Methodology: H.W., H.Z. (Haiping Zhang), H.L., S.N. and Y.Z.; Project administration: H.W., H.Z. Haiping Zhang), H.L. and Y.Z.; Resources: Y.Z.; Software: H.W., H.Z. (Haiping Zhang), H.L., G.L. and K.R.; Supervision: H.W. and Y.Z.; Validation: H.W., H.Z. (Haiping Zhang), G.L., K.R. and H.Z. (Haoxuan Zheng); Visualization: H.W., K.R., H.Z. (Haiping Zhang) and Y.Z.; Writing—original draft: H.W., K.R. and H.Z. (Haiping Zhang); Writing—review & editing: H.W., Y.Z. and J.P.

Funding: The authors gratefully thank the Natural Sciences and Engineering Research Council of Canada (NSERC strategic: 463140-2014STPGP, NSERC DISCOVERY: RGPIN-2015-03869), the Canada Foundation for Innovation (#31983), the Canada Research Chairs Program (950-228053), and Jiaxing University (#70518034) for financial assistance.

Conflicts of Interest: The authors declare no conflict of interest.

Abbreviations

WCO	waste cooking oil.
CoMoS	sulfided Cobalt and Molybdenum.
HDO	hydrodeoxygenation. In the HDO reaction, oxygen is removed in the form of water by adding hydrogen into oxygen-containing compounds.
HDCO	hydrodecarbonylation and decarboxylation.
DCO	the process in the hydrodecarbonylation reaction, in which oxygen is removed in the form of water and CO.
DCO_2	the process in the hydrodecarboxylation reaction, in which oxygen is removed in the form of CO_2.
SAPO-11	a medium-pore silicoaluminophosphate molecular sieve with tunable acidity.
A/B catalyst	A stands for the active metal of the catalyst, B stands for the support of the catalyst. For example, $CoMoS/SiO_2$ means that the active metal of the catalyst is sulfided Cobalt and Molybdenum, and the support of the catalyst is Silica.

References

1. Bezergianni, S.; Dimitriadis, A.; Kalogianni, A.; Pilavachi, P.A. Hydrotreating of waste cooking oil for biodiesel production. Part I: Effect of temperature on product yields and heteroatom removal. *Bioresour. Technol.* **2010**, *101*, 6651–6656. [CrossRef] [PubMed]

2. Bezergianni, S.; Dimitriadis, A.; Kalogianni, A.; Knudsen, K.G. Toward Hydrotreating of Waste Cooking Oil for Biodiesel Production. Effect of Pressure, H_2/Oil Ratio, and Liquid Hourly Space Velocity. *Ind. Eng. Chem. Res.* **2011**, *50*, 3874–3879. [CrossRef]

3. Veriansyah, B.; Han, J.Y.; Kim, S.K.; Hong, S.-A.; Kim, Y.J.; Lim, J.S.; Shu, Y.-W.; Oh, S.-G.; Kim, J. Production of renewable diesel by hydroprocessing of soybean oil: Effect of catalysts. *Fuel* **2012**, *94*, 578–585. [CrossRef]

4. Sankaranarayanan, T.M.; Banu, M.; Pandurangan, A.; Sivasanker, S. Hydroprocessing of sunflower oil–gas oil blends over sulfided Ni–Mo–Al–zeolite beta composites. *Bioresour. Technol.* **2011**, *102*, 10717–10723. [CrossRef] [PubMed]

5. Guzman, A.; Torres, J.E.; Prada, L.P.; Nuñez, M.L. Hydroprocessing of crude palm oil at pilot plant scale. *Catal. Today* **2010**, *156*, 38–43. [CrossRef]

6. Kubička, D.; Šimáček, P.; Žilková, N. Transformation of Vegetable Oils into Hydrocarbons over Mesoporous-Alumina-Supported CoMo Catalysts. *Top. Catal.* **2009**, *52*, 161–168. [CrossRef]

7. Wang, H.; Zhang, L.; Li, G.; Rogers, K.; Lin, H.; Seers, P.; Ledan, T.; Ng, S.; Zheng, Y. Application of uniform design experimental method in waste cooking oil (WCO) co-hydroprocessing parameter optimization and reaction route investigation. *Fuel* **2017**, *210*, 390–397. [CrossRef]

8. Wang, H.; Li, G.; Rogers, K.; Lin, H.; Zheng, Y.; Ng, S. Hydrotreating of waste cooking oil over supported CoMoS catalyst—Catalyst deactivation mechanism study. *Mol. Catal.* **2017**, *443*, 228–240. [CrossRef]

9. Zhang, H.; Lin, H.; Zheng, Y. The role of cobalt and nickel in deoxygenation of vegetable oils. *Appl. Catal. B Environ.* **2014**, *160*, 415–422. [CrossRef]

10. Bui, V.N.; Laurenti, D.; Afanasiev, P.; Geantet, C. Hydrodeoxygenation of guaiacol with CoMo catalysts. Part I: Promoting effect of cobalt on HDO selectivity and activity. *Appl. Catal. B Environ.* **2011**, *101*, 239–245. [CrossRef]

11. Shimada, H. Morphology and orientation of MoS_2 clusters on Al_2O_3 and TiO_2 supports and their effect on catalytic performance. *Catal. Today* **2003**, *86*, 17–29. [CrossRef]

12. Qiherima; Li, H.; Yuan, H.; Zhang, Y.; Xu, G. Effect of Alumina Support on the Formation of the Active Phase of Selective Hydrodesulfurization Catalysts Co-Mo/Al_2O_3. *Chin. J. Catal.* **2011**, *32*, 240–249. [CrossRef]

13. Sakashita, Y.; Araki, Y.; Shimada, H. Effects of surface orientation of alumina supports on the catalytic functionality of molybdenum sulfide catalysts. *Appl. Catal. A Gen.* **2001**, *215*, 101–110. [CrossRef]

14. Rana, M.S.; Ancheyta, J.; Maity, S.K.; Rayo, P. Maya crude hydrodemetallization and hydrodesulfurization catalysts: An effect of TiO_2 incorporation in Al_2O_3. *Catal. Today* **2005**, *109*, 61–68. [CrossRef]

15. De la Puente, G.; Gil, A.; Pis, J.J.; Grange, P. Effects of Support Surface Chemistry in Hydrodeoxygenation Reactions over CoMo/Activated Carbon Sulfided Catalysts. *Langmuir* **1999**, *15*, 5800–5806. [CrossRef]

16. Centeno, A.; Laurent, E.; Delmon, B. Influence of the Support of CoMo Sulfide Catalysts and of the Addition of Potassium and Platinum on the Catalytic Performances for the Hydrodeoxygenation of Carbonyl, Carboxyl, and Guaiacol-Type Molecules. *J. Catal.* **1995**, *154*, 288–298. [CrossRef]

17. Kubicka, D.; Horacek, J.; Setnicka, M.; Bulanek, R.; Zukal, A.; Kubickova, I. Effect of support-active phase interactions on the catalyst activity and selectivity in deoxygenation of triglycerides. *Appl. Catal. B Environ.* **2014**, *145*, 101–107. [CrossRef]

18. Wang, H.; Yan, S.; Salley, S.; Ng, K. Support effects on hydrotreating of soybean oil over NiMo carbide catalyst. *Fuel* **2013**, *111*, 81–87. [CrossRef]

19. Bui, V.N.; Laurenti, D.; Delichère, P.; Geantet, C. Hydrodeoxygenation of guaiacol: Part II: Support effect for CoMoS catalysts on HDO activity and selectivity. *Appl. Catal. B Environ.* **2011**, *101*, 246–255. [CrossRef]

20. Chen, H.; Wang, Q.; Zhang, X.; Wang, L. Effect of support on the NiMo phase and its catalytic hydrodeoxygenation of triglycerides. *Fuel* **2015**, *159*, 430–435. [CrossRef]

21. Gong, S.; Shinozaki, A.; Qian, W. Role of Support in Hydrotreatment of Jatropha Oil over Sulfided NiMo Catalysts. *Ind. Eng. Chem. Res.* **2012**, *51*, 13953–13960. [CrossRef]

22. Krár, M.; Kovács, S.; Kalló, D.; Hancsók, J. Fuel purpose hydrotreating of sunflower oil on CoMo/Al_2O_3 catalyst. *Bioresour. Technol.* **2010**, *101*, 9287–9293. [CrossRef] [PubMed]

23. Bezergianni, S.; Dimitriadis, A.; Sfetsas, T.; Kalogianni, A. Hydrotreating of waste cooking oil for biodiesel production. Part II: Effect of temperature on hydrocarbon composition. *Bioresour. Technol.* **2010**, *101*, 7658–7660. [CrossRef] [PubMed]

24. Toba, M.; Abe, Y.; Kuramochi, H.; Osako, M.; Mochizuki, T.; Yoshimura, Y. Hydrodeoxygenation of waste vegetable oil over sulfide catalysts. *Catal. Today* **2011**, *164*, 533–537. [CrossRef]

25. Calais, C.; Matsubayashi, N.; Geantet, C.; Yoshimura, Y.; Shimada, H.; Nishijima, A.; Lacroix, M.; Breysse, M. Crystallite Size Determination of Highly Dispersed Unsupported MoS_2 Catalysts. *J. Catal.* **1998**, *174*, 130–141. [CrossRef]

26. Pu, J.; Nishikado, K.; Wang, N.; Nguyen, T.T.; Maki, T.; Qian, E.W. Core-shell nickel catalysts for the steam reforming of acetic acid. *Appl. Catal. B Environ.* **2018**, *224*, 69–79. [CrossRef]

27. Chen, N.; Gong, S.; Qian, E.W. Effect of reduction temperature of $NiMoO_{3-x}$/SAPO-11 on its catalytic activity in hydrodeoxygenation of methyl laurate. *Appl. Catal. B Environ.* **2015**, *174*, 253–263. [CrossRef]

28. Šimáček, P.; Kubička, D.; Šebor, G.; Pospíšil, M. Hydroprocessed rapeseed oil as a source of hydrocarbon-based biodiesel. *Fuel* **2009**, *88*, 456–460. [CrossRef]

29. Zhang, H.; Lin, H.; Wang, W.; Zheng, Y.; Hu, P. Hydroprocessing of waste cooking oil over a dispersed nano catalyst: Kinetics study and temperature effect. *Appl. Catal. B Environ.* **2014**, *150*, 238–248. [CrossRef]

30. Pu, J.; Luo, Y.; Wang, N.; Bao, H.; Wang, X.; Qian, E.W. Ceria-promoted Ni@Al_2O_3 core-shell catalyst for steam reforming of acetic acid with enhanced activity and coke resistance. *Int. J. Hydrogen Energy* **2018**, *43*, 3142–3153. [CrossRef]

31. Pu, J.; Toyoda, T.; Qian, E.W. Evaluation of Reactivities of Various Compounds in Steam Reforming over RuNi/BaOAl_2O_3 Catalyst. *Energy Fuels* **2018**, *32*, 1804–1811. [CrossRef]

32. Donnis, B.; Egeberg, R.G.; Blom, P.; Knudsen, K.G. Hydroprocessing of Bio-Oils and Oxygenates to Hydrocarbons. Understanding the Reaction Routes. *Top. Catal.* **2009**, *52*, 229–240.

33. Huber, G.W.; O'Connor, P.; Corma, A. Processing biomass in conventional oil refineries: Production of high quality diesel by hydrotreating vegetable oils in heavy vacuum oil mixtures. *Appl. Catal. A Gen.* **2007**, *329*, 120–129. [CrossRef]

34. Gusmão, J.; Brodzki, D.; Djéga-Mariadassou, G.; Frety, R. Utilization of vegetable oils as an alternative source for diesel-type fuel: Hydrocracking on reduced Ni/SiO_2 and sulphided Ni-Mo/γ-Al_2O_3. *Catal. Today* **1989**, *5*, 533–544. [CrossRef]

35. Ryymin, E.-M.; Honkela, M.L.; Viljava, T.-R.; Krause, A.O.I. Competitive reactions and mechanisms in the simultaneous HDO of phenol and methyl heptanoate over sulphided NiMo/γ-Al_2O_3. *Appl. Catal. A Gen.* **2010**, *389*, 114–121. [CrossRef]

36. Brunet, S.; Mey, D.; Pérot, G.; Bouchy, C.; Diehl, F. On the hydrodesulfurization of FCC gasoline: A review. *Appl. Catal. A Gen.* **2005**, *278*, 143–172. [CrossRef]

37. Şenol, O.İ.; Ryymin, E.M.; Viljava, T.R.; Krause, A.O.I. Reactions of methyl heptanoate hydrodeoxygenation on sulphided catalysts. *J. Mol. Catal. A Chem.* **2007**, *268*, 1–8. [CrossRef]

38. Kubička, D.; Kaluža, L. Deoxygenation of vegetable oils over sulfided Ni, Mo and NiMo catalysts. *Appl. Catal. A Gen.* **2010**, *372*, 199–208. [CrossRef]

39. Boda, L.; Onyestyák, G.; Solt, H.; Lónyi, F.; Valyon, J.; Thernesz, A. Catalytic hydroconversion of tricaprylin and caprylic acid as model reaction for biofuel production from triglycerides. *Appl. Catal. A Gen.* **2010**, *374*, 158–169. [CrossRef]

40. Bezergianni, S.; Dimitriadis, A. Temperature effect on co-hydroprocessing of heavy gas oil–waste cooking oil mixtures for hybrid diesel production. *Fuel* **2013**, *103*, 579–584. [CrossRef]

41. Costa, R.; d'Acampora Zellner, B.; Crupi, M.L.; Fina, M.R.D.; Valentino, M.R.; Dugo, P.; Dugo, G.; Mondello, L. GC–MS, GC–O and enantio–GC investigation of the essential oil of *Tarchonanthus camphoratus* L. *Flavour Fragr. J.* **2008**, *23*, 40–48. [CrossRef]

42. Wang, H.; Lin, H.; Feng, P.; Han, X.; Zheng, Y. Integration of catalytic cracking and hydrotreating technology for triglyceride deoxygenation. *Catal. Today* **2017**, *291*, 172–179. [CrossRef]

43. Wang, H.; Lin, H.; Zheng, Y.; Ng, S.; Brown, H.; Xia, Y. Kaolin-based catalyst as a triglyceride FCC upgrading catalyst with high deoxygenation, mild cracking, and low dehydrogenation performances. *Catal. Today* **2019**, *319*, 164–171. [CrossRef]

 *catalysts*

Article

Artificial Intelligence Modelling Approach for the Prediction of CO-Rich Hydrogen Production Rate from Methane Dry Reforming

Bamidele Victor Ayodele [1,*], Siti Indati Mustapa [1], May Ali Alsaffar [2] and Chin Kui Cheng [3]

[1]   Institute of Energy Policy and Research, Universiti Tenaga Nasional, Putrajaya Campus, Jalan IKRAM-UNITEN, Kajang 43000, Selangor, Malaysia

[2]   Department of Chemical Engineering, University of Technology Iraq, Baghdad, Iraq

[3]   Faculty of Chemical and Natural Resources Engineering, Universiti Malaysia Pahang, Lebuhraya Tun Razak, Gambang 26300, Pahang, Malaysia

*   Correspondence: ayodelebv@gmail.com; Tel.: +60-3892-12020

Received: 10 July 2019; Accepted: 25 July 2019; Published: 31 August 2019

Abstract: This study investigates the applicability of the Leven–Marquardt algorithm, Bayesian regularization, and a scaled conjugate gradient algorithm as training algorithms for an artificial neural network (ANN) predictively modeling the rate of CO and H_2 production by methane dry reforming over a Co/Pr_2O_3 catalyst. The dataset employed for the ANN modeling was obtained using a central composite experimental design. The input parameters consisted of CH_4 partial pressure, CO_2 partial pressure, and reaction temperature, while the target parameters included the rate of CO and H_2 production. A neural network architecture of 3 13 2, 3 15 2, and 3 15 2 representing the input layer, hidden neuron layer, and target (output) layer were employed for the Leven–Marquardt, Bayesian regularization, and scaled conjugate gradient training algorithms, respectively. The ANN training with each of the algorithms resulted in an accurate prediction of the rate of CO and H_2 production. The best prediction was, however, obtained using the Bayesian regularization algorithm with the lowest standard error of estimates (SEE). The high values of coefficient of determination ($R^2 > 0.9$) obtained from the parity plots are an indication that the predicted rates of CO and H_2 production were strongly correlated with the observed values.

Keywords: artificial neural network; kinetic modeling; cobalt-praseodymium (III) oxide; CO-rich hydrogen; methane dry reforming

---

1. Introduction

Methane dry reforming is a thermo-catalytic process used for producing synthetic gas (syngas), a mixture of hydrogen (H_2) and carbon monoxide (CO), by utilizing methane (CH_4) and carbon dioxide (CO_2) as feedstocks [1]. Although there are several processes such as steam methane reforming [2], coal gasification [3], glycerol reforming [4], and partial oxidation reforming [5] that can be employed for syngas production, none of these processes have the advantages of mitigating greenhouse gas emission through the consumption of CH_4 and CO_2 [6]. Besides being a potential technical route for greenhouse gas emission reduction, methane dry reforming has the advantage of producing syngas with a H_2:CO ratio close to unity [7]. The syngas produced can in turn be used as an important building block for many industrial processes such as ammonia, methanol, and synthetic fuel production [8]. However, one of the key challenges of the methane dry reforming process is catalyst deactivation by carbon deposition and sintering which is caused due to the high temperature (>873 K) required for the reaction [9].

To overcome these challenges, several supported metal-based catalysts have been developed and tested. An extensive review by Abdullah et al. [10] revealed that supported nickel (Ni) catalysts have been mostly investigated for methane dry reforming due to its high catalytic performance. Nevertheless, the Ni-based catalysts are very prone to sintering and carbon deposition [11]. On the other hand, cobalt (Co)-based catalysts which have a comparative activity to Ni have been reported to show superior stability compare to Ni under the same process condition [12,13]. In our previous studies, the use of rare-earth metal oxide-supported Co catalysts for CO-rich hydrogen production showed considerable activity and stability [14–16]. However, one major challenge is understanding the kinetics of the methane dry reforming in terms of the rate of H_2 and CO production due to variations in the chemical composition of the various catalysts [17]. This challenge can be overcome by employing an artificial intelligence modeling approach for a better understanding of the process parameters [18,19]. Processes with non-linear and complex relationships between the input and the output parameters are often encountered in real life processes. The better understanding of the non-linear relationship between the input and the output parameters of the process can further be utilized to optimize the process operation and create the basis for the theoretical framework, process automation, and upscaling [20].

An artificial intelligence modeling approach using an artificial neural network (ANN) has been widely employed for different catalytic processes, such as hydrodesulfurization [20], methanol steam reforming, glycerol steam reforming [21,22], air gasification of biomass [23], water gas shift reaction [24], and steam gasification of palm oil waste [25]. Nasr et al. [26] reported the use of ANN for the predictive modeling of biohydrogen production using a back-propagation configuration and concluded that the experimental and the predicted biohydrogen production were strongly correlated. Zamaniyan et al. [27] employed a three-layer back-propagation feed-forward ANN for modeling industrial plant hydrogen. The study revealed that the ANN accurately predicted the temperature, pressure, and mole fraction of the hydrogen production in the plant. Ghasemzadeh et al. [28] predicted the performance of a silica membrane reactor during methanol steam reforming using a multilayer perceptron ANN. The study shows that the membrane pressure, temperature, and gas hourly space velocity were accurately predicted with a strong correlation between the actual and the predicted values. In a similar study by Ghasemzadeh et al. [22], ANN was also employed for the predictive modeling of hydrogen production by glycerol steam reforming over a Co/Al_2O_3 catalyst. The feed forward ANN accurately predicted the glycerol conversion, H_2 recovery, H_2 yield, H_2 selectivity, CO selectivity, and CO_2 selectivity with a high coefficient of determination (R^2) and low mean square error (MSE). In our previous study, ANN has been employed for the prediction of CH_4 conversion, CO_2 conversion, and syngas ratio from methane dry reforming over Sm_2O_3- and CeO_2- supported Co catalysts [19]. In all the above studies, the Leven–Marquardt algorithm was employed for the training of the ANN. In this study, the effect of employing three training algorithms, namely Leven–Marquardt, Bayesian regulation, and scaled conjugate gradient, on the predictability of the ANN model was investigated. The effectiveness of each of the trained ANN configurations was tested through the predicted rate of H_2 and CO production from the Co/Pr_2O_3-catalyzed methane dry reforming process.

2. Results and Discussions

2.1. Generated Data for the ANN Modeling

The data obtained from the experimental runs using a central composite design (CCD) are summarized in Table 1. The data consist of 50 experimental runs which are made up of treatment combinations of reaction temperature, CH_4 partial pressure, and CO_2 partial pressure as input parameters, while the target parameters include the rate of CO and H_2 production. The responses (target values) obtained from each of the experimental runs varies according to the treatment combinations of the reaction temperature, CH_4 partial pressure, and CO_2 partial pressure.

Table 1. Data obtained from central composite experimental design for artificial neural network (ANN) modeling.

S/N	Reaction Temperature (K)	CH_4 Partial Pressure (kPa)	CO_2 Partial Pressure (kPa)	Rate of CO Production (mmol/gcat/min)	Rate of H_2 Production (mmol/gcat/min)
1	973	27.5	27.5	0.2880	0.1032
2	1023	15.0	40.0	0.3085	0.1103
3	973	27.5	48.5	0.1736	0.0918
4	973	27.5	27.5	0.2878	0.1030
5	973	27.5	27.5	0.2879	0.1029
6	973	27.5	27.5	0.2878	0.1030
7	973	6.5	27.5	0.0013	0.0078
8	973	27.5	27.5	0.2881	0.1029
9	973	27.5	27.5	0.2878	0.1030
10	973	27.5	27.5	0.2879	0.1031
11	973	27.5	27.5	0.2881	0.1028
12	973	27.5	27.5	0.2880	0.1030
13	1023	40.0	15.0	0.3577	0.2601
14	973	27.5	27.5	0.2882	0.1031
15	923	40.0	40.0	0.0938	0.0422
16	1057	27.5	27.5	0.4623	0.3471
17	973	27.5	27.5	0.2878	0.1029
18	973	27.5	27.5	0.2880	0.1030
19	973	27.5	27.5	0.2881	0.1031
20	973	27.5	27.5	0.2879	0.1029
21	973	27.5	27.5	0.2878	0.1029
22	923	15.0	40.0	0.0381	0.0002
23	973	27.5	27.5	0.2877	0.1031
24	973	27.5	27.5	0.2880	0.1030
25	973	27.5	27.5	0.2876	0.1029
26	973	27.5	6.5	0.1581	0.0134
27	973	27.5	27.5	0.2874	0.1031
28	973	27.5	27.5	0.2877	0.1030
29	973	27.5	27.5	0.2880	0.1029
30	973	27.5	27.5	0.2878	0.1031
31	1023	15.0	15.0	0.3085	0.0113
32	973	27.5	27.5	0.2863	0.1029
33	1023	40.0	40.0	0.3624	0.2341
34	973	48.5	27.5	0.3495	0.1515
35	973	27.5	27.5	0.2878	0.1031
36	923	40.0	15.0	0.0728	0.0021
37	973	27.5	27.5	0.0877	0.013
38	973	27.5	27.5	0.0874	0.0129
39	923	15.0	15.0	0.0281	0.001
40	973	27.5	27.5	0.2881	0.1029
41	973	27.5	27.5	0.2880	0.1031
42	973	27.5	27.5	0.2878	0.1030
43	973	27.5	27.5	0.2880	0.1029
44	973	27.5	27.5	0.2879	0.1031
45	973	27.5	27.5	0.2878	0.1029
46	973	27.5	27.5	0.2880	0.1030
47	973	27.5	27.5	0.2881	0.1031
48	889	27.5	27.5	0.1379	0.0919
49	973	27.5	27.5	0.2880	0.1029
50	973	27.5	27.5	0.2878	0.1030

2.2. Interaction Effect of Process Parameters on the Rate of H_2 Production

Theoretically, a catalyzed methane dry reforming reaction as represented in Equation (1) involves the consumption of 1 mole of CH_4 and 1 mole CO_2 to produce 2 moles of CO and 2 moles of H_2 [29].

$$CH_4 + CO_2 \rightleftharpoons 2CO + 2H_2 \quad \Delta H_{298K} = +247 \, kJ/mol \tag{1}$$

The methane dry reforming process is highly endothermic [30]. Therefore, the reaction is favored by a high temperature >900 K [31]. Although, the mechanism of the methane dry reforming reaction is strongly dependent on the nature of the catalyst, it is generally believed that the reaction commences with the activation of CH_4 and CO_2 being adsorbed on the catalyst active sites [32]. The activation of the adsorbed CH_4 often leads to the formation of carbon and hydrogen. While the activation of the CO_2 often occurs at the interphase of the catalyst active site and the support often leads to the formation of CO and surface O_2, which is simultaneously utilized to gasify the carbon formed during the activation of CH_4 [10]. The partial pressure of CH_4 and CO_2 at varying temperatures are crucial in determining the rate of production of CO and H_2 during the methane dry reforming process [32].

Figure 1a–c shows the interaction effect of the CH_4 partial pressure, CO_2 partial pressure, and reaction temperature on the rate of H_2 production. As shown in Figure 1a, the rate of H_2 production was significantly influenced by the CH_4 partial pressure and reaction temperature. The rate of H_2 production increased steadily with an increase in the CH_4 partial pressure until 30 kPa and thereafter decreased. This phenomenon can be attributed to the dominance of methane cracking whereby the CH_4 is activated on the catalyst active site to give H_2 and carbon. The carbon formed is often gasified by the release of surface O_2 through the activation of CO_2. In the case where the rate of gasification of the carbon is not at equilibrium with the release of the surface, there would be net carbon deposition which often results in catalyst deactivation. The deactivation of the catalyst active site by the deposited carbon could be responsible for the decrease in the rate of H_2 production at CH_4 partial pressure >30 kPa. At a low CH_4 partial pressure, the rate of H_2 production was low and steady due to a high concentration of CO_2 present in the reactant mixture. However, as the CH_4 partial pressure increased to measure up with that of CO_2, an increase in the rate of H_2 was observed which is typical for the methane dry reforming reaction [33]. Similarly, the rate of H_2 production was found to steadily increase with an increase in the reaction temperature for all cases, which is consistent with Arrhenius' concept of temperature-dependent gas phase reactions [34]. Generally, the rate of H_2 production increased with an increase in both CH_4 partial pressure and reaction temperature, which is consistent with the work of Foo et al. [33], who reported an increase in the rate of production of H_2 with CH_4 partial pressure during methane dry reforming over an Al_2O_3-supported Co-Ni catalyst. In Figure 1b, it can be seen that both the CO_2 partial pressure and the reaction temperature had a significant influence on the rate of H_2 production. There was a steady increase in the rate of H_2 production between 5 and 30 kPa and thereafter a decline was observed. Again, within the CO_2 partial pressure range of 5–30 kPa, there was a steady release of surface O_2 from the activation of the CO_2. However, at a CO_2 partial pressure >30 kPa, there was no equilibrium between the rate of gasification of the carbon and the carbon deposition. Hence, there might be depletion in the catalyst active site which could be responsible for the decline in the rate of H_2 production. The interaction between the CO_2 partial pressure and the reaction temperature had a significant influence on the rate of H_2 production, as can be seen the yellow part of the mesh diagram. The interaction between the CO_2 partial pressure and the CH_4 partial pressure had a significant influence on the rate of H_2 production, as shown in Figure 1b, although at a lower CO_2 partial pressure, the rate of H_2 production was steady until 30 kPa. This can be attributed to the dominance of the methane decomposition reaction as stated earlier [35]. Although, there is a significant interaction between CO_2 and CH_4 partial pressure, the rate of H_2 production was greatly affected by the CH_4 partial pressure.

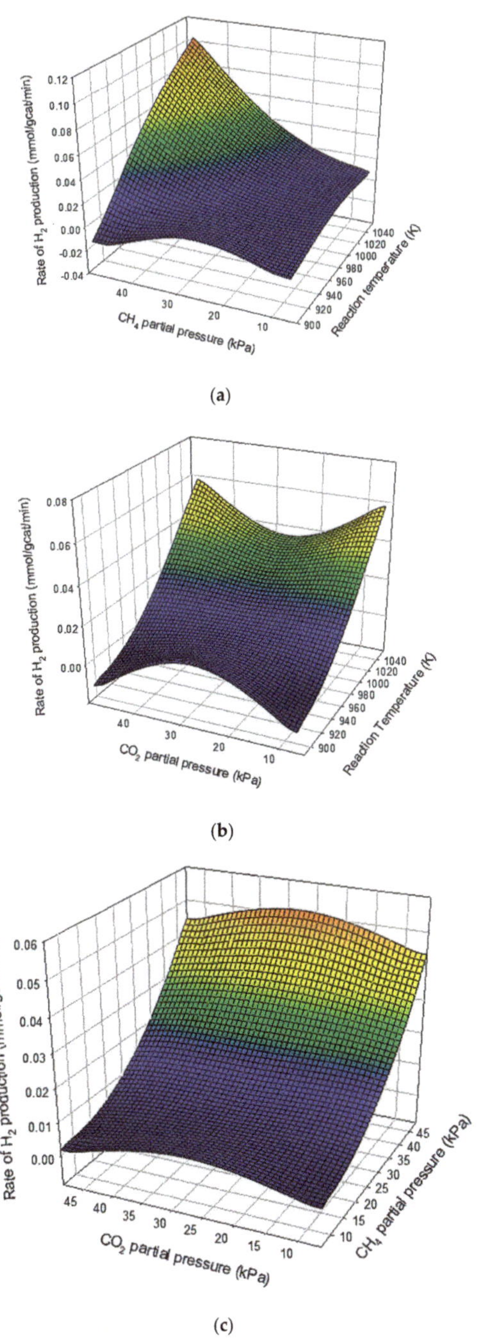

(a)

(b)

(c)

Figure 1. (**a**) Interaction effect of CH_4 partial pressure and reaction partial pressure on the rate of CO production; (**b**) Interaction effect of CO_2 partial pressure and reaction temperature on the rate of CO production; (**c**) Interaction effect of CH_4 partial pressure and CO_2 partial pressure on the rate of CO production.

2.3. Interaction Effect of Process Parameters on the Rate CO Production

The interaction effects of CH_4 partial pressure, CO_2 partial pressure, and reaction temperature on the rate of CO production are depicted in Figure 2. At a constant CH_4 partial pressure (Figure 2a), the rate of CO production was steady with increases in the CH_4 partial pressure, whereas a significant increase in the rate of CO production was observed with an increase in the reaction temperature, which agrees with the Arrhenius theory for temperature-dependent gas phase reactions. Based on Figure 2a, the CH_4 partial pressure did not have much influence on the rate of CO production. This is due to the fact that CO is solely produced from the activation of CO_2. At a lower CH_4 partial pressure, it can be inferred that the Bondouard reaction is dominant [36]. In this case, the CO produced was subsequently converted to CO_2 and carbon. However, as the CH_4 partial pressure increased, a state of equilibrium was attained with the CO_2 partial pressure, thereby resulting in an increase in the rate of CO production. A similar trend can be observed in Figure 2b, although there was a steady increase in the rate of CO production at a lower PCO_2 partial pressure, as reported by Foo et al. [33]. The interaction between CO_2 and the CH_4 partial pressure had a significant influence on the rate of CO production. However, the CO_2 partial pressure has the most significant influence on the rate of CO production, which is consistent with the fact that CO is produced during the activation of CO_2.

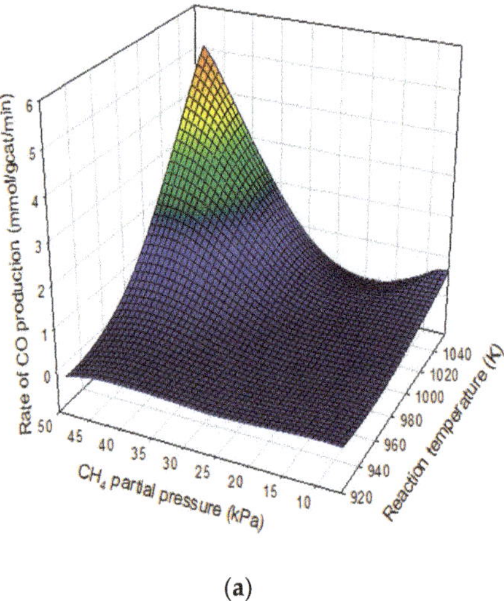

(**a**)

Figure 2. *Cont.*

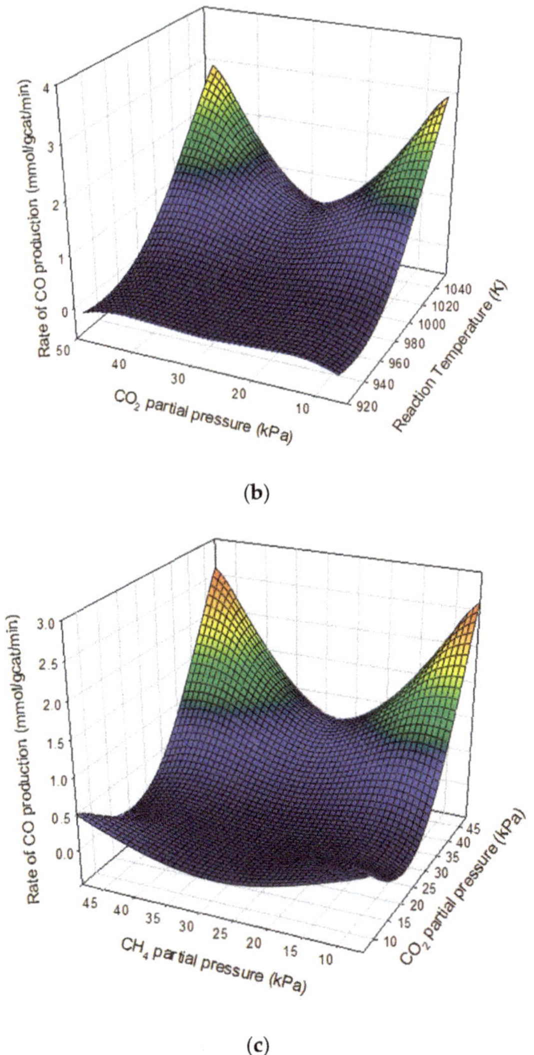

(b)

(c)

Figure 2. (**a**) Interaction effect of CH_4 partial pressure and reaction partial pressure on the rate of H_2 production; (**b**) Interaction effect of CO_2 partial pressure and reaction temperature on the rate of H_2 production; (**c**) Interaction effect of CH_4 partial pressure and CO_2 partial pressure on the rate of H_2 production.

2.4. Artificial Neural Network Modeling

Prior to the commencement of the network analysis, several ANN configurations were trained in other to determine the most suitable hidden neuron that minimized the MSE. As shown in Figures 3–5, the best ANN architecture for each of the training algorithms was obtained at the least MSE. The values of the MSE varied with changes in the number of hidden neurons. Hidden neuron ranges from 1 to 20 were tested for each of the algorithms, which resulted in the best hidden neuron of 13, 15, and 15 for Leven–Marquardt, Bayesian regularization, and scaled conjugate gradient algorithms, respectively. MSE values of 1.91×10^{-5}, 5.65×10^{-4}, and 9.34×10^{-4} were obtained for the ANN architecture using Leven–Marquardt, Bayesian regularization, and scaled conjugate gradient algorithms, respectively.

The high R values of 0.998, 0.977, and 0.956 revealed that the predicted rate of CO and H_2 at the obtained lowest MSE were very close to the actual values (Table 2).

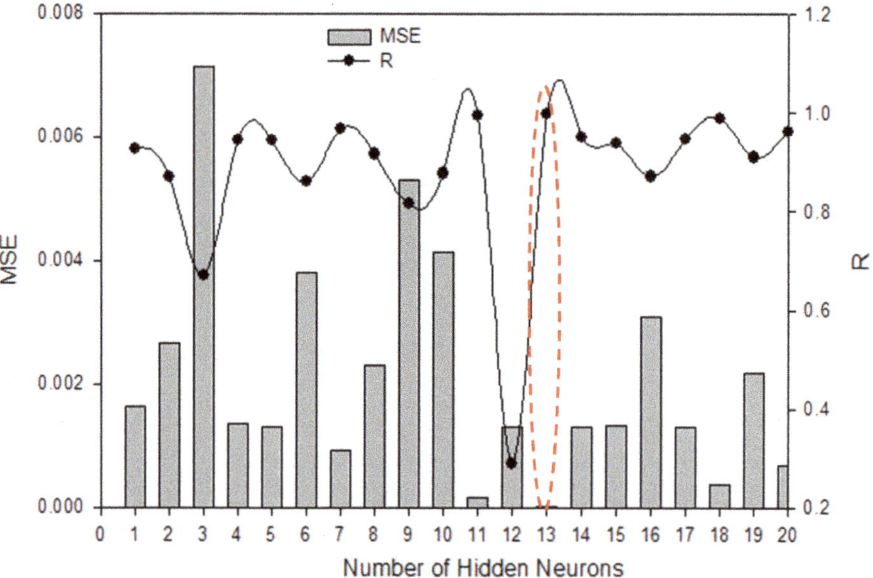

Figure 3. Determination of the optimized hidden neuron with the minimum mean square error (MSE) for ANN training using the Leven–Marquardt algorithm.

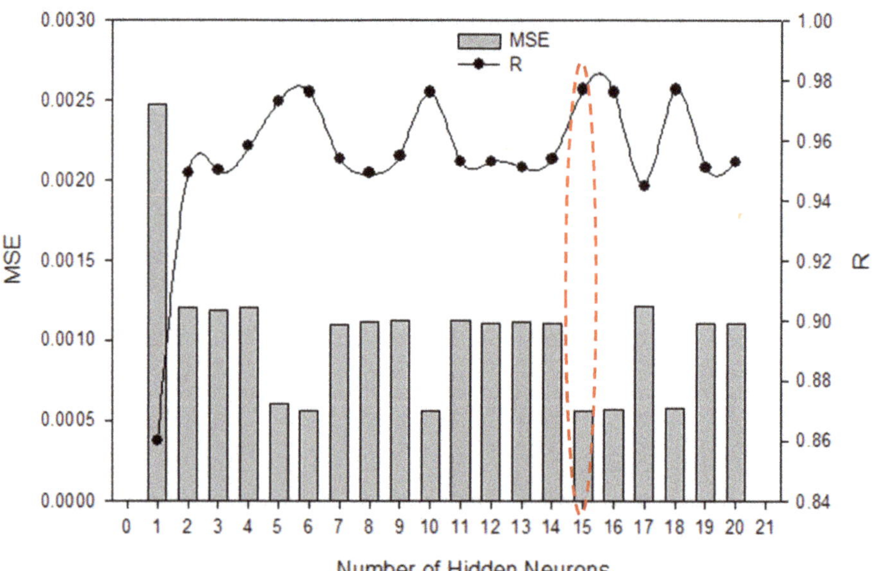

Figure 4. Determination of the optimized hidden neuron with the minimum MSE for ANN training using the Bayesian regularization algorithm.

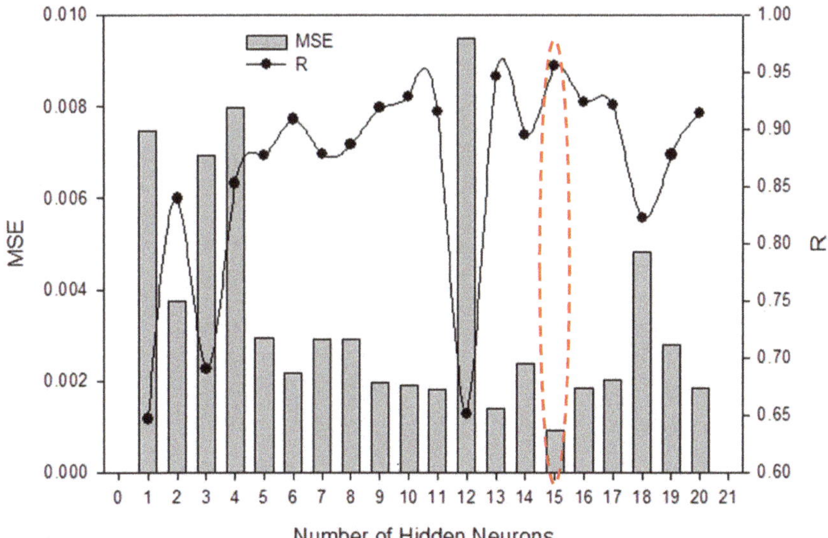

Figure 5. Determination of the optimized hidden neuron with the minimum MSE for ANN training using the scaled conjugate gradient algorithm.

Table 2. Determination of the best neuron for each of the training algorithms.

Hidden Neuron	Leven–Marquardt		Bayesian Regularization		Scaled Conjugate Gradient	
	MSE	**R**	**MSE**	**R**	**MSE**	**R**
1	1.63×10^{-3}	0.927	2.47×10^{-3}	0.860	7.47×10^{-3}	0.647
2	2.67×10^{-3}	0.870	1.21×10^{-3}	0.949	3.75×10^{-3}	0.840
3	7.14×10^{-3}	0.671	1.19×10^{-3}	0.950	6.93×10^{-3}	0.690
4	1.36×10^{-3}	0.945	1.21×10^{-3}	0.958	7.98×10^{-3}	0.853
5	1.32×10^{-3}	0.943	6.12×10^{-4}	0.973	2.94×10^{-3}	0.877
6	3.81×10^{-3}	0.861	5.67×10^{-4}	0.976	2.18×10^{-3}	0.909
7	9.39×10^{-4}	0.967	1.10×10^{-3}	0.954	2.92×10^{-3}	0.879
8	2.31×10^{-3}	0.916	1.12×10^{-3}	0.949	2.93×10^{-3}	0.887
9	5.31×10^{-3}	0.816	1.13×10^{-3}	0.955	1.97×10^{-3}	0.919
10	4.14×10^{-3}	0.877	5.66×10^{-4}	0.976	1.91×10^{-3}	0.929
11	1.57×10^{-4}	0.994	1.13×10^{-3}	0.953	1.81×10^{-3}	0.915
12	1.32×10^{-3}	0.290	1.11×10^{-3}	0.953	9.51×10^{-3}	0.651
13	$\mathbf{1.91 \times 10^{-5}}$	**0.998**	1.12×10^{-3}	0.951	1.39×10^{-3}	0.946
14	1.31×10^{-3}	0.949	1.11×10^{-3}	0.954	2.39×10^{-3}	0.895
15	1.33×10^{-3}	0.939	$\mathbf{5.65 \times 10^{-4}}$	**0.977**	$\mathbf{9.34 \times 10^{-4}}$	**0.956**
16	3.09×10^{-3}	0.871	5.68×10^{-4}	0.976	1.83×10^{-3}	0.924
17	1.31×10^{-3}	0.947	1.22×10^{-3}	0.945	2.01×10^{-3}	0.921
18	3.82×10^{-4}	0.989	5.84×10^{-4}	0.977	4.81×10^{-3}	0.823
19	2.19×10^{-3}	0.910	1.11×10^{-3}	0.951	2.81×10^{-3}	0.878
20	6.86×10^{-4}	0.963	1.11×10^{-3}	0.953	1.83×10^{-3}	0.914

2.5. The ANN Model Predictive Analysis

The performance of the ANN prediction of the rate of H_2 and CO production using the Leven–Marquardt, Bayesian Regularization, and scaled conjugate gradient algorithms are depicted in Figures 6–8. Figure 6 depicts the dispersion diagrams and the parity plots showing the actual and the ANN-predicted rates of CO and H_2 production using the Leven–Marquardt algorithm. The filled circles in the dispersion diagrams represent the actual rates of CO and H_2 production, while the spline curves depict the ANN-predicted rates of CO and H_2 production. It can be seen that the use of the Leven–Marquardt algorithm resulted in a good prediction of the rate of CO and H_2 production, as shown in the dispersion diagram (Figure 6a,c). The accuracy of the ANN prediction is further revealed from the parity plot. The actual values of the rate of CO and H_2 production are strongly correlated to the predicted values. Several authors have reported that the Leven–Marquardt algorithm is one of the most effective algorithms used for training ANN models. Its performance is hinged on the advantage of combining both the Gauss–Newton method and the steepest descent technique to attain convergence [37]. Furthermore, the use of the Leven–Marquardt algorithm enables the trained network to rapidly converge near the vicinity of the minimum error [38]. The good prediction of the ANN outputs in this study using the Leven–Marquardt algorithm is consistent with that reported in previous studies. Puig-Arnavat and Bruno [21] employed the Leven–Marquardt algorithm for the modeling of the thermochemical conversion of biomass. The application of the Leven–Marquardt algorithm for training the network resulted in an accurate prediction of H_2 in producer gas, CH_4 in producer gas, CO_2 in producer gas, and CO in producer gas. The predicted values were found to be in good agreement with the actual values based on the parity plots. In a similar study, George et al. [23] applied the Leven–Marquardt algorithm for the predictive modeling of producer gas composition during biomass gasification. The study revealed that the predicted values of the producer gas were in good agreement with the actual values with an R value of 0.987.

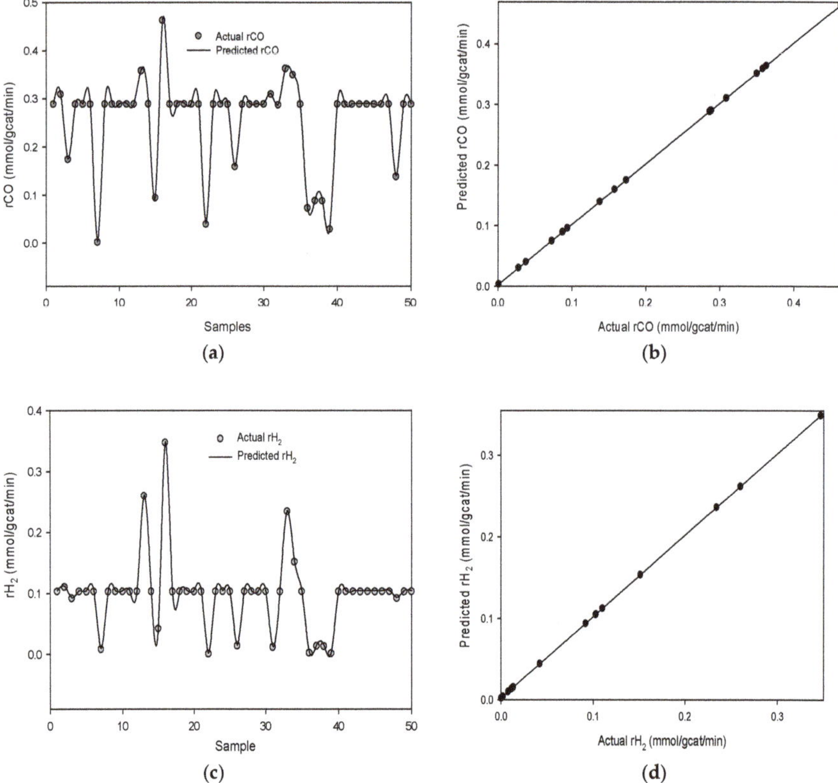

Figure 6. (**a**) Dispersion plot showing the comparison between actual and predicted rCO; (**b**) Parity plots showing the comparison between actual and predicted rCO; (**c**) Dispersion plot showing the comparison between actual and predicted rH2; (**d**) Parity plots the comparison between actual and predicted rH2 using Leven–Marquardt algorithm.

The ANN performance using the Bayesian regularization algorithm is represented in the dispersion and parity plots in Figure 7. The use of the Bayesian regularization algorithm for ANN training is founded on the probabilistic understanding of the network parameters [38]. It employs an optimum set of weights for the minimization of the error function [39]. As shown in Figure 7 the use of the Bayesian regularization also displayed a good prediction of the rate of CO and H_2 production. The dispersion diagrams in Figure 7a,c reveal the proximity between the predicted rate CO and H_2 production, while the parity plots (Figure 7b,d) show that both the predicted CO and H_2 production are in good agreement. Studies have shown that the use of the Bayesian regulation algorithm for ANN modeling results in a good prediction of the targets. George et al. [23] employed the Bayesian regularization algorithm for the ANN modeling of wheat output energy from a wheat production process. The study revealed that the use of the Bayesian regularization algorithm resulted in a good prediction of the wheat output energy which was in good agreement with the actual values. Shi et al. [40] applied the Bayesian regularization algorithm in the ANN modeling of explosion risk analysis of a fixed offshore platform. The Bayesian regulation-trained ANN accurately predicted the cumulative frequency of the maximum overpressure.

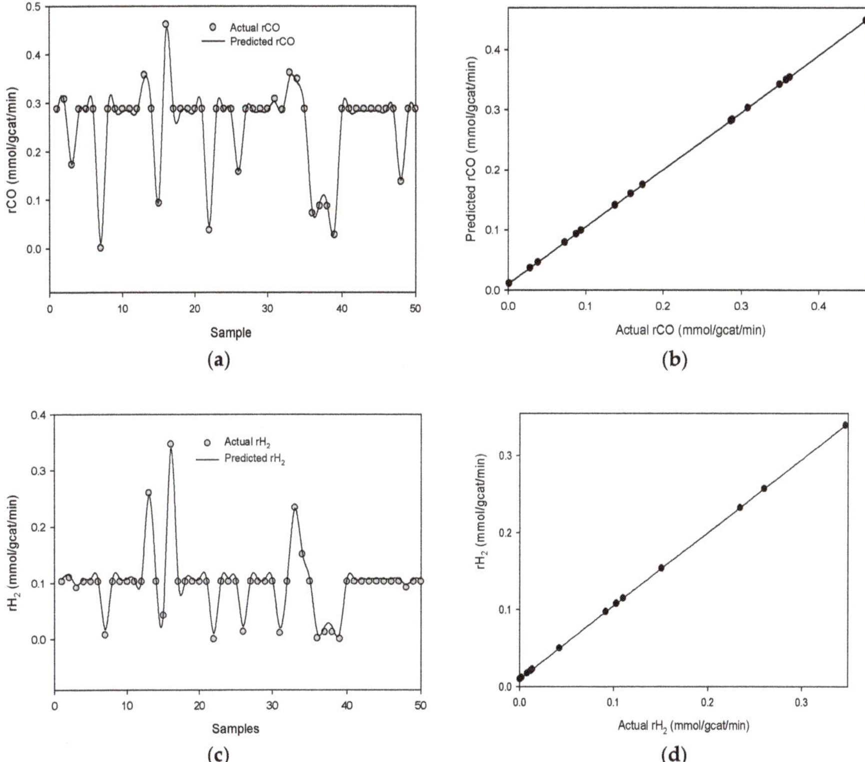

Figure 7. (**a**) Dispersion plot showing the comparison between actual and predicted rCO; (**b**) Parity plots showing the comparison between actual and predicted rCO; (**c**) Dispersion plot showing the comparison between actual and predicted rH_2; (**d**) Parity plots the comparison between actual and predicted rH_2 using Bayesian regularization algorithm.

The performance of the scaled conjugate gradient algorithm-trained ANN is represented in the dispersion diagrams and parity plots in Figure 8. The scaled conjugate gradient algorithm employed step size scaling mechanism which make it have a very fast iteration [41]. As depicted in Figure 8a,c, the use of the scaled conjugate gradient algorithm for ANN modeling resulted in a good prediction of the rate of CO and H_2 production. The predicted values of the rate of CO and H_2 production are in good agreement with the actual values as depicted by the parity plots in Figure 8b,d. The good prediction of the rate of CO and H_2 production obtained in this study is consistent with that reported by Khadse et al. [41] who employed the scaled conjugate gradient algorithm for the ANN modeling of an electromagnetic compatibility estimator. The authors revealed that the use of the scaled conjugate gradient algorithm for ANN modeling produced an accurate prediction of the output. Similarly, Mia and Dhar [39] also confirmed the robustness of the scaled conjugate gradient as a training algorithm for ANN predictive modeling of surface roughness in hard turning under high-pressure coolant. The prediction of the surface roughness using the scaled conjugate gradient-trained ANN model was in good agreement the actual values.

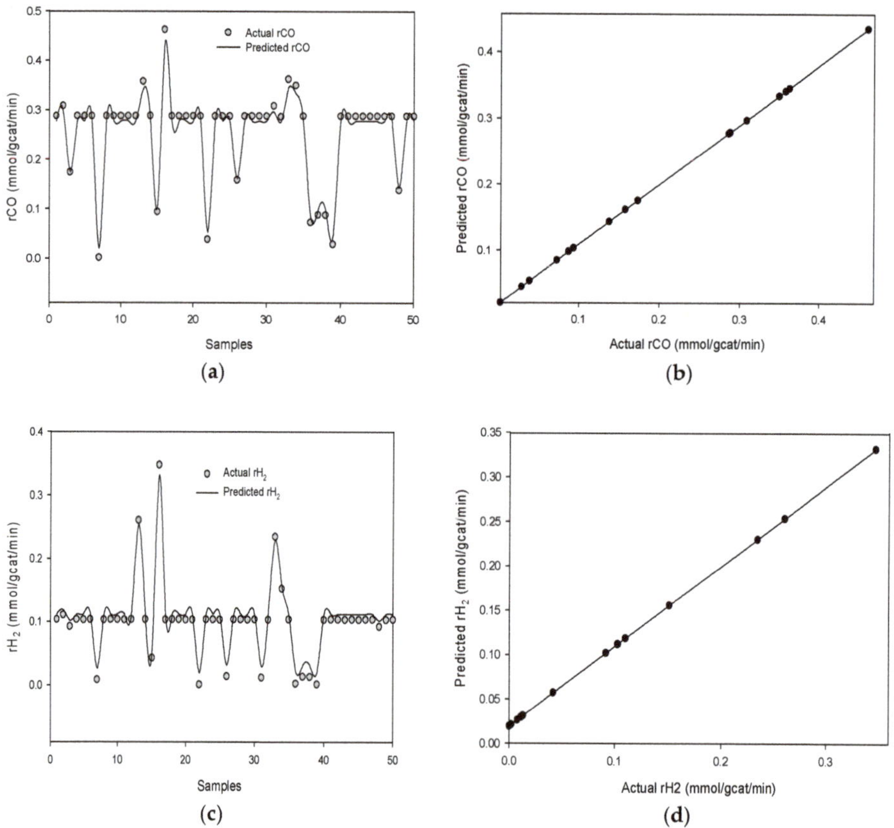

Figure 8. (a) Dispersion plot showing the comparison between actual and predicted rCO; (b) Parity plots showing the comparison between actual and predicted rCO; (c) Dispersion plot showing the comparison between actual and predicted rH_2; (d) Parity plots the comparison between actual and predicted rH_2 using scaled conjugate gradient algorithm.

2.6. Comparison of the Leven–Marquardt, Bayesian Regularization, and Scaled Conjugate Gradient Algorithms

The comparison of the ANN model using the Leven–Marquardt, Bayesian regularization, and scaled conjugate gradient algorithms using statistical parameters is depicted in Table 3. Statistical parameters, such as the standard error of estimates (SEE) and coefficient of determination (R^2) were employed to discriminate between the performance of the three algorithms. The ANN modeling using the Bayesian regularization algorithm resulted in the lowest SEE of 2.0526×10^{-17} obtained from the predicted and the actual rates of CO production compared to the ANN-trained using the Leven–Marquardt and scaled conjugate gradient algorithms. A high $R^2 > 0.9$ was obtained for the three algorithms indicating a strong agreement between the predicted rate of CO and the actual values. Nevertheless, the ANN trained with the Leven–Marquardt algorithm displayed the highest R^2 of 0.9992, which implies that the predicted rate of CO production is in closest agreement compared to the other two algorithms. On the contrary, the ANN trained with the scaled conjugate gradient algorithm produced the lowest SEE of 7.77×10^{-18} from the prediction of rate of H_2 production compared to Leven–Marquardt and scaled conjugate gradient algorithms. Although, all three algorithms used for the ANN training resulted in high $R^2 > 0.9$, the R^2 of 0.992 obtained using the Leven–Marquardt algorithm shows that the predicted and actual values of the rate of H_2 production are in closer agreement compared to the other two algorithms which have lower R^2 values. Mia and Dhar [39] compared the use of the Leven–Marquardt, Bayesian regularization, and scaled conjugate gradient algorithms for the predictive modeling of surface roughness in hard turning under high-pressure coolant using ANN. The results show that the Bayesian regularization-trained ANN presented the lowest root mean square errors with R^2 of 0.997.

Table 3. Statistical analysis of the ANN modeling using different algorithms.

	Leven–Marquardt		Bayesian Regularization		Scaled Conjugate Gradient	
	rCO	rH$_2$	rCO	rH$_2$	rCO	rH$_2$
SEE	2.54×10^{-17}	1.0607×10^{-17}	2.0526×10	9.9084×10^{-18}	2.80×10^{-17}	7.77×10^{-18}
R^2	0.9992	0.9992	0.9726	0.9726	0.9565	0.9565
Model Equation	Output = 1×Target + 0.0018	Output = 1×Target + 0.0018	Output = 0.95×Target + 0.0099	Output = 0.95×Target + 0.0099	Output = 0.9×Target + 0.019	Output = 0.9×Target + 0.019

3. Data Acquisition for ANN Modeling

Figure 9 shows the schematic representation of the stages involved in the data acquisition and the ANN modeling of the rate of H_2 and CO production. Basically, there are five stages involved, from the data acquisition for the ANN to the prediction of the output. The data used for the ANN modeling was obtained from experimental runs designed by employing a central composite design (CCD). The input variables for the experimental design include CH_4 partial pressure, CO_2 partial pressure, and reaction temperature while the rate of hydrogen (rH$_2$) and rate of CO (rCO) production were the output variables. Each of the output variables were obtained from the treatment combinations of the three input parameters.

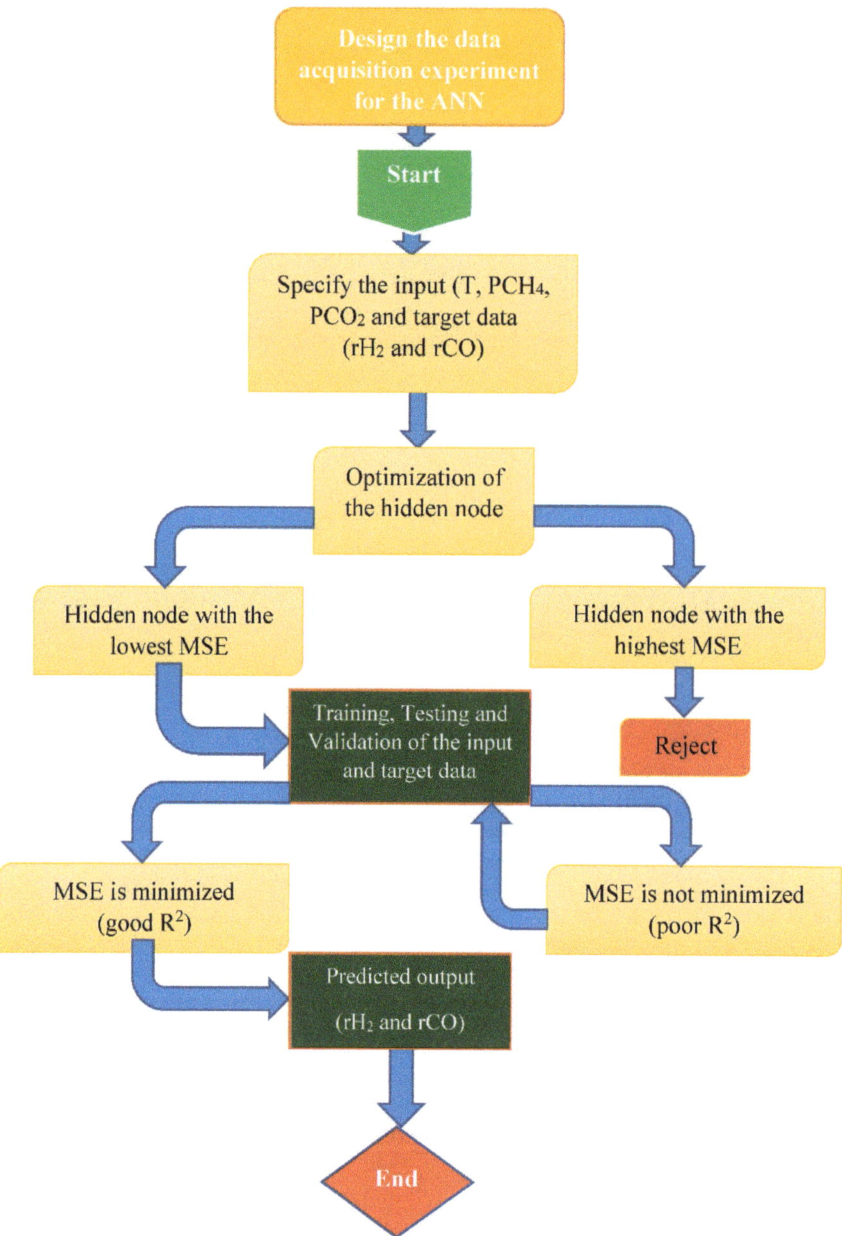

Figure 9. Flow diagram for the ANN modeling.

3.1. Artificial Neural Network Configurations

The ANN is an artificial intelligent model that is developed to mimic the pattern of processing information by the human brain [42]. The neural network configuration processes a large number of interlinked units arranged in layers. The interlinked units are patterned after the human neuron and they consist of the input layer, the hidden layers, and the output layers (also known as the target) [18].

Each of the interlinked units have varying connection strengths, known as weights. The ANN functions in such a way that each of the input signals is multiplied with the corresponding connection weights to obtain a combined weighted hidden layer [43]. The combined hidden layers are subsequently passed through an activation function which in turn generates the corresponding output. The Sigmoid function represented in Equation (2) is the most commonly use form of an activation function [20].

$$f(Z_i) = \frac{1}{1 + e^{-Z_i}} \tag{2}$$

where Z_i represents the summation of each of the hidden layers multiplied by an assigned weight plus bias from each neuron in the previous layer. Just like the human brain, the neural network model functions by exploring the non-linear relationship between the individual input and the target data. Subsequently, the network model creates a predicted output with minimized error. For an incorrect prediction, the weights are adjusted in a circle of iteration to produce an output with minimum error.

The manner of connection of the hidden neuron in an ANN is crucial to the performance of the network model. The neuron can either be connected in such a way to give a feedforward signals or a feedback signal. In this study, a feedforward ANN configuration is adopted due to its wide applicability in the process industries [44,45]. The feedforward ANN is a multilayer perceptron with 2 13 2 2, 2 15 2 2, and 2 15 2 2 architectures for the Leven–Marquardt, Bayesian regularization, and scaled conjugate gradient algorithms, respectively, as shown in Figure 10a–c. The input parameters to the neural network include CH_4 partial pressure, CO_2 partial pressure, and reaction temperature, while the target parameters are rate of H_2 and CO production (Figure 10d). The parameters employed for the ANN configuration are depicted in Table 4.

Table 4. Configuration parameters for the neural network architecture.

Configuration Parameters	Leven–Marquardt	Bayesian Regularization	Scaled Conjugate Gradient
Algorithm	Feed forward with 3 layers	Feed forward with 3 layers	Feed forward with 3 layers
Hidden layer size	1	1	1
Hidden neuron quantity	13	15	15
Output layer size	2	2	2
Output neuron quantity	2	2	2
Output layer neurons activation	Pure linear	Pure linear	Pure linear
Training ratio	0.01	0.01	0.01
Epochs	5	1000	21
Training target error	0.001	0.001	0.001

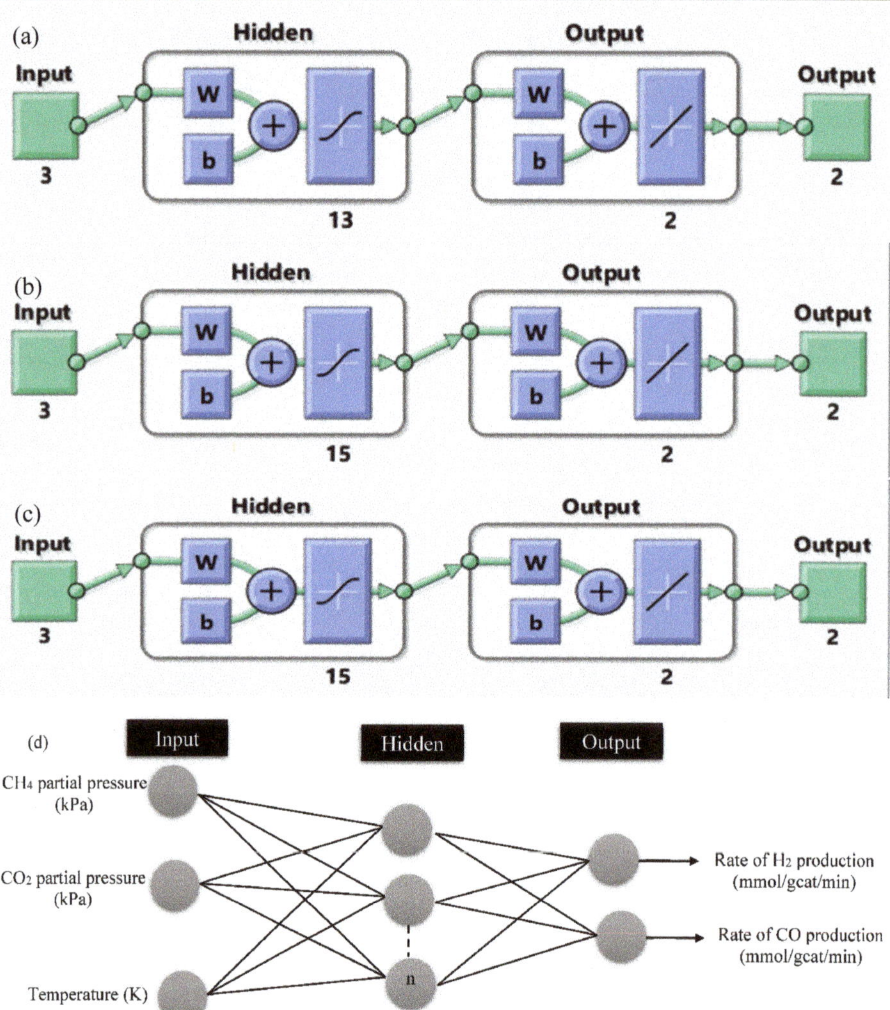

Figure 10. The Network architecture for the ANN modeling used. (**a**) Feed forward multi-layer perceptron architecture for prediction of CO-H$_2$ production; (**b**) Network configuration using the Leven–Marquardt training algorithm; (**c**) Network configuration using the Bayesian regularization training algorithm; (**d**) Network configuration using the scaled conjugate gradient training algorithm.

3.2. Network Training, Testing, and Validation

After applying the necessary configurations to the neural network and the data have been imputed, it is expedient to apply the necessary algorithm for the network training. During network training, the input data presented to the network are compared to the output unit, thereby adjusting the weight of all units based on their errors for an improved prediction. In this study, the network was trained using the Leven–Marquardt, Bayesian regularization, and scaled conjugate gradient algorithms [38,46,47]. The Leven–Marquardt training algorithm utilizes a damping factor which is self-adjusted during each iteration in order to obtain the least error between the predicted and the actual values [38]. The Leven–Marquardt training algorithm has the advantage of attaining a very fast convergence.

The Bayesian regularization training algorithm has the tendency to minimize the estimated errors through an inbuilt objective function that contains a residual sum of squares and the sum of squared weighs [40]. Hence, it is typical to obtain a good generalization model using the Bayesian regularization training algorithm [38]. To minimize the errors, the weights in scaled conjugate gradient algorithms are adjusted in the direction in which the network function performance is decreasing most rapidly. More iterations are required for convergence using scaled conjugate gradient algorithms compared to Leven–Marquardt and Bayesian regularization. Testing of the network provides an independent measure of its performance during and after the training. While validation is a form of measuring the network generalization and halting of training when generalization has stopped improving. The training, testing, and validation of the network were performed using the neural network toolbox in MATLAB 2019a (MathWorks Inc., Natick, MA, USA). The data were proportioned into 70%, 15%, and 15% for training, testing, and validation of the network, respectively. The detailed network architectures of the feedforward multilayer perceptron are depicted in Figure 3.

3.3. Evaluation of the ANN Performance

The accuracy of the ANN to predict the rate of CO and H_2 production were measured using parameters such as the mean square error (MSE) and the correlation coefficient (R) [27]. The MSE defined in Equation (3) was used to measure the average squared difference between the predicted rate of CO and H_2 production and the actual values. The lower the MSE values, the more accurate the ANN prediction.

$$MSE = \frac{1}{n} \sum_{i=1}^{n} \left(Y_p - Y_a \right)^2 \tag{3}$$

where n is the number of samples, Y_p and Y_a are the predicted and the actual values, respectively.

The correlation coefficient (R) defined in Equation (4) was employed to determine the strength of the linear relationship between the predicted rate of CO and H_2 production and the actual values.

$$R = \frac{1}{n-1} \left[\frac{\sum_x \sum_y (x - \bar{x})(y - \bar{y})}{S_x S_y} \right] \tag{4}$$

where n is the number of samples, $\bar{x}$ and $\bar{y}$ are the sample means of all the x and y values. S_x and S_y are the standard deviation of all the x and y values. An R value of 1 implies that a close relationship between the predicted rate of CO and H_2 production and the actual values exists, while an R value of 0 implies that a random relationship exists between the predicted rate of CO and H_2 production and the actual values.

4. Conclusions

In this study, the use of artificial neural network as a predictive model has been investigated. Three algorithms, namely Leven–Marquardt, Bayesian regularization, and scaled conjugate gradient, were employed to train the ANN for the prediction of the rate of CO and H_2 production from methane dry reforming catalyzed by Co/Pr_2O_3. The ANN predictive modeling was performed using datasets obtained from central composite experimental design. Several architectures of the ANN were tested using hidden neurons in the range of 1–20. The best ANN architectures were obtained using 13, 15, and 15 hidden neurons for Leven–Marquardt, Bayesian regularization, and scaled conjugate gradient algorithms, respectively. The training of the ANN model with the best neurons results in good predictions of the rates of CO and H_2 production using the three training algorithms. The R^2 values of 0.9992, 0.9726, and 0.9565 obtained using the Leven–Marquardt, Bayesian regularization, and scaled conjugate gradient algorithms, respectively, are an indication that the predicted rates of CO and H_2 production were in good agreement with the observed values. However, the best prediction was obtained using the Bayesian regularization training algorithm for the ANN model. This study has demonstrated the use of ANN predictive modeling to investigate the functional relationship that exists between the process parameters in the production of CO-rich hydrogen by a methane dry reforming reaction over a Co/Pr_2O_3 catalyst.

Author Contributions: B.V.A. conceptualized the idea, performed the analysis and wrote the manuscript, S.I.M. revised the manuscript for technical and language errors, M.A.A. revised the manuscript for technical and language errors. C.K.C. supervised the experimental design and technical analysis.

Acknowledgments: The financial support of Universiti Tenaga Nasional is well appreciated.

Funding: The authors would like to acknowledge the financial support of Universiti Tenaga Nasional, Malaysia through BOLD2025 researchers grant (10436494/B2019141).

Conflicts of Interest: The authors declare no conflict of interest.

References

1. Ayodele, B.V.; Khan, M.R.; Lam, S.S.; Cheng, C.K. Production of CO-rich hydrogen from methane dry reforming over lanthania-supported cobalt catalyst: Kinetic and mechanistic studies. *Int. J. Hydrogen Energy* **2016**, *41*, 4603–4615. [CrossRef]
2. Ashraf, M.A.; Sanz, O.; Montes, M.; Specchia, S. Insights into the effect of catalyst loading on methane steam reforming and controlling regime for metallic catalytic monoliths. *Int. J. Hydrogen Energy* **2018**, *43*, 11778–11792. [CrossRef]
3. Heng, L.; Xiao, R.; Zhang, H. Life cycle assessment of hydrogen production via iron-based chemical-looping process using non-aqueous phase bio-oil as fuel. *Int. J. Greenh. Gas Control* **2018**, *76*, 78–84. [CrossRef]
4. Bulutoglu, P.S.; Say, Z.; Bac, S.; Ozensoy, E.; Avci, A.K. Dry reforming of glycerol over Rh-based ceria and zirconia catalysts: New insights on catalyst activity and stability. *Appl. Catal. A Gen.* **2018**, *564*, 157–171. [CrossRef]
5. Sengodan, S.; Lan, R.; Humphreys, J.; Du, D.; Xu, W.; Wang, H.; Tao, S. Advances in reforming and partial oxidation of hydrocarbons for hydrogen production and fuel cell applications. *Renew. Sustain. Energy Rev.* **2018**, *82*, 761–780. [CrossRef]
6. Ayodele, B.V.; Khan, M.R.; Cheng, C.K. Greenhouse gases mitigation by CO2 reforming of methane to hydrogen-rich syngas using praseodymium oxide supported cobalt catalyst. *Clean Technol. Environ. Policy* **2016**, *19*, 795–807. [CrossRef]
7. Ayodele, B.V.; Khan, M.R.; Nooruddin, S.S.; Cheng, C.K. Modelling and optimization of syngas production by methane dry reforming over samarium oxide supported cobalt catalyst: Response surface methodology and artificial neural networks approach. *Clean Technol. Environ. Policy* **2016**, *19*, 1181–1193. [CrossRef]
8. Abatzoglou, N.; Fauteux-lefebvre, C. Review of catalytic syngas production through steam or dry reforming and partial oxidation of studied liquid compunds. *WIREs Energy Environ.* **2016**, *5*, 169–187. [CrossRef]
9. Shah, Y.T.; Gardner, T.H. Dry Reforming of Hydrocarbon Feedstocks. *Catal. Rev.* **2014**, *56*, 476–536. [CrossRef]
10. Abdullah, B.; Abd Ghani, N.A.; Vo, D.V.N. Recent advances in dry reforming of methane over Ni-based catalysts. *J. Clean. Prod.* **2017**, *162*, 170–185. [CrossRef]
11. Sehested, J. Four challenges for nickel steam-reforming catalysts. *Catal. Today* **2006**, *111*, 103–110. [CrossRef]
12. San-José-Alonso, D.; Juan-Juan, J.; Illán-Gómez, M.J.; Román-Martínez, M.C. Ni, Co and bimetallic Ni-Co catalysts for the dry reforming of methane. *Appl. Catal. A Gen.* **2009**, *371*, 54–59. [CrossRef]
13. Huang, X.; Ji, C.; Wang, C.; Xiao, F.; Zhao, N.; Sun, N.; Wei, W.; Sun, Y. Ordered mesoporous CoO-NiO-Al2O3 bimetallic catalysts with dual confinement effects for CO2 reforming of CH4. *Catal. Today* **2016**, *281*, 241–249. [CrossRef]
14. Ayodele, B.V.; Khan, M.R.; Cheng, C.K. Production of CO-rich hydrogen gas from methane dry reforming over Co/CeO2 Catalyst. *Bull. Chem. React. Eng. Catal.* **2016**, *11*, 210–219. [CrossRef]
15. Ayodele, B.V.; Khan, M.R.; Cheng, C.K. Syngas production from CO2 reforming of methane over ceria supported cobalt catalyst: Effects of reactants partial pressure. *J. Nat. Gas Sci. Eng.* **2015**, *27*, 1016–1023. [CrossRef]
16. Ayodele, B.V.; Khan, M.R.; Cheng, C.K. Catalytic performance of ceria-supported cobalt catalyst for CO-rich hydrogen production from dry reforming of methane. *Int. J. Hydrogen Energy* **2015**, *41*, 198–207. [CrossRef]
17. Kathiraser, Y.; Oemar, U.; Saw, E.T.; Li, Z.; Kawi, S. Kinetic and mechanistic aspects for CO 2 reforming of methane over Ni based catalysts. *Chem. Eng. J.* **2015**, *278*, 62–78. [CrossRef]

18. Hossain, M.A.; Ayodele, B.V.; Cheng, C.K.; Khan, M.R. Artificial neural network modeling of hydrogen-rich syngas production from methane dry reforming over novel Ni/CaFe2O4 catalysts. *Int. J. Hydrogen Energy* **2016**, *41*, 11119–11130. [CrossRef]

19. Ayodele, B.V.; Cheng, C.K. Modelling and optimization of syngas production from methane dry reforming over ceria-supported cobalt catalyst using artificial neural networks and Box-Behnken design. *J. Ind. Eng. Chem.* **2015**, *32*, 246–258. [CrossRef]

20. Arce-Medina, E.; Paz-Paredes, J.I. Artificial neural network modeling techniques applied to the hydrodesulfurization process. *Math. Comput. Model.* **2009**, *49*, 207–214. [CrossRef]

21. Puig-Arnavat, M.; Bruno, J.C. Artificial Neural Networks for Thermochemical Conversion of Biomass. *Recent Adv. Thermochem. Conver. Biomass* **2015**, 133–156. [CrossRef]

22. Ghasemzadeh, K.; Ahmadnejad, F.; Aghaeinejad-Meybodi, A.; Basile, A. Hydrogen production by a Pd—Ag membrane reactor during glycerol steam reforming: ANN modeling study. *Int. J. Hydrogen Energy* **2018**, *43*, 7722–7730. [CrossRef]

23. George, J.; Arun, P.; Muraleedharan, C. Assessment of producer gas composition in air gasification of biomass using artificial neural network model. *Int. J. Hydrogen Energy* **2018**, *43*, 9558–9568. [CrossRef]

24. Basile, A.; Curcio, S.; Bagnato, G.; Liguori, S.; Jokar, S.M.; Iulianelli, A. Water gas shift reaction in membrane reactors: Theoretical investigation by artificial neural networks model and experimental validation. *Int. J. Hydrogen Energy* **2015**, *40*, 5897–5906. [CrossRef]

25. Shahbaz, M.; Taqvi, S.A.; Minh Loy, A.C.; Inayat, A.; Uddin, F.; Bokhari, A.; Naqvi, S.R. Artificial neural network approach for the steam gasification of palm oil waste using bottom ash and CaO. *Renew. Energy* **2019**, *132*, 243–254. [CrossRef]

26. Nasr, N.; Hafez, H.; El, M.H.; Nakhla, G. Application of artificial neural networks for modeling of biohydrogen production. *Int. J. Hydrogen Energy* **2013**, *38*, 3189–3195. [CrossRef]

27. Zamaniyan, A.; Joda, F.; Behroozsarand, A.; Ebrahimi, H. Application of artificial neural networks (ANN) for modeling of industrial hydrogen plant. *Int. J. Hydrogen Energy* **2013**, *38*, 6289–6297. [CrossRef]

28. Ghasemzadeh, K.; Aghaeinejad-Meybodi, A.; Basile, A. Hydrogen production as a green fuel in silica membrane reactor: Experimental analysis and artificial neural network modeling. *Fuel* **2018**, *222*, 114–124. [CrossRef]

29. Usman, M.; Daud, W.M.A.W.; Abbas, H.F. Dry reforming of methane: Influence of process parameters—A review. *Renew. Sustain. Energy Rev.* **2015**, *45*, 710–744. [CrossRef]

30. Sun, Y.; Ritchie, T.; Hla, S.S.; McEvoy, S. Thermodynamic analysis of mixed and dry reforming of methane for solar thermal applications. *J. Nat. Gas Chem.* **2011**, *20*, 568–576. [CrossRef]

31. Ayodele, B.V.; Cheng, C.K. Process modelling, thermodynamic analysis and optimization of dry reforming, partial oxidation and auto-thermal methane reforming for hydrogen and syngas production. *Chem. Prod. Process Model.* **2015**, *10*, 211–220. [CrossRef]

32. Pakhare, D.; Schwartz, V.; Abdelsayed, V.; Haynes, D.; Shekhawat, D.; Poston, J.; Spivey, J. Kinetic and mechanistic study of dry (CO2) reforming of methane over Rh-substituted La2Zr2O7 pyrochlores. *J. Catal.* **2014**, *316*, 78–92. [CrossRef]

33. Foo, S.Y.; Cheng, C.K.; Nguyen, T.-H.; Adesina, A.A. Kinetic study of methane CO2 reforming on Co–Ni/Al2O3 and Ce–Co–Ni/Al2O3 catalysts. *Catal. Today* **2011**, *164*, 221–226. [CrossRef]

34. Aquilanti, V.; Mundim, K.C.; Elango, M.; Kleijn, S.; Kasai, T. Temperature dependence of chemical and biophysical rate processes: Phenomenological approach to deviations from Arrhenius law. *Chem. Phys. Lett.* **2010**, *498*, 209–213. [CrossRef]

35. Maneerung, T.; Hidajat, K.; Kawi, S. Co-production of hydrogen and carbon nanofibers from catalytic decomposition of methane over LaNi(1-x)Mx O3-α perovskite (where M = Co, Fe and X = 0, 0.2, 0.5, 0.8, 1). *Int. J. Hydrogen Energy* **2015**, *40*, 13399–13411. [CrossRef]

36. Chen, D.; Lødeng, R.; Anundskås, A.; Olsvik, O.; Holmen, A. Deactivation during carbon dioxide reforming of methane over Ni catalyst: Microkinetic analysis. *Chem. Eng. Sci.* **2001**, *56*, 1371–1379. [CrossRef]

37. Cui, M.; Yang, K.; Xu, X.L.; Wang, S.D.; Gao, X.W. A modified Levenberg-Marquardt algorithm for simultaneous estimation of multi-parameters of boundary heat flux by solving transient nonlinear inverse heat conduction problems. *Int. J. Heat Mass Transf.* **2016**, *97*, 908–916. [CrossRef]

38. Kayri, M. Predictive Abilities of Bayesian Regularization and Levenberg–Marquardt Algorithms in Artificial Neural Networks: A Comparative Empirical Study on Social Data. *Math. Comput. Appl.* **2016**, *21*, 20. [CrossRef]

39. Mia, M.; Dhar, N.R. Prediction of surface roughness in hard turning under high pressure coolant using Artificial Neural Network. *Meas. J. Int. Meas. Confed.* **2016**, *92*, 464–474. [CrossRef]

40. Shi, J.; Zhu, Y.; Khan, F.; Chen, G. Application of Bayesian Regularization Artificial Neural Network in explosion risk analysis of fixed offshore platform. *J. Loss Prev. Process Ind.* **2019**, *57*, 131–141. [CrossRef]

41. Khadse, C.B.; Chaudhari, M.A.; Borghate, V.B. Electromagnetic Compatibility Estimator Using Scaled Conjugate Gradient Backpropagation Based Artificial Neural Network. *IEEE Trans. Ind. Inform.* **2017**, *13*, 1036–1045. [CrossRef]

42. Li, H.; Zhang, Z.; Liu, Z. Application of Artificial Neural Networks for Catalysis: A Review. *Catalysts* **2017**, *7*, 306. [CrossRef]

43. Bustillo, A.; Pimenov, D.Y.; Matuszewski, M.; Mikolajczyk, T. Using artificial intelligence models for the prediction of surface wear based on surface isotropy levels. *Robot. Comput. Integr. Manuf.* **2018**, *53*, 215–227. [CrossRef]

44. Benardos, P.G.; Vosniakos, G.C. Optimizing feedforward artificial neural network architecture. *Eng. Appl. Artif. Intell.* **2007**, *20*, 365–382. [CrossRef]

45. Onalo, D.; Adedigba, S.; Khan, F.; James, L.A.; Butt, S. Data driven model for sonic well log prediction. *J. Pet. Sci. Eng.* **2018**, *170*, 1022–1037. [CrossRef]

46. Du, Y.C.; Stephanus, A. Levenberg-marquardt neural network algorithm for degree of arteriovenous fistula stenosis classification using a dual optical photoplethysmography sensor. *Sensors (Switzerland)* **2018**, *18*, 2322. [CrossRef]

47. Sharma, B.; Venugopalan, K. Comparison of Neural Network Training Functions for Hematoma Classification in Brain CT Images. *IOSR J. Comput. Eng.* **2014**, *16*, 31–35. [CrossRef]

 *catalysts*

Article

Effect of Surface Composition and Structure of the Mesoporous Ni/KIT-6 Catalyst on Catalytic Hydrodeoxygenation Performance

Xianming Zhang [1,*,†], Shuang Chen [1,†], Fengjiao Wang [2], Lidan Deng [2], Jianmin Ren [2], Zhaojie Jiao [1] and Guilin Zhou [1,2,*]

1   Engineering Research Center for Waste Oil Recovery Technology and Equipment, Ministry of Education, Chongqing Technology and Business University, Chongqing 400067, China; chenshuang023@gmail.com (S.C.); jiaozhaojie2001@ctbu.edu.cn (Z.J.)
2   Department of Materials Science and Engineering, Chongqing Technology and Business University, Chongqing 400067, China; wfj1122345@163.com (F.W.); dld2006lidan@163.com (L.D.); renjianmin123@sohu.com (J.R.)
*   Correspondence: zxm215@126.com (X.Z.); dicpglzhou@ctbu.edu.cn (G.Z.); Tel.: +86-23-62769-785 (G.Z.)
†   These authors contributed equally to this work and should be considered co-first authors.

Received: 13 October 2019; Accepted: 22 October 2019; Published: 25 October 2019

Abstract: A series of Ni/KIT6 catalyst precursors with 25 wt.% Ni loading amount were reduced in H_2 at 400, 450, 500, and 550 °C, respectively. The studied catalysts were investigated by XRD, Quasi in-situ XPS, BET, TEM, and H_2-TPD/Ranalysis methods. It was found that reduction temperature is an important factor affecting the hydrodeoxygenation (HDO) performance of the studied catalysts because of the Strong Metal Support Interaction Effect (SMSI). The reduction temperature influences mainly the content of active components, crystal size, and the abilityfor adsorbing and activating H_2. The developed pore structure and large specific surface area of the KIT-6 support favored the Ni dispersion. The RT450 catalyst, which was prepared in H_2 atmosphere at 450 °C, has the best HDO performance. Ethyl acetate can be completely transformed and maintain 96.8% ethane selectivity and 3.2% methane selectivity at 300 °C. The calculated apparent activation energies of the prepared catalysts increased in the following order: RT550 > RT400 > RT500 > RT450.

Keywords: hydrodeoxygenation; Ni/KIT-6; ethyl acetate

---

1. Introduction

With the process of economic globalization, the energy demand is increasing in human society [1]. The development of non-renewable energyhasnot fully met the developmentneeds of industry and society [2,3]. Therefore, finding a reliable alternative energy product is imminent. In order to cope with the above energy crisis, the new green energy concept and energy conversion technology have been put forward. The use of fed oil to prepare biodiesel is one of the most effective ways to reduce the supply pressure of petrochemical diesel [4]. The edible oil mainly includes vegetable oil and animal oil, which are mainly composed of fatty acid or ester with long chain. Fatty acid or ester usually contains rich carbon resources with C_{12}–C_{24} bonds [5,6]. Therefore, the successful transformation of fed oil to biodiesel may effectively mitigate the current tense energy situation. At present, the first-generation biodiesel cannot completely be deoxygenated due to technical limitations. The products contain high oxygen content, resulting in low calorific value and poor adaptability [7–9]. In contrast, the second-generation biodiesel can be completely deoxygenated because of technical advantages, and the prepared product has excellent performance. The second-generation biodiesel produced by HDO technology is closer to the petrochemical diesel in physical and chemical properties than the first-generation biodiesel.

Therefore, HDO technology has attracted wide attention because of the important applicable value. Of course, it also will play adominant role in promoting the chemical industry in the future.

The research and development of high performance catalytic materials is the core of the HDO technology. At present, the most widely used catalytic materials are the supported noble metal catalysts, such as Pt [10], Pd [11], Ru [12], and so on. Pd/C catalyst has high yield of alkane products, which are used in the stearic acid HDO reaction. The yield of n-C_{17} alkane is over 60%, olefins and polymers are also produced with these products [13]. Yang et al. [14] prepared Pt/ZIF-67/zeolite 5A catalyst for oleic acid HDO and found that the oleic acid is completely transformed at 320 °C/2.0 MPa when Pt loading is only 0.5% due to the special structure of the studied catalyst. Good at stearic acid HDO reaction is 2% Ru/TiO_2. The results show that the product is composed of C_{17} and octadecanol whose selectivities are 18.1% and 78.2%, respectively, which indicates the product is the hydrolyzate product, and the hydrolysis rate is greater than the HDO rate [15]. Chen et al. [16] reports that the Ru/zeolite catalyst is used in the HDO study of fatty acid methyl ester. It is found that the fed oil mainly goes through decarbonylation pathway. At 300 °C/3.0 MPa, the fed oil conversion reaches 90.8%, and the selectivity of C_{17} reaches 64.3%. As is well known, the main advantage of the noble metal catalyst is the high catalytic activity, that is, the HDO reaction can be achieved at a relatively low reaction temperature. Nevertheless, noble metal catalysts are expensive. Therefore, transition metal catalyst has attracted much attention in the research and application in the HDO field. HDO and hydrodesulfurization have many similar characteristics; therefore, HDO catalysts are originally developed primarily as hydrodesulfurization catalysts. Kubicka et al. [17] use the CoMoS/MCM-41 to catalyze the HDO reaction of rapeseed oil in a fluidized bed, mainly obtaining the diesel components such as C_{17} and C_{18}, under the condition of 300~320 °C, 2.0–11.0 MPa, and 1.0–4.0 h^{-1}. The introduction of S element is conducive to preventing catalyst from being poisoned and even reducing some negative factors such as carbon deposition [18]. However, the S element may also be introduced into the product to reduce the environmental protection of the corresponding product [19]. Therefore, sulfur-free catalysts are getting more and more attention. Imane Hachemi et al. [20] use the sulfur-free Ni/γ-Al_2O_3 catalyst for the fatty acids HDO. Under 300 °C/3.0 MPa, 6.0 h is running stably. The selectivity of C_{17} reaches 96.2%, while the conversion of fatty acid reaches 100%. Mesoporous silicon-based materials have developed mesoporous structure and large specific surface area, which facilitates the active component in high dispersion on the support surface [21–23]. Therefore, replacing traditional support with mesoporous silicon-based material is a good method to solve the above problems we have met at the moment. At the same time, the use of ethyl acetate as a model compound can even effectively eliminate the influence of long chain alkyl structures over the studied catalytic HDO performance, thus obtaining more accurate information on HDO of oxygen containing functional groups [24,25]. As we all know, the reduction temperature can affect the valence state of the variable valence transition metal species, then influence the activation ability of the corresponding catalysts for oxygen containing functional groups in the ester molecules and H_2 molecules in the reaction system. Zs. Ferencz et al. [26] studied the reduction of Co/Al_2O_3 catalyst at different reduction temperatures. It was found that the composition of the corresponding catalyst is obviously affected by the reduction temperature. The higher the reduction temperature is, the more the Co content is. Furthermore, the higher the temperature is, the narrower the particle size distribution is. Therefore, the reduction temperature of the studied catalyst is a very important factor in the hydrogen reaction. However, the impact of the reduction temperature of Ni/KIT-6 catalyst on its HDO properties has rarely been reported.

In this paper, a series of 25 wt.% Ni/KIT-6 catalysts at different reduction temperatures. Ethyl acetate was used as model compound to investigate the catalytic HDO performance of the studied catalysts. The morphology structure, chemical composition, and physicochemical properties of the prepared catalysts were studied by means of XRD, Quasi in situ XPS, BET, TEM, and H_2-TPD.

2. Results and Discussion

2.1. H₂-TPR Studies

Figure 1 is the H_2-TPR spectrum of calcined precursor and pure NiO. It is known from Figure 1 that there is an obvious hydrogen consumption phenomenon when the reduction temperature reaches 100 °C. With the increase of the temperature, the catalyst precursor, at 185, 345, 390, and 530 °C, forms distinct hydrogen consumption peaks which are labeled as α, β, γ, and δ. Interestingly, the hydrogen consumption peak of the pure phase NiO sample appears at 410 °C, and the pure phase NiO sample has only one peak.

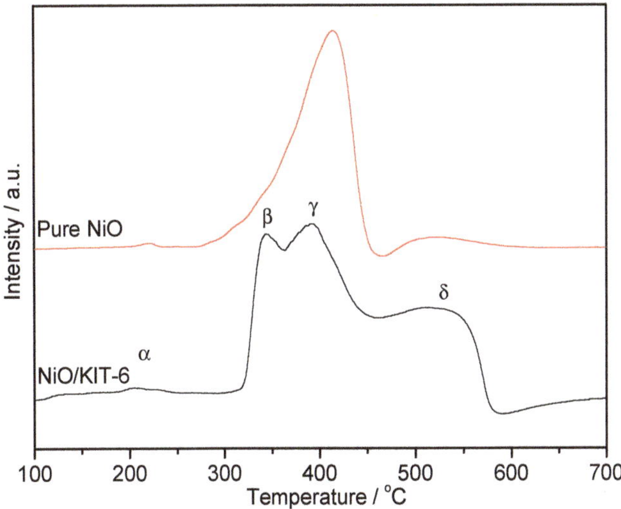

Figure 1. H_2-TPR patterns of catalyst precursor and pure NiO.

KIT-6 support has large specific surface area and developed pore structure, which can effectively promote the dispersion of NiO species. The H_2 reduction of highly dispersed NiO species forms α hydrogen consumption peak because of the preferred reducibility. The β hydrogen consumption peak is mainly attributed to the reduction of the dispersed NiO species on the surface. The interaction between this partial NiO and support SiO_2 is weak, thus showing good low temperature reducibility. The subsurface NiO species in the precursor is covered by the surface NiO species and has a certain interaction with the surface SiO_2. Therefore, the subsurface NiO species show a poor reducibility, and the reduction hydrogen consumption peak γ is formed in a higher temperature region than that of the surface NiO species. At the same time, the reduction peak at 530 °C may be attributed to the reduction of the NiO species in the support pore structure or the $NiSiO_3$ species formed at high temperature. The confinement and attachment effects of the nanopore structure hinder the reduction of partial NiO species in the channel. Moreover, this partial NiO species even may block the pore structure to restrict the diffusion and migration of H_2 flow in the pore structure, resulting in the reduction of NiO species in the higher temperature region. The confinement effect of KIT-6 pore structure enhances the interaction between NiO species and SiO_2 in the channel, leading to forming more stable NiO or $NiSiO_3$ species, which increase the difficulty of reduction. As a result, the large specific surface area and developed pore structure of the KIT-6 support can promote the NiO species to be highly dispersed, thus promoting the reduction of NiO species in the NiO/KIT-6 precursor. In addition, the results indicate that the NiO/KIT-6 precursor can be effectively reduced to prepare Ni/KIT-6 catalyst when the reduction temperature is 450 °C. However, there may still be some Ni^{2+} species (such as, NiO and $NiSiO_3$) that cannot be reduced by hydrogen at 450 °C.

2.2. XRD Characterization

The XRD spectra of the catalysts, which were produced by H_2 reduction at different temperatures, are shown in Figure 2. RT400, RT450, RT500, and RT550 catalysts form a series of XRD diffraction peaks at 2θ = 44.51, 51.81, and 76.41°, which corresponds to the face-centered cubic phase of metallic Ni (JCPDS NO. 04-0850) with {111}, {200}, and {220} planes, respectively [27]. In addition, the intensity of the metallic Ni diffraction peak is related to the reduction temperature. There is no other characteristic diffraction peak in the prepared catalysts expect for RT400 obtained by H_2 reduction at 400 °C for 2.0 h. In addition to the diffraction peak of the metallic Ni, the RT400 catalyst forms NiO peaks corresponding to the {111} and {220} planes at 2θ = 37.28 and 62.62°, using standard data (JCPDS No. 44-1159) [28]. The diffraction peaks of NiO phase have not been observed in RT450, RT500, and RT550 catalysts.

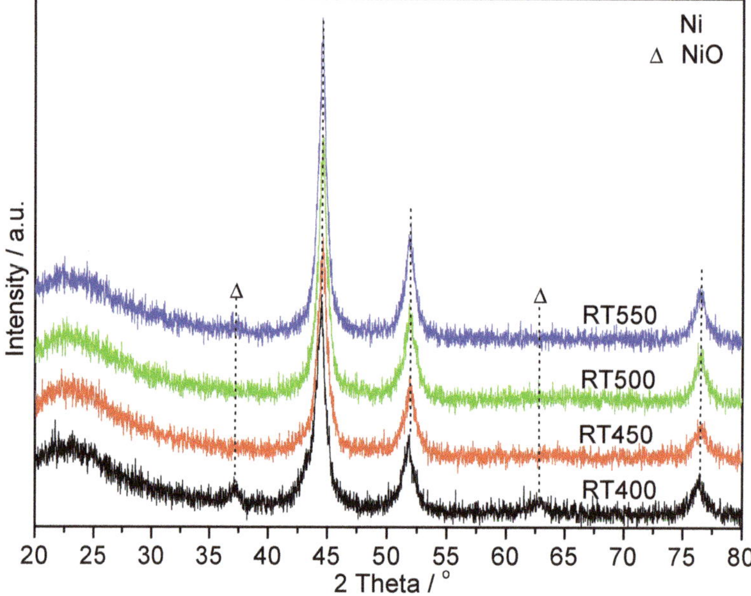

Figure 2. XRD patterns of RTx catalysts. (x = 400, 450, 500, and 550).

After reducing precursors at different temperatures, NiO species in precursors can be effectively reduced by H_2 to form metallic Ni species. The reduction temperature directly affects the reduction degree of NiO species, crystallinity, and dispersion of the metallic Ni species. Non reductive NiO species in the RT400 catalyst may belong to the NiO species that exist in the KIT-6 support pore structure. Meanwhile, the intensity of XRD diffraction peak of metallic Ni increases with increasing reduction temperature. The high reduction temperature can promote the migration and aggregation of metallic Ni to form stable Ni particles with large crystal size. Calculated by the Scherrer equation, the metallic Ni crystal sizes of the RT400, RT450, RT500, and RT550 catalysts are 5.9, 7.4, 9.1, and 10.0 nm, respectively (see Table 1). Under the heat dissipation action of H_2 flow, the migration and accumulation rate of metallic Ni during the reduction process of NiO/KIT-6 precursor can be decreased, thus, the high dispersion of metallic Ni crystal is maintained when the complete reduction is achieved. The lower reduction temperature is, the slower reduction rate of NiO species is. This can also be beneficial to the refinement of the metallic Ni crystal size. By contrary, at high reduction temperature, the effect of H_2 flow is not enough to prevent the high temperature sintering of metallic Ni, which reduces the dispersion and increases the crystal size. The reduction rate of NiO species is accelerated when the reduction temperature is too high, so that the formed metallic Ni particles with high surface energy can overcome the limitation of the support structure and migrate to the surface of the support [24]. At this time, the metallic Ni exists on the support surface with supersaturated state; unfortunately, the crystal size increases gradually [29]. Therefore, too high reduction temperature will lead to the aggregation of metallic Ni, which may cause excessive crystallization of metallic Ni active components and even decrease the number of exposed active sites to affect the catalytic performance. At the high reduction temperature, the metallic Ni component may even produce a crystal surface that is not conducive to the HDO reaction. The XRD results are also confirmed that the NiO species can be effectively reduced by H_2 at 450 °C for 2.0 h to form metallic Ni. This result is consistent with the previous H_2-TPR results and previous reports [24,30]. However, there may still be some amorphous NiO or/and $NiSiO_3$ species, which formed at high-temperature calcination. What is more, it is also possible that the amount of NiO or/and $NiSiO_3$ is lower than the detection limit of XRD. This is probably the reason why NiO or/and $NiSiO_3$ has not been observed in RT450, RT500, and RT550 catalysts in XRD spectra.

Table 1. Physicochemical properties of RTx catalysts.

Catalyst	Crystal Size from XRD (nm) [a]	H_2-TPD (α Region) Quantity (μmol/g)	Surface Area (m^2/g) [b]	Average Pore Diameter (nm) [c]	Total Pore Volume (cm^3/g) [d]	Particle Size from TEM (nm) [e]
RT400	5.9	27.6	349.2	6.08	0.55	6.3
RT450	7.4	65.8	348.9	6.35	0.56	7.7
RT500	9.1	52.4	347.7	6.17	0.54	9.7
RT550	10	49.9	343.4	6.03	0.56	10.3
KIT-6	—	—	562.5	6.56	1.22	—

[a] Average Ni crystal diameter calculated from Ni(111) plane using Scherrer equation from XRD. [b] Calculated by the BET equation. [c] BJH adsorption average pore diameter. [d] BJH adsorption pore volume. [e] Average Ni particle size observed from TEM images.

2.3. Quasi In-Situ XPS Studies

Quasi in-situ XPS technology is mainly used to analyze surface elements and the state of the surface elements of the studied RTx catalysts. The XPS spectra of Ni 2p for RTx catalysts are shown in Figure 3. The peaks of Ni 2p at 856.5 and 852.7 eV are attributed to Ni^{2+} and Ni^0 species, respectively [29,31]. From the results of in situ XPS, it is found that the RTx catalysts have rich metallic Ni species, which indicates that Ni^{2+} in the catalyst precursors can be effectively reduced to metallic Ni active species. The results are similar to that of the XRD studies. Ni^{2+} species may originate from the amorphous NiO species in the pore structure or the $NiSiO_3$ species formed at high-temperature calcination or/and reduction. However, the peak area and intensity corresponding to Ni^0 are different in the RTx catalysts, which can be attributed to the interactions between the active component and the KIT-6 support having significant difference in the function of the different reduction temperatures.

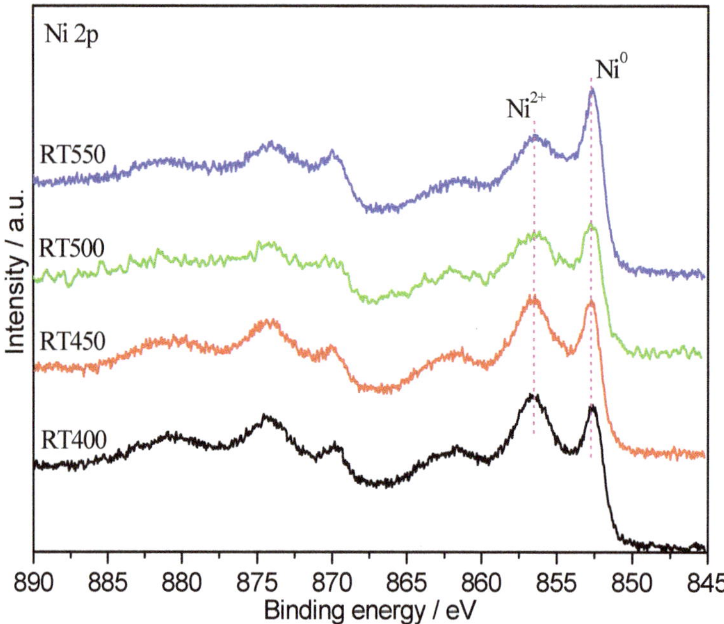

Figure 3. Spectra of the RTx catalysts. (x = 400, 450, 500, and 550).

2.4. H_2-TPD Studies

The H_2-TPD spectra of the studied RT400, RT450, RT500, and RT550 catalysts, which were reduced at different temperatures, are shown in Figure 4. It is known from Figure 4 that the prepared Ni based catalysts have good hydrogen desorption capacities and mainly form two hydrogen desorption regions, which are recorded as α and β, respectively. There are significant differences in the peak area and strength on the H_2-TPD patterns of the studied Ni/KIT-6 catalysts prepared at different reduction temperatures. The area of α peak increases with the increase of the reduction temperature and then decreases, and RT450 catalyst has the largest area of α peak among that of RTx catalysts. Conversely, the area and desorption temperature of β peak increase with the increase of the reduction temperature. The RT400 catalyst has a significant γ peak, mainly for the desorption of the bulk hydrogen species, which can store and supply the reactive hydrogen species for the reaction.

H_2-TPD can be used to effectively study the adsorption of H_2 molecules by corresponding catalyst and the ability to provide reactive hydrogen species for the HDO reaction [32]. Hydrogen species exist on the surface and in the bulk of the prepared Ni-based catalyst after reduction, and the physically adsorbed hydrogen species on the catalyst surface have been removed by Ar purge. In contrast, the chemisorbed hydrogen species are difficult to remove at room temperature by Ar purge because of the formation of a chemical bond between chemisorbed hydrogen species and metallic Ni active center. Therefore, the α and β hydrogen desorption regions formed by the catalysts can be attributed to the desorption of surface chemisorbed hydrogen species and metal hydride [24,32]. As is well known, the chemisorbed hydrogen species can also be effectively desorbed to form the reactive hydrogen species, which is required for the reaction by increasing the system temperature. Moreover, the bond energy of the chemisorbed hydrogen species is relatively weak, so it can be desorbed at low temperature (<400 °C). Unfortunately, compared with the chemisorbed hydrogen species, metal hydrides need to be decomposed at a higher temperature to release the hydrogen species. KIT-6 support has developed mesoporous structure, so Ni active components can penetrate into the inner surface, which helps to form active centers with different strength. H_2 molecules can also be adsorbed on the inner surface of the studied Ni/KIT-6 catalyst, which is concluded as weakly and strongly adsorbed hydrogen. According to the XRD and quasi in situ XPS results, a small amount of non-reductive NiO species exist in the RT400 catalyst, which is not conducive to the formation of the active center in the catalyst. As a result, the adsorption capacity of H_2 decreases on the RT400 catalyst, resulting in a poor low temperature desorption ability of H_2. After elevating the reduction temperature, α region of the RT450 catalyst has the maximum peak area among the studied RTx catalysts. These findings suggest that the RT450 catalyst has the best hydrogen adsorption storage capacity, which can provide more reactive hydrogen species for the HDO reaction. By continuing to increase the reduction temperature of the prepared catalyst, the area of the α hydrogen desorption region gradually decreases, which is attributed to the gradual weakening of the hydrogen adsorption and storage capacity of the corresponding catalysts. The XRD results also confirm that the crystal size of Ni increases with the increase of the reduction temperature. However, the large crystal size leads to reducing the exposed active crystal surface. Moreover, the large crystal size is not conducive to the formation or fracture of the "Ni-H" species, thus weakening the adsorption and activation ability of the catalyst for the H_2 molecules. The area of the β region increases with the increase of the reduction temperature; the desorption peak temperature increases gradually. This find may suggest that the formed hydrogen species in the high temperature region are mainly due to the decomposition of the metal hydride; the stable metal hydride is more easily formed at higher reduction temperature.

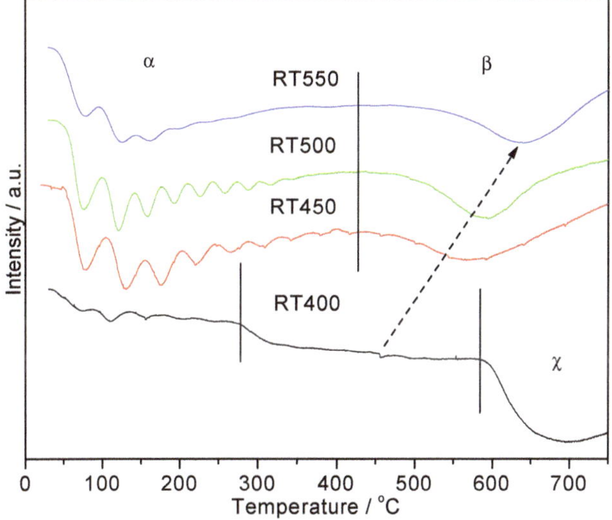

Figure 4. H_2-TPD patterns of RTx catalysts. (x = 400, 450, 500, and 550).

2.5. BET Characteristics

The porosities of the prepared RT400, RT450, RT500, and RT550 catalysts and KIT-6 support were studied by BET, as shown in Figure 5. The pore size distributions of the corresponding catalysts and KIT-6 support, calculated from the adsorption isotherm, were shown in the inset in Figure 5. All prepared RTx catalysts have a type IV isotherms with clear H1-type hysteresis loop, which indicates the corresponding catalysts have mesoporous structure. The specific surface area, pore volume, and pore diameter of the prepared catalysts are showed in Table 1. The prepared catalysts have inherited the developed pore structure of the KIT-6 support. Unfortunately, compared with the KIT-6 support, the specific surface area, pore volume, and pore diameter of the corresponding catalysts obviously decrease. These findings can be attributed to the loading of metallic Ni species entering the pore structure of KIT-6, resulting in the blockage of the pore structure. At the same time, the loading of

Ni species also reduces the support content in the unit volume or unit mass catalyst, which leads to decreasing the specific surface area of the corresponding catalyst. The results of Table 1 show that the RTx catalysts obtained at different reduction temperatures have maintained a specific surface area of about 340 m^2/g, the pore size also around 6 nm. It can be known that the change of the reduction temperature does not have a significant effect on the structure of the prepared catalysts. Because the high temperature accelerates the agglomeration of crystal and the increase of crystal size, the specific surface area has a slightly decrease by elevating the reduction temperature. However, the average pore diameter and total pore volume have not been affected by elevating the reduction temperature; that is to say, in the studied temperature region, the average pore diameter and total pore volume have no obvious rule to be observed by increasing reduction temperature. The pore diameter range is 6.03–6.35 nm, which is close to the pore diameter of KIT-6 support as showing in the Table 1. Based on the BET results, the developed pore structure and large specific surface area of KIT-6 contribute to support metallic Ni on its inner and outer surface, which leads to the metallic Ni in a high dispersion state and may even reach monolayer dispersion. With the filling and attachment of the metallic Ni on the KIT-6 inner surface, the micropore may be blocked, while the macropore structure can be filled to form mesopore structure. The reduction temperature can only change the chemical state and crystal size of Ni species; fortunately, it cannot change the structure of KIT-6 support. Therefore, the pore structure can maintain the pore size distribution and pore volume because of its confinement and attachment effects [24]. The slight changes of average pore diameter and total pore volume are mainly due to the growth of metal Ni crystal in the pore structure.

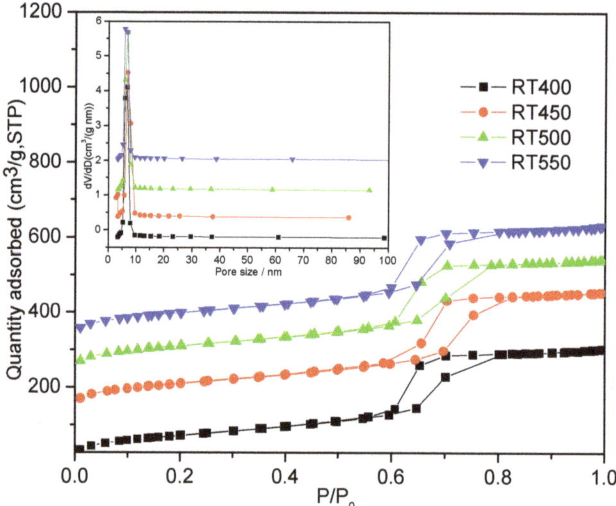

Figure 5. N$_2$ adsorption-desorption isotherms and pore diameter distributions of the prepared RTx catalysts. (x = 400, 450, 500, and 550).

2.6. TEM Characteristics

The prepared Ni/KIT-6 catalysts were studied by TEM, as shown in Figure 6. The prepared catalysts have obvious mesoporous structure, which is similar to that of the BET results. The metal Ni mainly presents two states. Some of the metal Ni successfully enters the pore of the support KIT-6, and some of the metal Ni mainly disperses on the KIT-6 surface. As the reduction temperature increases, the Ni particle size gradually increases, which is similar to that of XRD study. This phenomenon is mainly due to the high temperature accelerating the accumulation and migration of metal Ni particles, leading them gradually to increase at high temperature. The particle size of the metal Ni is shown in Table 1.

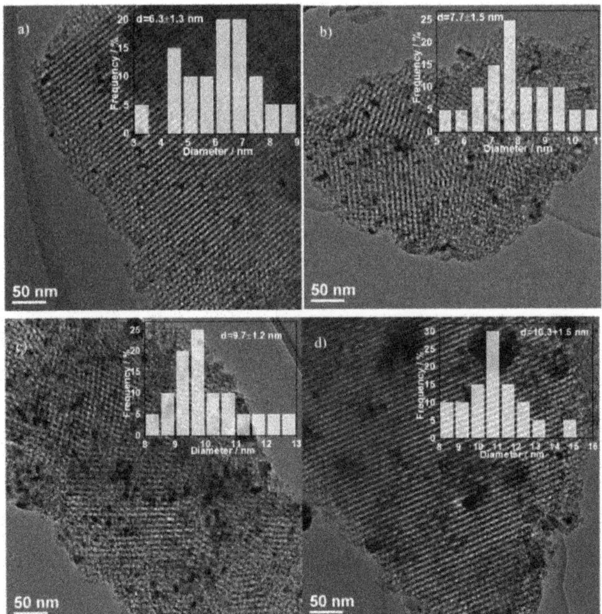

Figure 6. TEM patterns of the studied catalysts: (**a**) 400, (**b**) 450, (**c**) 500, and (**d**) 550.

2.7. Catalytic Activity Test

The prepared RT400, RT450, RT500, and RT550 catalysts exhibit good catalytic HDO performance and ethyl acetate (EA) catalytic HDO activity and product selectivity, as shown in Figure 7a–d. The catalytic HDO performances of the studied catalysts are obviously influenced by the reduction temperature, which are mainly reflected in the complete conversion temperature (T_{100}) of reactant and selectivity of the product (S). The reduction temperature has a significant effect on the catalytic performance. The T_{100} is 300 °C on the RT450 catalyst; the T_{100} of RT400 and RT500 catalysts rise to 320 °C, while the T_{100} of RT550 catalyst further increases to 340 °C. When the reaction temperature is 220 °C, the RT450 catalyst has the highest S_{ethane}, which is 86.9%; among the RTx catalysts at this point, a small amount of ethanol is detected, $S_{ethanol}$ is 6.7%. However, the intermediate product ethanol can be completely transformed at 240 °C. S_{ethane} can reach 96.8%, but $S_{methane}$ is only 3.2% at 300 °C over the RT450 catalyst which shows the best catalytic performance. For RT400 catalyst, at the initial reaction temperature, $S_{ethanol}$ reaches 37%, and the complete conversion temperature of the ethanol is higher by 40 °C than that of the RT450 catalyst. Finally, S_{ethane} is 97.3%, while $S_{methane}$ is just only 2.7% at 320 °C. For RT500 catalyst, $S_{ethanol}$ is more than 13% at the initial temperature of 220 °C, and the complete conversion temperature of the ethanol is the same as that of the RT400 catalyst. At 320 °C, S_{ethane} and $S_{methane}$ are 97.1% and 2.9%, respectively. However, the activity of RT550 catalyst is the poorest. When ethyl acetate completely transforms, the reaction system also has a little ethanol and methane, which are up to 1.5% and 1.3%, respectively. At the beginning of the reaction, in the RT550 catalyst, $S_{ethanol}$ is 30.5 percent points higher than that of RT450 catalyst; conversely, S_{ethane} is 24.1 percent points lower than that of RT450 catalyst at 220 °C.

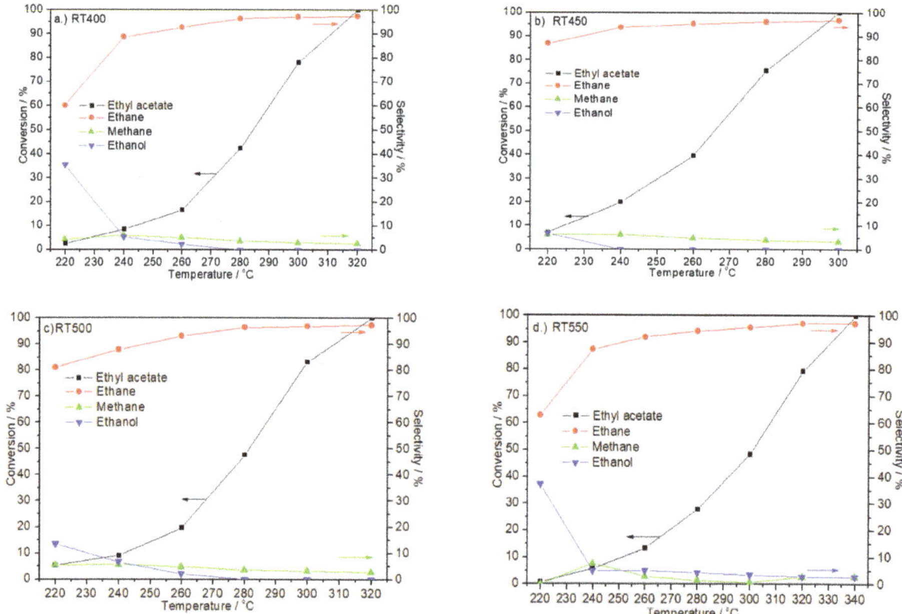

Figure 7. Influence of reduction temperature on the ethyl acetate hydrodeoxygenation conversion and products selectivity over the RTx catalysts. (**a**) RT400, (**b**) RT450, (**c**) RT500, and (**d**) RT550.

In chemical reaction system, the molecules of the reactant can be effectively activated to promote the chemical reaction [33]. Following the general catalytic reaction law, increasing the reaction temperature helps to elevate the energy of the reaction system to accelerate the chemical reaction [34]. Therefore, H_2 molecules are difficult to effectively activate at low reaction temperature, which is not conducive to the formation of reactive H species. When the reaction system energy is low, the activation ability of the corresponding catalyst for the reactants and intermediates is limited. At this time, the intermediate product ethanol hardly achieves further HDO, so the reaction maintains high $S_{ethanol}$ at relative low reaction temperature. The energy of the reaction system increases by enhancing reaction temperature, which accelerates the shift rate of the reactant molecules and the effective collision frequency between activated reactant molecules. Therefore, the HDO reaction of ethyl acetate and intermediate product ethanol can be promoted, and the conversion of ethyl acetate and S_{ethane} can also be significantly improved. Under the studied conditions, ethyl acetate may hydrolyze to produce two intermediate products: acetic acid and ethanol, which can be further deoxygenated to produce ethane [34]. Acetic acid can also react with metallic Ni to produce nickel acetate, and nickel acetate can be further hydrogenated to produce ethane and methane [24,25]. At the same time, the ethane molecules can be cracked to methane at a high reaction temperature [25]. In the ester HDO reaction, decarboxylation or decarboxylation can also happen to generate CO_2/CO. The generated CO_2/CO can react with H_2 over the metallic Ni catalyst to produce methane, which is also called methanation reaction [35]. Therefore, the corresponding catalysts show high $S_{methane}$ at studied reaction temperatures.

Based on the results of XRD, NiO species in RT400 catalyst cannot be completely reduced to form metallic Ni species; the result is similar to that of the quasi in situ XPS studies. The content of active centers in the prepared Ni/KIT-6 catalyst is relatively low, which may weaken the ability of the corresponding catalyst to activate H_2 molecules; H_2-TPD can also well confirm the results. In addition, water formed during the HDO reaction could be competing for the active sites together with ethyl acetate as the conversion increases. Moreover, these effects could be enhanced if the Ni particles are not completely reduced during reaction as water can bind strongly on oxidized metal surfaces. Then, the performance of the RT400 catalyst is weakened. RT450 catalyst, which was obtained by H_2 reduction at 450 °C, shows good HDO performances, and its catalytic performance is obviously higher than that of RT400 catalyst. Compared with RT400, this interesting result can be attributed to the effective reduction of NiO in the precursor of RT450 catalyst. At the same time, quasi in situ XPS confirmed the formation of active metallic Ni in the RT450 catalyst. The crystallinity of the metallic Ni active component is relatively low, and the crystal size is relatively small in RT450 catalyst, which is more conducive to the formation of the catalytic active center. Meanwhile, the smaller crystal size of metallic Ni can also promote the exposure of more active planes to a certain extent, which are conducive to the activation of H_2 molecules and also promote the HDO of reactants and intermediates. Therefore, the RT450 catalyst shows wonderful reactant conversion and S_{ethane}. The catalytic activities of RT500 and RT550 catalysts are significantly poorer than that of RT450 catalyst, and the conversions of ethyl acetate and S_{ethane} continue to decline at low temperature (220–260 °C) by increasing the reduction temperature. This can be attributed to the high reduction temperature resulting in a change in the chemical composition of the catalyst. The samples may even generate new inert crystal planes under high temperature, resulting in a decrease in the ratio of the active crystal planes, which weakens the ability of the corresponding catalyst to activate the reactants. The crystallinities and crystal sizes of metallic Ni in RT500 and RT550 catalysts obviously increase, and the crystal sizes of two catalysts reach 9.1 and 10 nm, respectively. The high crystallinity and large crystal size limit the formation and exposure of the active crystal plane. At first, this weakens the adsorption and activation ability of the active crystal to H_2 molecules; in addition, this hinders the formation or fracture of the "Ni-H" bond. What is more, this weakens the activation capacity of the corresponding catalyst for the "C = O" bond and the "C-O" bond. Therefore, the HDO performance is influenced. According to the results of BET and TEM, the prepared catalysts have large specific surface areas and developed mesoporous structures, which effectively promote the migration and diffusion of gaseous reactant molecules in the pore structure. Therefore, the studied catalysts have perfect catalytic HDO performances.

2.8. Activation Energy Studies

Under the condition of atmospheric pressure and 12,000 mL/h·g_{cat} space velocity, the influence of internal and external diffusion is eliminated as is using the fixed bed reactor to study the mechanism of the prepared RTx catalysts for HDO performance by activation energy experiments. According to the Arrhenius equation, $k = k_0 \cdot \exp(-E_a/RT)$; this equation can be obtained: $\ln k = \ln k_0 - E_a/RT$ (where k_0 is the pre-exponential factor, E_a is the activation energy, R is the gas constant, and T is the thermodynamic temperature). Map the -ln$k$ to T^{-1} over RTx catalysts with different reduction temperature, as shown in Figure 8. From Figure 8, the activation energies of ethyl acetate HDO on RT400, RT450, RT500, and RT550 catalysts are 81.0, 67.7, 79.5, and 90.9 kJ/mol, respectively, according to the slope of the obtained straight line, as shown in Table 2.

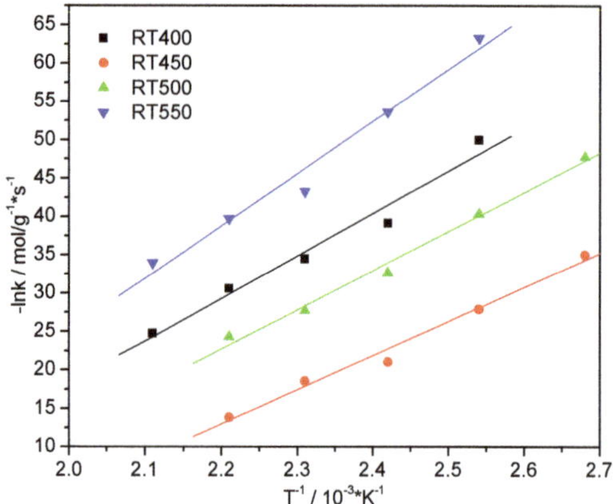

Figure 8. The relationship of ln$k$ and T^{-1}.

Table 2. Activation energy studies of RTx catalysts (x = 400, 450, 500, and 550).

Catalysts	Equation	Activation Energy (kJ/mol)
RT400	Y = 9.208X − 12.44	81.0
RT450	Y = 8.066X − 23.17	67.7
RT500	Y = 9.264X − 16.14	79.5
RT550	Y = 11.33X − 9.65	90.9

There are significant differences in the activation energies of the HDO reaction on the studied RTx catalysts, indicating that the changes of the reduction temperatures can further alter the energy barriers over the corresponding catalysts for HDO reaction. In the gas–solid heterogeneous catalytic hydrogenation reaction, the purposes of the adsorption and activation of the reactant molecules on the catalyst surface are to reduce the reaction activation energy. Therefore, the activation energy represents the energy required by the reactant molecules from the normal state to the active state in the chemical reaction under the function of the corresponding catalyst. In other words, under the function of the catalyst, the activation energy, which reflects the difficulty of the catalytic reaction, is the minimum energy needed for the chemical reaction. Meanwhile, chemical reaction rate is closely related to the activation energy. The lower the activation energy is, the faster the reaction rate is; thus, reducing activation energy will effectively promote the reaction rate. Moreover, the reaction system with lower activation energy can ensure that more reactant molecules are activated to form activated molecules, which increases the number of activated molecules, and further, to increase the effective collision probability between the activated molecules, thus accelerating the corresponding chemical reaction rate. According to the XRD and quasi in situ XPS results, the RT400 catalyst obtained by lower temperature reduction has lesser active component, which cannot effectively activate the reactant molecules, thereby elevating the activation energy and the difficulty of corresponding reaction. With the increase of the reduction temperature, even though the content of the active metallic Ni increases, the crystal size of the metallic Ni gradually increases. The large crystal size reduces the number of the exposed active planes on the metallic Ni; then, this action weakens the interaction between catalytic active centers and reactant molecules, which is not conducive to further decreasing the activation energy of ethyl acetate HDO reaction. The increase of crystal size may also lead to the formation of inert crystal plane, which reduces the number of active centers of the corresponding catalyst and affects the activation ability of the studied catalyst to the reactant molecules. Therefore, these adverse factors result in the number of activated molecules in the reaction system being lower and the activation energy of the corresponding reaction being higher. The activation energy of ethyl acetate HDO reaction under the function of RT450 catalyst is lowest among that of the RTx catalysts, and the energy required for effective collision is lowest. This find shows that the effective activation among the reactant molecules can be realized in the relative lower energy system, thus forming more activated molecules, which is beneficial to the catalytic hydrogenation reaction. At the same time, the lower reaction activation energy leads the formation of active molecules more likely to have effective collisions in the lower energy system, which is beneficial to improve the HDO reaction performance. Hence, decreasing the reaction activation energy can provide an easy way for the ethyl acetate HDO reaction, thus accelerating the reaction to thermodynamic equilibrium. H_2-TPD results also strongly confirm that under the same conditions, RT450 catalyst reaction system has the best adsorption and activation ability for reactant H_2 molecules, thus forming a large amount of reactivated hydrogen species. Effective collisions among activated molecules will help reduce reaction activation energy. This is mainly attributed to the sufficient metallic Ni content, large specific surface area, and suitable crystal size for the RT450 catalyst. The activation energies of the RT500 and RT550 catalysts obtained by enhancing the reduction temperature significantly increase, indicating that the higher the reduction temperature is, the larger the activation energy is. Simultaneously, the BET and TEM results show that the specific surface area of the studied catalyst slightly decreased with the increase of the reduction temperature. High reduction temperature may accelerate the shrinkage or even collapse of mesorpous structure, which weakens the adsorption and activation ability of reactant molecules in the mesorpous structure of corresponding catalyst, as well as further affects the migration and effective collision of activated molecules in mesorpous structure. The effective collision frequency between activated molecules is reduced, which leads to the difficult ethyl acetate HDO reaction increase, as well as to the increase of activation energy. H_2-TPD also confirms that a series of changes in the chemical composition of the studied catalysts have a significant effect on the adsorption and activation ability for the H_2 molecules. The results show that the chemical composition and particle structure of the studied catalyst have a conspicuous influence on the ethyl acetate HDO reaction. In our previous study, it was also found that ethyl acetate HDO reaction is a structural sensitive reaction [30]. To sum up, the amount of active metallic Ni and its crystal size, as well as other chemical compositions may be important reasons for the change of activation energy of RTx series reaction system. The activation energy of the Ni/KIT-6 catalysts prepared at different reduction temperatures for ethyl acetate HDO reaction has the lowest value. The activation energies of the prepared catalysts obey the following order: RT550 > RT400 > RT500 > RT450.

3. Materials and Methods

3.1. Catalyst Preparation

The Ni/KIT-6 catalyst was prepared by impregnation method. KIT-6, which is prepared following the procedure described by our subject group [36,37], was used as the support for the Ni/KIT-6 catalysts. 0.62 g Ni(NO$_3$)$_2$·6H$_2$O (Chengdu Kelong Chemical Reagent Factory, China) was dissolved in deionized water with HNO$_3$. KIT-6 (0.38 g) was added. The sample continued to dry overnight in an oven and then calcined at 550 °C for 4.0 h in air. Finally, the obtained catalyst precursor was subjected to hydrogen reduction 2.0 h to produce the Ni/KIT-6 catalyst at 400, 450, 500, and 550 °C, which were labeled as RT400, RT450, RT500, and RT550, respectively.

3.2. Catalyst Characterization

3.2.1. XRD Characterization

X-ray diffraction (XRD) pattern were record on a Shimadzu-6100 diffractometer (Tokoyo, Japan) with a rotating anode using Ni filtered Cu Kα radiation at 40 kV of a tube voltage and 30 mA of a tube current. The data of 2θ from 20° to 80° range were collected with the step size of 0.02 at the rate of 3°/min, according to our preview work [24].

3.2.2. Quasi In-Situ XPS Studies

The prepared RTx catalyst samples (x = 400, 450, 500, and 550) were in situ reduced in a pretreatment chamber in high purity hydrogen at 400 °C for 60 min to obtain RTx catalysts, and then, obtained RTx catalysts were cooled down to room temperature. Then, the RTx catalysts were directly moved to analysis chamber without being exposed to the environment to analyze the Ni 2p message. The chemical states of the RTx (x = 400, 450, 500, and 550) samples were determined using in situ X-ray photoelectron spectra (in situ XPS). The XPS signals were collected using an ESCALAB 250Xi analyzer (Thermo Fisher Scientific, MA, USA).

3.2.3. BET Characterization

The Brunauer–Emmett–Teller (BET) analysis of the catalysts was determined by N_2 adsorption–desorption using an ASAP 3020 (Micromeritics, GA, USA) instrument. The catalysts were outgassed at 300 °C for 3.0 h before being subjected to N_2 adsorption, according to our preview work [24].

3.2.4. H_2-TPD/R Studies

Temperature-programmed desorption of H_2 (H_2-TPD) was tested on a self-made instrument, the steps follows: 50 mg of catalyst precursor was placed in a quartz U-tube reactor. The catalyst precursors were reduced with a 5% H_2/Ar mixture (25 mL/min) at 400, 450, 500, and 550 °C for 2.0 h, respectively. Afterward, the catalyst is flushed with Ar flow (25 mL/min) to remove all physically adsorbed molecules. The studied catalysts were then heated at 10 °C/min to 700 °C with pure Ar at a flow rate of 25 mL/min. The hydrogen desorption signals were monitored using a thermal conductivity detector (TCD), according to our previous work [24].

3.2.5. TEM Characterization

The transmission electron microscopy (TEM) was performed using a FEI Tecnai G^2 F30 microscope (Oregon, OR, USA) operating at an acceleration voltage of 300 KV.

3.2.6. Catalytic Activity Measurement

Catalytic tests of all the studied catalysts were carried out in a tubular, continuous-flow, fixed-bed reactor under atmospheric pressure. The quartz tube reactor containing 50 mg of the supported nickel oxide precursors were placed inside the tubular furnace. High-purity hydrogen was flowed downward through the reactor containing the catalyst bed, while an electronic mass flow controller was used to control the hydrogen gas flow rate. The precursors were reduced at 400, 450, 500, and 550 °C at hydrogen flow for 2.0 h, followed by cooling under a H_2 flow to the studied temperature. Catalytic activity tests were performed at temperature range of 220 to 340 °C. The actual reduction and reaction temperatures were measured using a thermocouple which was directly inserted into the catalyst bed and were monitored by a temperature programmable controller at a heating rate of 10 °C/min. The fed gas was obtained by bubbling hydrogen through a saturator, which contains the studied model bio-oil compound ethyl acetate in liquid phase kept at constant temperature to achieve the fed gas mixture consisting of gaseous acetate ester (3.3%) and H_2 (96.7%). The fed gas was led over the catalyst at a flow rate of 10 mL·min^{-1}, which is equivalent to a gas mass space velocity of 12,000 mL/g·h. The gas products were monitored using an on-line gas chromatograph (SC-3000B, Chuanyi Automation CO. LTD, Chongiqng, China), equipped with a FID, according to our preview work [24].

The conversion of reactant and the selectivity of products were calculated according to our previous study [24].

4. Conclusions

A series of Ni/KIT-6 catalysts with a metallic Ni loading of 25 wt.% were prepared at different reduction temperatures. They show good catalytic HDO performance to the model compound ethyl acetate. Ethyl acetate is completely converted, while high S_{ethane}, all greater than 95%, can be maintained. The catalytic performance of RT450 catalyst is the best among the prepared Ni/KIT-6 catalysts. The complete conversion of ethyl acetate can be achieved at 300 °C, S_{ethane} is up to 96.8%, $S_{methane}$ is only 3.2%, and the intermediate product ethanol is completely converted at 240 °C. This is mainly attributed to the abundant metallic Ni content in the RT450 catalyst, the small crystal size, only 7.4 nm. KIT-6 support promotes the highly dispersed Ni components and also promotes the migration and activation of reactant molecules in the pore structure, and the Strong Metal Support Interaction Effect (SMSI). In the range of studied reduction temperature, the mesoporous structure does not change significantly, but the reduction temperature affects the HDO performances of the corresponding catalyst. The adsorption and desorption properties of the prepared catalysts for H_2 molecules are in the following order: RT450 > RT500 > RT550 > RT400, and the order of HDOactivities of the prepared catalysts is consistent with that of H_2 desorption. By calculation, the activation energy of the prepared catalysts obeys the following order: RT550 > RT400 > RT500 > RT450.

Author Contributions: X.Z., S.C., and G.Z. conceived and designed the experiments. X.Z. and G.Z. supervised the study. S.C., F.W., L.D., and Z.J. carried out materials syntheses. F.W., L.D., and J.R. performed all kinetic degradation tests and related measurements and interpretation of results. All authors discussed the results and co-wrote and edited the manuscript. All authors reviewed and approved the manuscript.

Funding: This research is funded by Scientific and Technological Key Program of Chongqing Municipal Education Commission (KJZD-K201800801, KJZD-M201900802); Science and Technology of Chongqing Municipal Education Commission Funded Research Projects (KJQN20190081).

Acknowledgments: Thank you for the support of Engineering Research Center for Waste Oil Recovery Technology and Equipment (Ministry of Education) in the experimental site and equipment.

Conflicts of Interest: The authors declare no conflict of interest.

References

1. Wang, H.; Li, G.L.; Rogers, K.; Lin, H.F.; Zheng, Y. Hydrotreating of waste cooking oil over supported CoMoS catalyst—Catalyst deactivation mechanism study. *Mol. Catal.* **2017**, *443*, 228–240. [CrossRef]
2. Ameen, M.; Azizan, M.T.; Yusup, S.; Ramli, A.; Yasir, M. Catalytic hydrodeoxygenation of triglycerides: An approach to clean diesel fuel production. *Renew. Sustain. Energy Rev.* **2017**, *80*, 1072–1088. [CrossRef]
3. Arun, N.; Sharma, R.V.; Dalai, A.K. Green diesel synthesis by hydrodeoxygenation of bio-based feedstocks: Strategies for catalyst design and development. *Renew. Sustain. Energy Rev.* **2015**, *48*, 240–255. [CrossRef]
4. Ullah, Z.; Bustam, M.A.; Man, Z.; Khan, A.S.; Muhammad, N.; Sarwono, A. Preparation and kinetics study of biodiesel production from waste cooking oil using new functionalized ionic liquids as catalysts. *Renew. Energy* **2017**, *114*, 755–765. [CrossRef]
5. Farid, M.A.A.; Hassan, M.A.; Taufiq-Yap, Y.H.; Shirai, Y.; Hasan, M.Y.; Zakaria, M.R. Waterless purification using oil palm biomass-derived bioadsorbent improved the quality of biodiesel from waste cooking oil. *J. Clean. Prod.* **2017**, *165*, 262–272. [CrossRef]
6. Jung, J.M.; Lee, S.R.; Lee, J.; Lee, T.; Tsang, D.C.W.; Kwon, E.E. Biodiesel synthesis using chicken manure biochar and waste cooking oil. *Bioresour. Technol.* **2017**, *244*, 810–815. [CrossRef]
7. Tan, Y.H.; Abdullah, M.O.; Nolasco-Hipolito, C.; Zauzi, N.S.A. Application of RSM and Taguchi methods for optimizing the transesterification of waste cooking oil catalyzed by solid ostrich and chicken-eggshell derived CaO. *Renew. Energy* **2017**, *114*, 437–447. [CrossRef]
8. Ling, T.R.; Chang, J.S.; Chiou, Y.J.; Chern, J.M.; Chou, T.C. Characterization of high acid value waste cottonseed oil by temperature programmed pyrolysis in a batch reactor. *J. Anal. Appl. Pyrolysis* **2016**, *120*, 222–230. [CrossRef]
9. Wang, X.M.; Qin, X.L.; Li, D.M.; Yang, B.; Wang, Y.H. One-step synthesis of high-yield biodiesel from waste cooking oils by a novel and highly methanol-tolerant immobilized lipase. *Bioresour. Technol.* **2017**, *235*, 18–24. [CrossRef]

10. Horácek, J.; St'ávová, G.; Kelbichová, V.; Kubicka, D. Zeolite-Beta-supported platinum catalysts for hydrogenation/hydrodeoxygenation of pyrolysis oil model compounds. *Catal. Today* **2013**, *204*, 38–45. [CrossRef]

11. Zeng, Y.; Wang, Z.; Lin, W.G.; Song, W.L.; Christensen, J.M.; Jensen, A.D. Hydrodeoxygenation of phenol over Pd catalysts by in-situ generated hydrogen from aqueous reforming of formic acid. *Catal. Commun.* **2016**, *82*, 46–49. [CrossRef]

12. Lu, J.M.; Faheem, B.M.S.; Heyden, A. Theoretical investigation of the decarboxylation and decarbonylation mechanism of propanoic acid over a Ru(0001) model surface. *J. Catal.* **2015**, *324*, 14–24. [CrossRef]

13. Kubičková, I.; Kubička, D. Utilization of triglycerides and related feedstocks for production of clean hydrocarbon fuels and petrochemicals: A review. *Waste Biomass Valorization* **2010**, *1*, 293–308. [CrossRef]

14. Yang, L.Q.; Carreon, M.A. Effect of reaction parameters on the decarboxylation of oleic acid over Pt/ZIF-67membrane/zeolite 5A bead catalysts. *J. Chem. Technol. Biotechnol.* **2016**, *92*, 52–58. [CrossRef]

15. Di, L.; Yao, S.K.; Guang, S.S.; Wu, J.; Dai, W.L.; Guan, N.J.; Li, L.D. Robust ruthenium catalysts for the selective conversion of stearic acid to diesel-range alkanes. *Appl. Catal. B* **2016**, *201*, 137–149. [CrossRef]

16. Chen, J.Z.; Xu, Q.Y. Hydrodeoxygenation of biodiesel-related fatty acid methyl esters to diesel-range alkanes over zeolite-supported ruthenium catalysts. *Catal. Sci. Technol.* **2016**, *6*, 7239–7251. [CrossRef]

17. Kubicka, D.; Bejblov, M.; Vlk, J. Conversion of vegetable oils into hydrocarbons over CoMo/MCM-41 catalysts. *Top. Catal.* **2010**, *53*, 168–178. [CrossRef]

18. Kovács, S.; Kasza, T.; Thernesz, A.; Horváth, I.W.; Hancsók, J. Fuel production by hydrotreating of triglycerides on NiMo/Al$_2$O$_3$/F catalyst. *Chem. Eng. J.* **2011**, *176*, 237–243. [CrossRef]

19. Hachemi, I.; Kumar, N.; Mäki-Arvela, P.; Roine, J.; Peurla, M.; Hemming, J.; Salonen, J.; Murzin, D.Y. Sulfur-free Ni catalyst for production of green diesel by hydrodeoxygenation. *J. Catal.* **2017**, *347*, 205–221. [CrossRef]

20. Hachemi, I.; Jeništová, K.; Mäki-Arvela, P.; Kumar, N.; Eränen, K.; Hemming, J.; Murzin, D.Y. Comparative study of sulfur-free nickel and palladium catalysts in hydrodeoxygenation of different fatty acid feedstocks for production of biofuels. *Catal. Sci. Technol.* **2016**, *6*, 1476–1487. [CrossRef]

21. Aziz, M.A.A.; Jalil, A.A.; Triwahyono, S.; Saad, M.W.A. CO$_2$ methanation over Ni-promoted mesostructured silica nanoparticles: Influence of Ni loading and water vapor on activity and response surface methodology studies. *Chem. Eng. J.* **2015**, *260*, 757–764. [CrossRef]

22. Zhou, G.L.; Wu, T.; Xie, H.M.; Zheng, X.X. Effects of structure on the carbon dioxide methanation performance of Co-based catalysts. *Int. J. Hydrogen Energy* **2013**, *38*, 10012–10018. [CrossRef]

23. Zhou, G.L.; Wu, T.; Zhang, H.B.; Xie, H.M.; Feng, Y.C. Carbon dioxide methanation on ordered mesoporous Co/KIT-6 catalyst. *Chem. Eng. Commun.* **2014**, *201*, 233–240. [CrossRef]

24. Chen, S.; Pan, X.Y.; Miao, C.X.; Xie, H.M.; Zhou, G.L.; Jiao, Z.J.; Zhang, X.M. Study of catalytic hydrodeoxygenation performance for the Ni/KIT-6 catalysts. *J. Saudi Chem. Soc.* **2018**, *22*, 614–627. [CrossRef]

25. Chen, S.; Miao, C.X.; Luo, Y.; Zhou, G.L.; Xiong, K.; Jiao, Z.J.; Zhang, X.M. Study of catalytic hydrodeoxygenation performance of Ni catalysts: Effects of prepared method. *Renew. Energy* **2018**, *115*, 1109–1117. [CrossRef]

26. Ferencz, Z.; Varga, E.; Puskás, R.; Kónya, Z.; Baán, K.; Oszkó, A.; Erdohelyi, A. Reforming of ethanol on Co/Al$_2$O$_3$ catalysts reduced at different temperatures. *J. Catal.* **2018**, *358*, 118–130. [CrossRef]

27. Zhou, G.L.; Liu, H.R.; Cui, K.K.; Xie, H.M.; Jiao, Z.J.; Zhang, G.Z.; Xiong, K.; Zheng, X.X. Methanation of carbon dioxide over Ni/CeO$_2$ catalysts: Effects of support CeO$_2$ structure. *Int. J. Hydrogen Energy* **2017**, *42*, 16108–16117. [CrossRef]

28. Zhou, G.L.; Xie, H.M.; Gui, B.G.; Zhang, G.Z.; Zheng, X.X. Influence of NiO on the performance of CoO-based catalysts for the selective oxidation of CO in H$_2$-rich gas. *Catal. Commun.* **2012**, *19*, 42–45. [CrossRef]

29. Zhou, G.L.; Liu, H.R.; Cui, K.K.; Jia, A.P.; Hu, G.S.; Jiao, Z.J.; Liu, Y.Q.; Zhang, X.M. Role of surface Ni and Ce species of Ni/CeO$_2$ catalyst in CO$_2$ methanation. *Appl. Surf. Sci.* **2016**, *383*, 248–252. [CrossRef]

30. Arslan, A.; Dogu, T. Effect of calcination/reduction temperature of Ni impregnated CeO$_2$-ZrO$_2$ catalysts on hydrogen yield and coke minimization in low temperature reforming of ethanol. *Int. J. Hydrogen Energy* **2013**, *38*, 7268–7279. [CrossRef]

31. Pauly, N.; Yubero, F.; García-García, F.J.; Tougaard, S. Quantitative analysis of Ni2p photoemission in NiO and Ni diluted in a SiO$_2$ matrix. *Surf. Sci.* **2016**, *644*, 46–52. [CrossRef]

32. Boudjahem, A.G.; Bettahar, M.M. Effect of oxidative pre-treatment on hydrogen spillover for a Ni/SiO$_2$ catalyst. *J. Mol. Catal. A* **2017**, *426*, 190–197. [CrossRef]

33. Zhou, G.L.; Gui, B.G.; Xie, H.M.; Yang, F.; Chen, Y.; Chen, S.M.; Zheng, X.X. Influence of CeO$_2$ morphology on the catalytic oxidation of ethanol in air. *J. Indus. Eng. Chem.* **2014**, *20*, 160–165. [CrossRef]

34. Bie, Y.; Lehtonen, J.; Kanervo, J. Hydrodeoxygenation (HDO) of methyl palmitate over bifunctional Rh/ZrO$_2$ catalyst: Insights into reaction mechanism via kinetic modeling. *Appl. Catal. A* **2016**, *526*, 183–190. [CrossRef]

35. Kellenberger, A.; Jahne, E.; Adler, H.J.; Khandelwal, T.; Dunscha, L. In situ FTIR spectroelectrochemistry of poly [2-(3-thienyl) ethyl acetate] and its hydrolyzed derivatives. *Electrochim. Acta* **2008**, *53*, 7054–7060. [CrossRef]

36. Tan, X.; Lan, H.; Xie, H.M.; Zhou, G.L.; Jiang, Y. Role of surface oxygen species of mesoporous CeCu oxide catalyst in OVOCs catalytic combustion. *J. Environ. Chem. Eng.* **2017**, *5*, 2068–2076. [CrossRef]

37. Zhou, G.L.; Dai, B.C.; Xie, H.M.; Zhang, G.Z.; Xiong, K.; Zheng, X.X. CeCu composite catalyst for CO synthesis by reverse water–gas shift reaction: Effect of Ce/Cu mole ratio. *J. CO$_2$ Util.* **2017**, *21*, 292–301. [CrossRef]

 *catalysts*

Article

Atomic Layer Deposition ZnO Over-Coated Cu/SiO$_2$ Catalysts for Methanol Synthesis from CO$_2$ Hydrogenation

Jinglin Gao [1], Philip Effah Boahene [1], Yongfeng Hu [2], Ajay Dalai [1] and Hui Wang [1,*]

[1] Department of Chemical and Biological Engineering, University of Saskatchewan, Saskatoon, SK S7N 5A9, Canada; jig367@usask.ca (J.G.); peb225@mail.usask.ca (P.E.B.); akd983@mail.usask.ca (A.D.)

[2] Canadian Light Source, University of Saskatchewan, Saskatoon, SK S7N 2V3, Canada; yongfeng.hu@lightsource.ca

* Correspondence: hui.wang@usask.ca

Received: 17 October 2019; Accepted: 3 November 2019; Published: 6 November 2019

Abstract: Cu-ZnO-based catalysts are of importance for CO$_2$ utilization to synthesize methanol. However, the mechanisms of CO$_2$ activation, the split of the C=O double bond, and the formation of C-H and O-H bonds are still debatable. To understand this mechanism and to improve the selectivity of methanol formation, the combination of strong electronic adsorption (SEA) and atomic layer deposition (ALD) was used to form catalysts with Cu nanoparticles surrounded by a non-uniform ZnO layer, uniform atomic layer of ZnO, or multiple layers of ZnO on porous SiO$_2$. N$_2$ adsorption, H$_2$ temperature-programmed reduction (H$_2$-TPR) X-ray diffraction (XRD), transmission electron microscope (TEM), energy-dispersive X-ray spectroscopy (EDX), CO-chemisorption, CO$_2$ temperature-programmed desorption (CO$_2$-TPD), X-ray adsorption near edge structure (XANES), and extended X-ray absorption fine structure (EXAFS) were used to characterize the catalysts. The catalyst activity was correlated to the number of metallic sites. The catalyst of 5 wt% Cu over-coated with a single atomic layer of ZnO exhibited higher methanol selectivity. This catalyst has comparatively more metallic sites (smaller Cu particles with good distribution) and basic site (uniform ZnO layer) formation, and a stronger interaction between them, which provided necessary synergy for the CO$_2$ activation and hydrogenation to form methanol.

Keywords: CO$_2$ activation; methanol synthesis; atomic layer deposition; copper nanoparticles; zinc oxide atomic layer

---

1. Introduction

Carbon dioxide (CO$_2$) is considered to be the most severe greenhouse gas by the amount of anthropogenic emission (36 billion tons per year in 2017 [1]), and the prominent free carbon source in the future, on the other hand. Currently, only 0.5% of emitted CO$_2$ can be used for industrial purposes [1]. However, many research efforts have been devoted to converting CO$_2$ into value-added products of commercial importance. CO$_2$ is thermodynamically stable [2] and barely reacts with other chemicals without catalysts and driving energy. Thus, it is important to understand the mechanism of catalytic activation of CO$_2$, especially the relationship between catalysts' structural properties and catalytic performance such that effective catalyst materials can be developed for a specific reaction, such as CO$_2$ hydrogenation for methanol (MeOH or CH$_3$OH). MeOH is a clean liquid fuel. It is also an important commodity used as feedstock for the production of many other chemicals. Although the hydrogenation of CO$_2$ for MeOH:

$$CO_2 + 3H_2 \rightarrow CH_3OH + H_2O, \ \Delta H_{298K} = -49.5 \ kJ \ mol^{-1}, \tag{1}$$

is an exothermal reaction, high reaction temperatures favor the activation of CO_2 and the formation of methanol. However, high temperatures also contribute to the formation of by-products of CO and hydrocarbons during the reaction [3]. Catalysts are developed in order to minimize or prevent the occurrence of side reactions, among which the Cu-ZnO catalyst for CO_2 hydrogenation has been predominantly studied [2].

A literature review observed that some catalysts exhibited high conversion but poor selectivity to MeOH while others showed a prominent selectivity of MeOH but lower CO_2 conversion. However, there is no catalyst that yet offers both high CO_2 conversion and high MeOH selectivity. On the high conversion and low selectivity side, Duan and coworkers [4] prepared a Cu–ZnO catalyst supported on mesoporous carbon. The catalyst showed a relatively high CO_2 conversion of 27%, and high MeOH selectivity of 69% due to the good contact between the Cu/CuO and ZnO particles. The Cu–ZnO catalyst [5] supported by reduced graphene oxide nanosheets also showed a CO_2 conversion of 26%, which resulted from the good dispersion of Cu–Zn particles enhanced by the support, but a much lower MeOH selectivity of 5%. $Cu/ZnO/ZrO_2$ catalyst [6] prepared by the precipitation–reduction method showed a CO_2 conversion of 23% and selectivity to MeOH of 57%. The study found that CO_2 conversion was connected to the surface area of the exposed Cu, and MeOH selectivity was related to the number of basic sites. On the high selectivity and low conversion side, $Cu@ZnO_x$ core-shell catalyst [7] prepared by precipitated ZnO on the surface of Cu powders showed 100% selectivity to MeOH and a CO_2 conversion of 2%. The excellent selectivity was due to the alloy formation and migration of Zn. Cu–Zn/Al foam catalyst [8] prepared via the hydrothermal method showed a selectivity to MeOH of 83% and CO_2 conversion of 10%. Compared with other Cu-based catalysts, Cu–Zn/Al foam catalyst possessed great heat and mass transfer properties and suitable ratio of strong basic sites.

Cu or ZnO alone had little effect on methanol synthesis by an experimental study, as shown by Karelovic and coworkers [9]. However, the combination of Cu and ZnO enhanced the catalyst performance dramatically, as revealed by Gesmanee and Koo-Amornpattana [10]. ZnO was regarded as a structure-directing support to improve the dispersion of metallic Cu particles during reduction and acts as a spacer between the Cu particles [11,12]. ZnO was able to provide the active sites for the spillover of hydrogen [10,13–15]. According to the study of Cu-based catalysts modified by ZnO, ZrO_2, and MgO, Ren et al. [16] observed that Cu^0 sites were the catalytic active centers for hydrogenation of CO_2 to MeOH. However, Choi et al. [17] discovered that Cu–Zn sites were the active sites for CO_2 hydrogenation to MeOH. Kanai and coworkers [18] found that parts of ZnO migrated onto the surface to form Cu–Zn alloy, and that Cu^+ species formed in the vicinity of ZnO_x were regarded as active sites for MeOH synthesis. The group of Tisseraud and Le Valant [7,19–21] reported that the active sites for CO and MeOH synthesis were different. CO was produced on ZnO sites and MeOH was produced on ZnO_x, which were formed on the interface of Cu and ZnO through cross-diffusion of Cu and ZnO. The activity of the catalysts was found to be related to the number of "contact points" of Cu and ZnO.

Although a variety of mechanisms have been proposed and some are debatable, one recognizes that CO_2 is activated by one type of site, say ZnO related, and H_2 by another type, say metallic Cu or Cu_2O/CuO related. Each type alone cannot have the activity to either H_2 and CO_2; however, they must be next to each other to form the right structures to facilitate the formation of methanol or to form the wrong structures to help the formation of CO. The activated CO_2, still in the molecular form, needs assistance from the activated H to pull one O away and to insert H in at the same time to form C-H and O-H bonds. In other words, the right structures must be able to perform two functions at the same time, pulling one O away to form H_2O and inserting H atoms in to form methanol. Based on this thinking, we planned to make two extremely different catalysts with the help of the atomic layer deposition (ALD) method. One catalyst contains isolated Cu-related particles, specifically metallic Cu nanoparticles, distributed in a layer of ZnO atoms, on which the activated CO_2 lacks surrounding H atoms, thus tending to form CO not methanol during hydrogenation. The other catalyst contains isolated ZnO phases distributed in a layer of Cu or surrounded by sufficient Cu nanoparticles, on which the activated CO_2 has enough H to access. In this paper, we prepared the first type of extreme catalysts. Cu was loaded onto a SiO_2 support by strong electronic adsorption (SEA) and ZnO was over-coated by ALD. The catalysts were characterized by N_2 adsorption, H_2 temperature-programmed reduction (H_2-TPR) X-ray diffraction (XRD), transmission electron microscope (TEM), energy-dispersive X-ray spectroscopy (EDX), CO-chemisorption, CO_2 temperature-programmed desorption (CO_2-TPD), X-ray adsorption near edge structure (XANES), and extended X-ray absorption fine structure (EXAFS) and their performance for CO_2 hydrogenation to form methanol was tested using a fixed bed reactor.

2. Results and Discussion

2.1. Formation of Cu Particles with a Uniform Distribution

The samples were prepared by impregnation (Imp) and strong electronic adsorption (SEA) methods over SiO_2 with certain amount of Cu loading.

2.1.1. Reduction Properties of Samples from H_2-TPR

The H_2-TPR results of Imp-Cu and SEA-Cu samples with different Cu contents are shown in Figure 1. Two hydrogen consumption peaks were observed in the range of 186 to 194 °C (referred to as the first peak) and 206 to 240 °C (second peak) in the H_2-TPR profile. The first peak was ascribed to the reduction of the highly dispersed CuO phases [16,22], while the second peak was attributed to the reduction of the CuO species having relatively stronger interactions with the support [16]. Imp-5Cu sample (5 means 5 wt% target Cu loading) prepared by impregnation resulted in greater formation of the dispersed CuO species, as shown by the higher intensity of the first peak. SEA-5Cu and SEA-10Cu, prepared by the SEA method with 5 wt% and 10 wt% Cu loading, exhibited a larger second peak. During the preparation of these two samples, the silica gel was first pretreated with ammonia solution at a certain pH to adjust the surface charge prior to mixing with the Cu complex of the opposite charge to form the CuO species. So, the catalysts prepared by the SEA method show more CuO species, which have stronger interactions with the support. To ensure full reduction, 500 °C was used as the reduction temperature in later research.

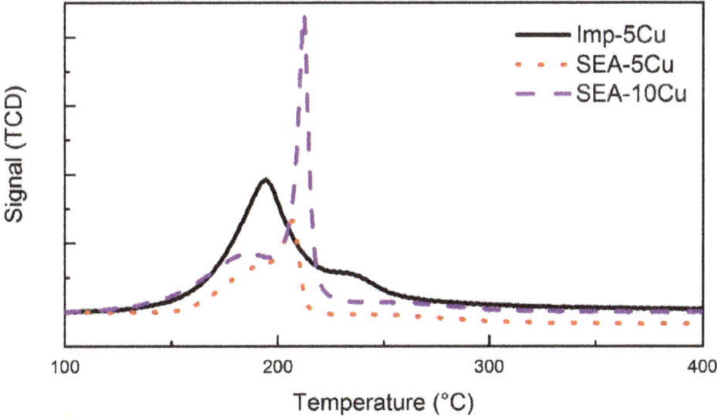

Figure 1. H_2-TPR profiles of samples prepared by the Imp or SEA method with different Cu contents.

2.1.2. Samples Screened by TEM Images

These three samples reduced at 500 °C were used for TEM analysis. Figure 2 shows the TEM images and the histograms of the samples. The CuO species prepared by the SEA method showed relatively stronger interactions with the support, which was revealed by TPR. The pretreatment in SEA to adjust the surface charge also resulted in a better distribution of the Cu particles. The average particle sizes of catalyst SEA-5Cu and SEA-10Cu were 8.4 and 8.6 nm, respectively. Further comparison of the particle size distribution between SEA-5Cu and SEA-10Cu shows that SEA-5Cu should be used for further study. The final catalysts were prepared by ALD to deposit ZnO with different pulse times or cycles on SEA-5Cu. The catalysts were denoted as 5Cu-1CyZn-5s, 5Cu-1CyZn-30s, and 5Cu-5CyZn-5s (called ZnO over-coated Cu/SiO$_2$ catalysts). 1Cy and 5Cy represent the number of cycles of ALD performed. Furthermore, 5 and 30 s represent the exposure time of the SEA-5Cu to the Zn precursor in each cycle.

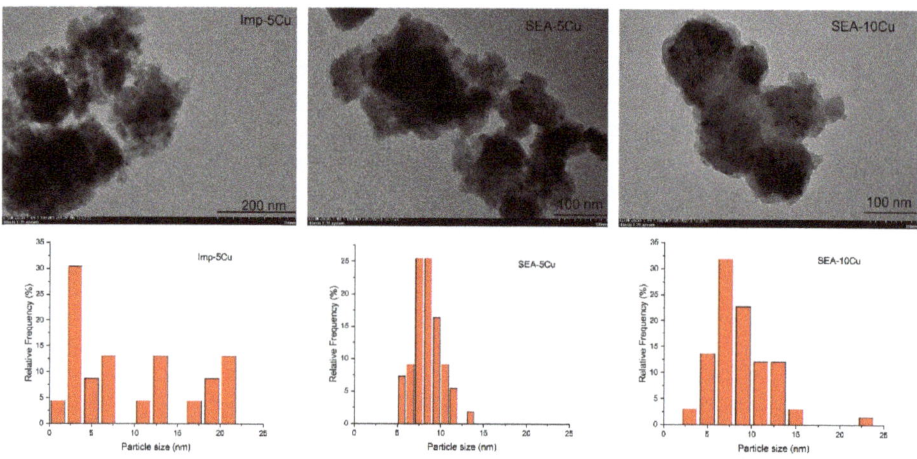

Figure 2. TEM images and particle size distribution of samples Imp-5Cu, SEA-5Cu, and SEA-10Cu.

2.2. Catalysts and Support Characterization

2.2.1. Results of ICP, BET, and XRD

Table 1 shows the ICP and BET results of the substrate silica gel (SiO$_2$), SEA-5Cu, and the ZnO over-coated Cu/SiO$_2$ catalysts. The Cu contents in these samples were approximately 4.4 wt% as expected; however, the Zn content varied depending on the pulse time and ALD cycles. With a single-cycle deposition, the content of Zn increased from 0.3 (pulse time of 5 s) to 1.5 wt% (pulse time of 30 s). With the same pulse time (5 s), the Zn content increased from 0.3 (1 cycle of ALD) to 1.3 wt% (5 cycles of ALD). The loading of Cu by the SEA method dramatically decreased the specific surface area and pore volume of SiO$_2$. The average pore size increased from 3.0 to 4.5 nm, indicating blockage of smaller mesopores. The surface area and pore volume were decreased further by ZnO ALD, but to a smaller extent on the basis of the Cu loading.

Table 1. Bulk properties of SiO$_2$, SEA-5Cu, and ZnO over-coated Cu/SiO$_2$ catalysts.

Samples	Cu [1] Loading (wt%)	Zn [1] Loading (wt%)	BET Surface Area [2] (m^{2}g^{-1})	Pore Volume [2] (cm^{3}g^{-1})	Average Pore Size [2] (nm)	Cu^0 Lattice Constant [3] (Å)
SiO$_2$	-	-	718	0.41	3.0	-
SEA-5Cu	4.4	-	264	0.30	4.5	3.62
5Cu-1CyZn-5s	4.4	0.3	200	0.21	4.2	3.62
5Cu-1CyZn-30s	4.5	1.5	210	0.24	4.3	3.64
5Cu-5CyZn-5s	4.3	1.3	238	0.28	4.6	3.65

[1] Tested by ICP analysis; [2] Tested by N$_2$ adsorption–desorption; [3] Calculated from XRD patterns.

The XRD patterns of the SiO$_2$ and reduced SEA-Cu, ZnO over-coated Cu/SiO$_2$ catalysts are shown in Figure 3. The diffraction peak relative to amorphous SiO$_2$ can be seen at 2θ = 22 ° [23]. After reduction, most of the CuO is reduced to metallic Cu. For the reduced SEA-5Cu, the Cu phase is assigned to a face-centered cubic copper structure with characteristic diffraction peaks 2θ at 43 °, 50 °, and 74 ° [21]. The ALD over-coating of ZnO did not change much of the Cu signals. No new peaks were formed for the ZnO species, which indicated that the structure of ZnO prepared by ALD was not large enough to be detected by XRD. The value of the calculated Cu^0 lattice constant (Table 1) increased with the increase in ZnO deposition, from 3.62 to 3.64 Å. Kanai et al. [18] believed that the increasing of the lattice constant from 3.62 to 3.67 Å in their study on the Cu-ZnO catalysts was attributable to the formation of a Cu–Zn alloy. In our case, the increase in the lattice constant upon the deposition of ZnO may also be associated with the formation of the Cu–ZnO surface alloy.

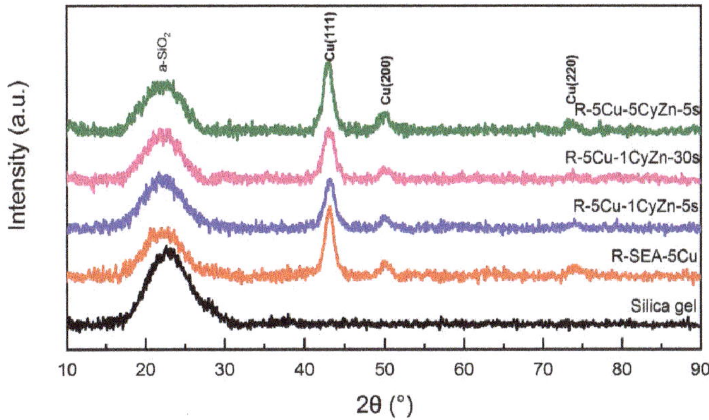

Figure 3. XRD patterns corresponding to SiO_2, SEA-5Cu, and ZnO over-coated Cu/SiO_2 catalysts.

2.2.2. Reduction Properties, Cu Particle Size, and Structure of ZnO

The H_2-TPR results of SEA-5Cu and ZnO over-coated Cu/SiO_2 catalysts are shown in Figure 4. Two hydrogen consumption peaks are seen in the range of 148 to 194 °C (referred to as the first peak) and 206 to 254 °C (second peak) in the TPR profile. The two peaks are ascribed to the reduction of the highly dispersed CuO phases and the reduction of CuO species having a strong interaction with SiO_2 [16,22]. With the increase in the amount of ZnO by either lengthening the pulse time or having more ALD cycles, the second peak clearly shifted to a higher temperature zone. This right shift could indicate that the over-coating of ZnO on the CuO species may cause them to be harder to reduce. The H_2-TPR profiles also verify that using 500 °C as the reduction temperature is sufficient to activate the catalysts.

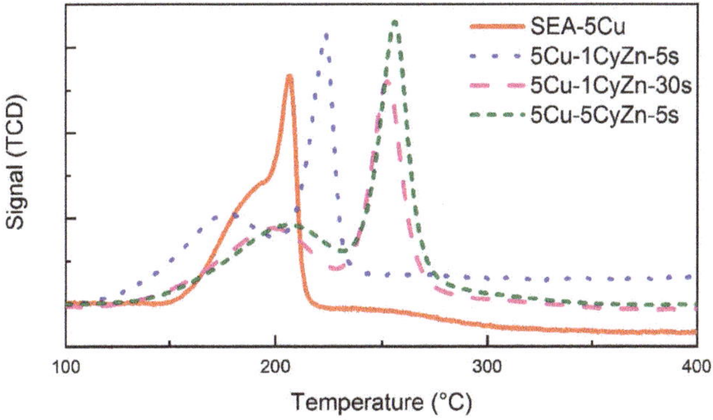

Figure 4. H_2-TPR profiles of SEA-5Cu and ZnO over-coated Cu/SiO_2 catalysts.

When imaging the three reduced catalysts, 5Cu-1CyZn-5s, 5Cu-1CyZn-30s, and 5Cu-5CyZn-5s, by TEM, four spots were randomly selected to measure the distribution of the elements by the EDX technique (Table 2). The diameter of the spot was 1 μm. On 5Cu-1CyZn-5s, with 25% probability, a Cu particle was detected (87 wt% of Cu), where the Zn content was also relatively high (1.9 wt% vs. 0.2–0.3 wt% on the other three spots). The Cu content on the other three spots varied from 7.8 to 22.6 wt%. On 5Cu-1CyZn-30s, the chance to detect a big change in the Cu and Zn contents become smaller. For all four spots, the Zn content changed from 1.1 to 1.9 wt% while the Cu content changed from 8.3 to 12.9 wt%. On 5Cu-5CyZn-5s, the Zn content varied from 0.1 to 4.1 wt% with four different values for four spots. With a 25% chance, a relatively high Cu spot was detected (26.5 wt% of Cu). The uniformity of the ZnO layer over-coating depended on the number of ALD cycles used and the length of the pulse time of each cycle. One cycle of ZnO over-coating with a pulse time of 5 s possibly led to some uncoated areas. 5 cycles and 5 s possibly led to multiple layers of ZnO as well as uncoated areas. Only one cycle of 30 s led to a relatively uniform ZnO layer and a small chance of having uncoated areas. The layer of ZnO "disciplined" the growth of Cu particles during reduction. Under a uniform single layer of ZnO, 5Cu-1CyZn-30s grew well-distributed Cu particles. However, 5Cu-1CyZn-5s and 5Cu-5CyZn-5s, both possibly having an uncovered surface and specifically having uncovered CuO species, grew larger Cu particles [24].

Table 2. The content of elements on the reduced ZnO over-coated Cu/SiO_2 catalysts detected by TEM-EDX.

Catalysts	5Cu-1CyZn-5s Element Content (wt%)				5Cu-1CyZn-30s Element Content (wt%)				5Cu-5CyZn-5s Elements Content (wt%)			
Spots	Zn	Cu	Si	O	Zn	Cu	Si	O	Zn	Cu	Si	O
1	1.9	87.0	1.8	9.3	1.1	10.9	43.7	44.3	0.1	5.0	60.7	34.2
2	0.2	22.6	42.0	35.2	1.9	12.3	48.6	37.2	4.1	8.0	43.1	44.8
3	0.3	17.7	46.8	35.2	1.1	8.5	52.8	37.6	2.4	9.0	43.2	45.4
4	0.2	7.8	45.2	46.8	1.8	8.3	50.1	39.8	1.0	26.5	34.6	37.9

This argument has immediate support from the TEM images shown in Figure 5. The TEM images and histogram show that 5Cu-1CyZn-30s had the largest number of Cu particles and most uniform particle size from 4 to 10 nm. The average Cu particle sizes of the three ZnO over-coated Cu/SiO_2 catalysts as well as their precursors, Cu/SiO_2 (SEA-5Cu), as measured from the TEM images are shown in Table 3. The particle size calculated from the XRD peaks by Scherrer's equation coincided with the TEM results (Table 3). A greater regulation effect can be observed when a longer pulse time or more ALD layers were used.

The CO chemisorption results are also shown in Table 3. Using 5Cu-1CyZn-5s as a benchmark, lengthening the pulse time from 5 to 30 s did not change or decrease too much of the Cu dispersion, metallic Cu surface area, and thus, the number of Cu metallic sites per gram of catalysts. However, using 5 cycles of ALD, each having 5 s of pulse time, significantly changed or decreased the Cu dispersion, metallic Cu surface area, and number of metallic sites per gram of catalysts. A certain position of multiple layers of ZnO may hinder the Cu reduction of the CuO species under multiple layers.

Table 3. Metallic Cu particle size, dispersion, and metallic surface area of reduced catalysts.

Samples	Cu Particle Size (nm)		Cu Dispersion (%)	Cu Surface Area ($m^2 \cdot g^{-1}$)	The Number of Metallic Sites ($\mu mol \cdot g^{-1}$)
	TEM	XRD			
SEA-5Cu	8.4	7.1	-	-	-
5Cu-1CyZn-5s	8.1	6.5	1.9	0.69	304
5Cu-1CyZn-30s	6.7	5.3	1.7	0.65	286
5Cu-5CyZn-5s	6.9	6.4	1.3	0.53	228

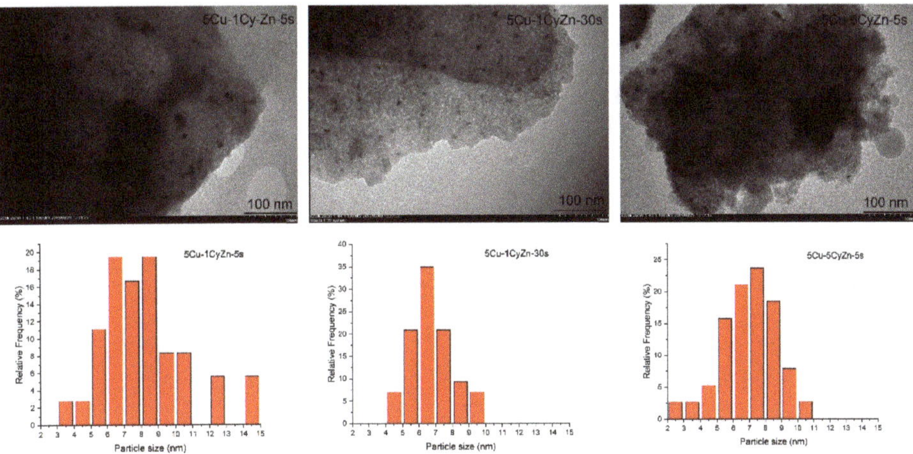

Figure 5. TEM images, particle size, and distribution of ZnO over-coated Cu/SiO$_2$ catalysts.

2.2.3. Basicity of the Catalysts from CO$_2$-TPD

The basicity of the SEA-5Cu and ZnO over-coated Cu/SiO$_2$ catalysts was detected using CO$_2$-TPD (Figure 6), which revealed the species and number of basic sites. There are three not totally separated peaks shown in the profiles. The adsorption temperatures of each peak are centered around 140, 340, and 570 °C. All three peaks of 5Cu-1CyZn-30s shifted towards the higher temperature. These peaks can be assigned to the weak, moderate, and strong basic sites. According to the literature, the weak basic sites are ascribed to the OH$^-$ groups. The moderate basic sites are related to the M-oxygen pairs (M = metal). The strong basic sites are associated with the coordinatively unsaturated O^{2-} ions, which is a low coordination oxygen atom [25–28].

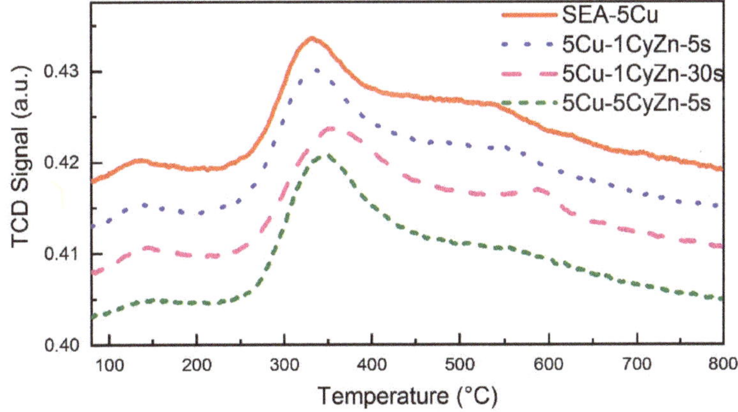

Figure 6. CO$_2$-TPD curves of the SEA-5Cu and ZnO over-coated Cu/SiO$_2$ catalysts.

The peaks of CO$_2$ desorption were then separated and quantified, and the results are shown in Table 4. The comparison of SEA-5Cu and ZnO over-coated Cu/SiO$_2$ catalysts indicated that ZnO over-coating increased the number of total basic sites mainly via the formation of more moderate and strong basic sites. Compared to the catalyst 5Cu-1CyZn-5s, a longer pulse time (5Cu-1CyZn-30s) or multiple cycles of ZnO over-coated (5Cu-5CyZn-5s) led to more basic site formation, which is from 19 (weak) to 42 or 21 μmol·g^{-1}, 301 (moderate) to 369 or 405 μmol·g^{-1}, 838 (strong) to 1017 or 884 μmol·g^{-1}, and 1158 (total) to 1428 or 1310 $\mu mol \cdot g^{-1}$, respectively. It is worth mentioning that a longer pulse time of the ZnO precursor ended in the formation of more strong basic sites, and multiple cycles of ZnO over-coating resulted in more moderated basic sites being formed. Comparing the number of basic sites of the ZnO over-coated Cu/SiO_2 catalysts, there are more basic sites on the catalyst 5Cu-1CyZn-30s. The basic sites are where the CO_2 adsorption and activation happen.

Table 4. Basic properties of the SEA-5Cu and ZnO over-coated Cu/SiO_2 catalysts.

Samples	Number of Basic Sites ($\mu mol \cdot g^{-1}$)			Total Number of Basic Sites
	Weak	**Moderate**	**Strong**	
SEA-5Cu	13	193	578	784
5Cu-1CyZn-5s	19	301	838	1158
5Cu-1CyZn-30s	42	369	1017	1428
5Cu-5CyZn-5s	21	405	884	1310

2.3. Chemical Structure from XAS Spectra

2.3.1. XANES for Cu K-Edge of SEA-5Cu and ZnO Over-Coated Cu/SiO_2 Catalysts

The local structure and electronic state of Cu and Zn in the catalysts were investigated by XAS. The Cu K-edge XANES spectra and their first derivative spectra of SEA-5Cu and ZnO over-coated Cu/SiO_2 catalysts, along with those of the Cu, Cu_2O, and CuO standards are shown in Figure 7. The comparison of the spectra of SEA-5Cu and ZnO over-coated Cu/SiO_2 catalysts and those of the standards indicated that the copper in all the calcined samples took the oxidation state of Cu^{2+}. A pre-edge at approximately 8981 eV, which was the characteristic Cu^{1+} 1s–4d transition feature [29], was only observed in the spectrum of Cu_2O. The absence of this peak in the spectra of samples also demonstrated there only Cu^{2+} species were present in the calcined samples.

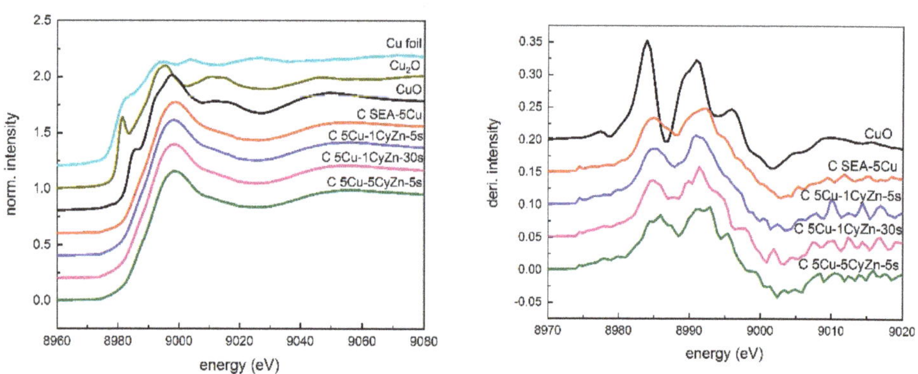

Figure 7. Experimental XANES spectra registered Cu K-edge of SEA-5Cu and ZnO over-coated Cu/SiO_2 catalysts (**left**). The first derivative XANES spectra (**right**).

Further comparison of the first derivative spectra showed that Cu in the prepared samples maintained the two CuO peaks, 8984 and 8991 eV, which represented the 1s–4p transition shakedown and the main transition, respectively [29]. The notable peak shift toward higher energy (1–3 eV) may suggest a noncentrosymmetric CuO attributable to the ligand donor effect [29]. The spectra of the samples did not show the pre-edge peak (1s–3d quadrupole-allowed transition [29]), which was frequently observed in the Cu^{2+} compounds.

The Cu K-edge XANES spectra and the first derivative spectra of the reduced SEA-5Cu and ZnO over-coated Cu/SiO_2 catalysts, as well as the spectra of Cu, Cu_2O, and CuO as the reference, are shown in Figure 8. Compare to the spectra of the standards, those of SEA-5Cu and ZnO over-coated Cu/SiO$_2$ catalysts showed the same shape with the Cu foil and a similar absorption edge, meaning that the chemical environment of Cu species in SEA-5Cu and Cu-ZnO catalysts was close to that of the Cu foil standard. The catalyst 5Cu-1CyZn-30s showed more obviously separated peaks corresponding to resonance features at $E_0 + 15$ eV and $E_0 + 24$ eV [30], which indicated the Cu particles on the catalyst had better dispersion. This conclusion was also corroborated by the TEM images of the reduced catalysts. To further confirm the oxidation state of Cu, linear combination fitting of XANES was then used for reduced SEA-5Cu and ZnO over-coated Cu/SiO$_2$ catalysts (Table 5). There was approximately 75% of metallic Cu formation in SEA-5Cu and the catalysts. With multiple layers of ZnO being over-coated, more Cu was reduced on catalyst 5Cu-5CyZn-5s.

There were two main peaks at approximately 8979 and 8990 eV in the first derivative spectra of the reduced SEA-5Cu and ZnO over-coated Cu/SiO$_2$ catalysts, which were both assigned to 1s–4p transitions [30]. According to the first derivative spectra of the reduced 5Cu-1CyZn-5s and 5Cu-1CyZn-30s catalysts, the peaks assigned to the 1s–4p transition shifted towards higher energy compared to that of SEA-5Cu. This demonstrated that the chemical environment of the Cu species on these two catalysts was affected by the ZnO over-coated, and metallic Cu may be partially oxidized by ZnO. Higher energy was needed to make the electron transition from the 1s to 4p orbital. So, there may be an interaction between Cu and ZnO, and possibly a dissolution of Cu atoms in the ZnO lattice. However, the spectrum of the catalyst R 5Cu-5CyZn-5s showed no shift compared to that of SEA-5Cu, which means that with multiple layers of ZnO, the interaction of Cu and ZnO becomes weaker. The energy shift was not that much because not all the metallic Cu had an interaction with the ZnO, and the spectra can only provide the average feature of the bulk catalysts.

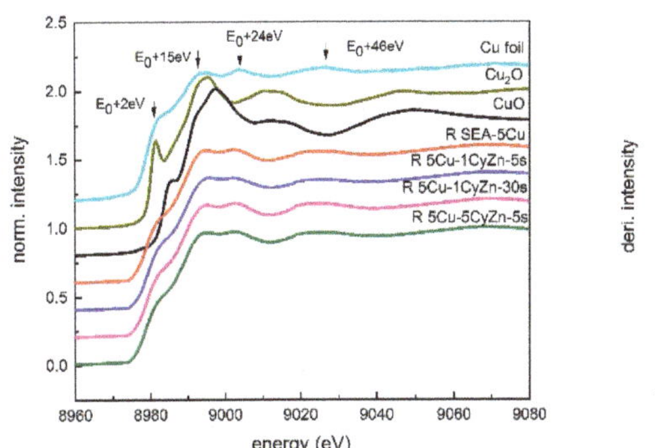

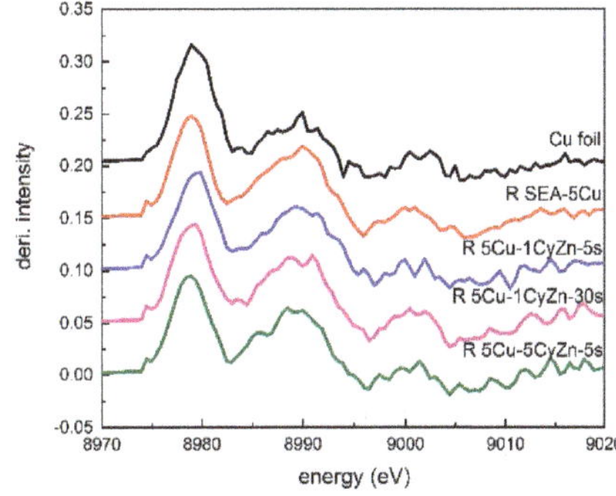

Figure 8. Experimental XANES spectra registered the Cu K-edge of reduced SEA-5Cu and ZnO over-coated Cu/SiO$_2$ catalysts (**left**). The first derivative XANES spectra (**right**).

Table 5. Cu species in the reduced SEA-5Cu and ZnO over-coated Cu/SiO$_2$ catalysts based on linear combination fitting.

Samples	Configuration		
	Cu	Cu$_2$O	CuO
SEA-5Cu	0.75	0.01	0.24
5Cu-1CyZn-5s	0.75	0	0.25
5Cu-1CyZn-30s	0.75	0	0.25
5Cu-5CyZn-5s	0.79	0	0.21

2.3.2. EXAFS for Cu K-Edge of SEA-5Cu and ZnO Over-Coated Cu/SiO$_2$ Catalysts

The EXAFS spectra of the Cu foil, Cu$_2$O, and CuO standards in Figure 9 (left) gave us the information of the bond species and their lengths, which are marked out. The EXAFS spectra of

SEA-5Cu and ZnO over-coated Cu/SiO$_2$ catalysts displayed two peaks at around 1.47 and 2.58 Å, which were assigned to the lengths of the Cu-O bond and Cu-(O)-Cu bond. The lengths of these two bonds were both in the middle of those of Cu$_2$O and CuO. Compared to the spectra of SEA-5Cu, the peaks of calcined ZnO over-coated Cu/SiO$_2$ catalysts both shifted right, which indicated that with ZnO over-coated the bonds of Cu-O and Cu-(O)-Cu became longer.

The EXAFS spectra of reduced SEA-5Cu and ZnO over-coated Cu/SiO$_2$ catalysts are shown in Figure 9 (right). There were two peaks in the spectra of catalysts on behalf of the two bonds, respectively; the Cu-O bond whose length was in the range from 1.41 to 1.47 Å, and the Cu-Cu bond whose length was in the range from 2.22 to 2.24 Å. The appearance of the peak representing the Cu-O bond means not all the CuO species were reduced, and this result corresponded to the result of linear combination fitting. The absorption peaks of 5Cu-1CyZn-5s were at 1.42 and 2.24 Å. The first peak went right to 1.47 Å with a longer pulse time or multiple cycles of ZnO over-coated, revealing the length of the Cu-O bond became longer due to more ZnO over-coated. The intensity of this peak became less with the increase of the pulse time and cycles of ZnO deposition, which disclosed that ZnO over-coated helped the reduction of Cu. The Cu-Cu bond length of ZnO over-coated Cu/SiO$_2$ catalysts was similar to that of SEA-5Cu.

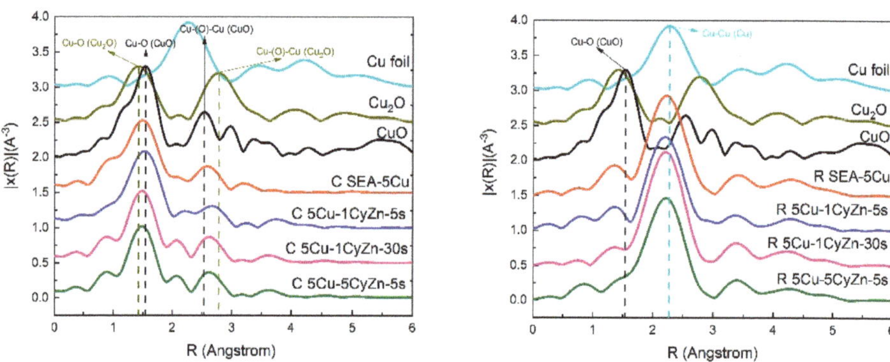

Figure 9. Experimental EXAFS spectra registered Cu K-edge of calcined (**left**) and reduced (**right**) SEA-5Cu and ZnO over-coated Cu/SiO$_2$ catalysts.

2.3.3. XANES for Zn-K Edge of ZnO Over-Coated Cu/SiO$_2$ Catalysts

The Zn XANES spectra of calcined ZnO over-coated Cu/SiO$_2$ catalysts are shown in Figure 10 (left). The absorption edge was at 9662 eV, disclosing the existence of bivalent Zn. However, the spectra of ZnO over-coated Cu/SiO$_2$ catalysts were considerably different from that of ZnO standard. Compared to the spectrum of ZnO, the peaks at 9663.8 and 9668.4 eV assigned to 1s–4sp and 1s–4p transitions, were merged to one peak at 9667.5 eV in the spectra of the catalysts. The peaks at 9679.3 and 9683.4 eV were smoothed and became a shoulder (9681.1 eV), which was assigned to multiple scattering (midrange order and short-range order) [31]. This peak was only shown in the spectrum of 5Cu-5CyZn-5s instead of the catalysts with one cycle of ZnO deposition, since there was a single layer of ZnO appearance with an uncoated area on the catalyst 5Cu-1CyZn-5s and a relatively uniform ZnO layer on the catalyst 5Cu-1CyZn-30s. The difference between the spectra of Cu-ZnO and the spectrum of ZnO was also because the Zn^{2+} on the catalysts, which existed as molecules or much smaller particles, were mostly influenced by the substrate, SiO$_2$.

The Zn XANES spectra of reduced ZnO over-coated Cu/SiO$_2$ catalysts are shown in Figure 10 (right). These spectra showed no difference with those before reduction. The Zn XANES spectra of ZnO over-coated Cu/SiO$_2$ catalysts revealed a disordered ZnO material with an embryonic short-range order. The nearest neighbor of Zn, O, stayed at the expected angles and distances for the wurtzite structure [16,28].

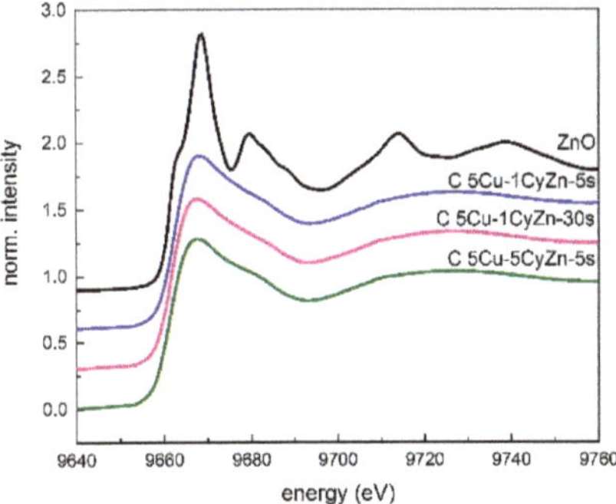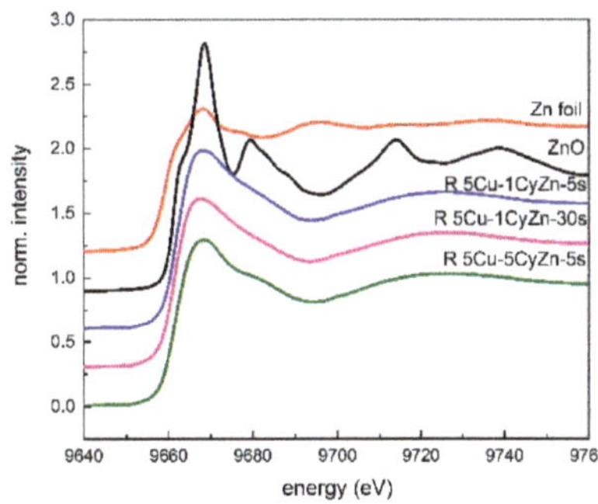

Figure 10. Experimental XANES spectra registered Zn K-edge of calcined (**left**) reduced (**right**) SEA-5Cu and ZnO over-coated Cu/SiO$_2$ catalysts (**left**).

2.3.4. EXAFS for Zn-K Edge of ZnO Over-Coated Cu/SiO$_2$ Catalysts

The Zn EXAFS spectra of calcined and reduced ZnO over-coated Cu/SiO$_2$ catalysts are shown in Figure 11. In the EXAFS spectra of ZnO standard, the bond length between the central zinc atom and the first shell atom, which was an oxygen atom, was 1.60 Å. The second peak was at 2.85 Å, which was attributed to Zn-(O)-Zn interaction. According to the EXAFS spectra of Zn foil, the bond length of Zn-Zn was 2.28 Å. There was a Zn-O bond in the spectra of calcined Cu-ZnO, and not much difference was shown in the spectra of reduced Cu-ZnO. It is worth noting that there was no peak shown to represent the bond of Zn-(O)-Zn, which means no bigger ZnO particles or many layers of ZnO formed.

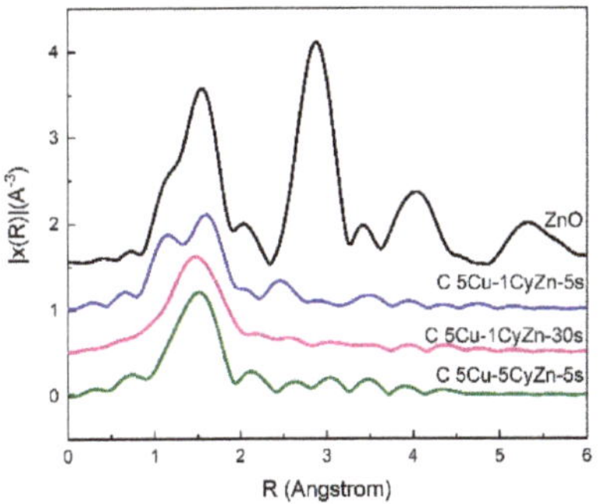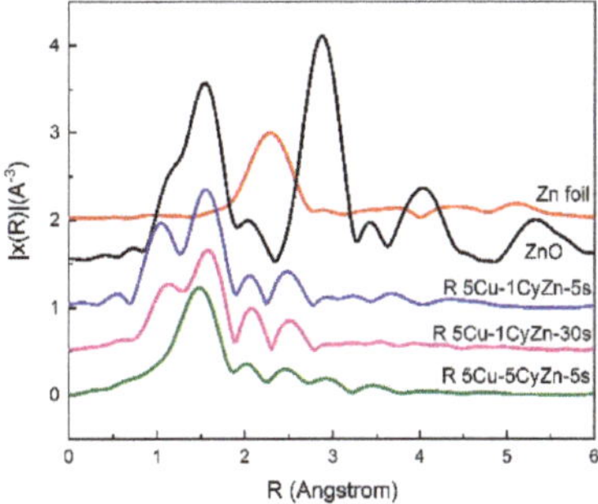

Figure 11. Experimental EXAFS spectra registered Zn K-edge of calcined (**left**) and reduced (**right**) SEA-5Cu and ZnO over-coated Cu/SiO$_2$ catalysts.

2.4. CO$_2$ Conversion and MeOH Synthesis

Catalytic evaluations of the ZnO over-coated Cu/SiO$_2$ catalysts are shown in Table 6. When the catalysts were evaluated by the CO$_2$ hydrogenation reaction, 1 g of catalyst was packed in a packed-bed reactor. The mixture of gas reactants, H$_2$ and CO$_2$, were fed to the reactor at a flow rate of 9.6 L·h-1. The unreacted gases and gaseous products were analyzed by an off-line GC every 6 h. The CO$_2$ reaction rate and CO formation rate barely changed during a run (acted for 24 h), and the average values are shown in the table. The liquid products were analyzed at the end of the run. No other product besides MeOH and CO was formed. The CO$_2$ reaction rate of catalyst 5Cu-1CyZn-5s is 4.9 mmol·g^{-1}h^{-1}, which decreases to 4.77 mmol·g^{-1}h^{-1} with a pulse time of 30 s, and becomes even smaller (4.06 mmol·g^{-1}h^{-1}) with five cycles of ZnO over-coated. The formation rate of CO has the same trend with the CO$_2$ reaction rate, which is from 4.57 (5Cu-1CyZn-5s) to 4.32 (5Cu-1CyZn-30s) and 3.89 mmol·g^{-1}h^{-1} (5Cu-5CyZn-5s), respectively. The MeOH formation rate of catalyst 5Cu-1CyZn-30s is 0.48 mmol·g^{-1}h^{-1}, which is higher than that of catalyst 5Cu-1CyZn-5s (0.33 mmol·g^{-1}h^{-1}) and 5Cu-5CyZn-5s (0.21 mmol·g^{-1}h^{-1}). The MeOH selectivity of catalyst 5Cu-1CyZn-30s is 10.1%, which is higher than 6.7% for 5Cu-1CyZn-5s, and 5.2% for 5Cu-5CyZn-5s. The ratios of the number of metallic sites to the number of basic sites also are also shown in the table, which are 0.26, 0.19, and 0.17 on the catalysts 5Cu-1CyZn-5s, 5Cu-1CyZn-30s, and 5Cu-5CyZn-5s, respectively. The TOF of CO_2 was calculated based on the number of basic sites from the results of CO_2-TPD. The TOF of H_2 was calculated based on the number of metallic sites from the results of CO chemisorption. 5Cu-1CyZn-5s has higher TOF$_{CO2}$ of 4.2 h^{-1}, and lower TOF$_{H2}$ of 18.3 h^{-1}. 5Cu-1CyZn-30s and 5Cu-5CyZn-5s show similar TOF$_{CO2}$ of 3.3 and 3.1 h^{-1}, respectively, and the similar TOF$_{H2}$ of 20.0 and 19.7 h^{-1}.

Table 6. Catalytic performance of the ZnO over-coated Cu/SiO$_2$ catalysts.

Catalysts [1]	CO$_2$ Reaction Rate	MeOH Formation Rate	CO Formation Rate	MeOH Selectivity	Number of Basic Sites	Number of Metallic Sites	Ratio of Metallic to Basic Sites	TOF$_{CO2}$ by ZnO	TOF$_{H2}$ by Cu
	(mmol·g^{-1}h^{-1})			(%)	(μmol·g^{-1})			(h^{-1})	
5Cu-1CyZn-5s	4.90	0.33	4.57	6.7	1158	304	0.26	4.2	18.3
5Cu-1CyZn-30s	4.77	0.48	4.32	10.1	1428	286	0.19	3.3	20.0
5Cu-5CyZn-5s	4.06	0.21	3.89	5.2	1310	228	0.17	3.1	19.7

[1] Reaction conditions: T = 250 °C, P = 4.0 MPa, F = 9.6 L/h, 1.0 g catalyst.

According to our discussion in the introduction, the CO_2 was adsorbed and activated by basic sites, ZnO related, and H_2 by metallic sites, Cu related. They must be close to each other to form the right structures to facilitate the formation of methanol or to form the wrong structures to help the formation of CO. During the reaction, the activated CO_2 needs assistance from the activated H to pull one O away and to insert H in at the same time to form C-H and O-H bonds. When the activated CO_2 lacks surrounding H atoms, only O can be pulled away from the activated CO_2, resulting in the formation of CO. To form MeOH only, three metallic sites should be next to one basic site to supply three moles of activated H_2 to react with the activated CO_2 in the molecular form. So, the ratio of the number of metallic sites to the number of basic sites on the catalyst, on which only MeOH forms, should be at least 3. This value should be 1 on the catalyst producing only CO. This conclusion is based on an ideal situation, which is that all the active sites are involved in the reaction. In the actual reaction, not all of the active sites are working during the reaction. Only the metallic and basic sites, who are next to each other, have the opportunity to provide the activated reactants and participate in the reaction.

In our case, many more basic sites are formed than the metallic sites on the ZnO over-coated Cu/SiO$_2$ catalysts. The ratios of the number of metallic sites to the number of basic sites are much smaller than 3. During the reaction, the basic sites are enough to adsorb and activate CO_2. It is more important if there are enough metallic sites to activate H_2 and provide atom H to the basic sites. Since more metallic sites are formed on the 5Cu-1CyZn-5s, it shows a higher CO_2 reaction rate as observed by other researchers [6,11,14]. It indicates that the CO_2 reaction rate is related to the number of metallic sites.

When it comes to the selectivity of the catalyst, another factor should be considered, which is whether the metallic and basic sites are close enough to each other. To understand this, the characterization of XRD patterns, XANES, and EXAFS spectra were studied on the ZnO over-coated Cu/SiO$_2$ catalysts. The value of the Cu^0 lattice constant of 5Cu-1CyZn-30s is 3.64 Å, but this value for catalyst 5Cu-1CyZn-5s is 3.62 Å. The increase in the lattice constant upon the deposition of ZnO with a longer pulse time may be attributable to the formation of the Cu-ZnO surface alloy [18]. From the XANES spectra registered the Cu K edge of reduced 5Cu-1CyZn-5s and 5Cu-1CyZn-30s, the peaks assigned to the 1s–4p transition drift towards a higher energy compared to that of SEA-5Cu, which demonstrates that the chemical environment of the Cu species on these two catalysts is affected by the ZnO over-coated, and metallic Cu may be oxidized by the ZnO. So, there may be an interaction between Cu and ZnO, and possibly a dissolution of Cu atoms in the ZnO lattice. Longer ZnO pulse times help this dissolution, but multiple layers do not. Compared to the EXAFS spectrum of calcined SEA-5Cu, the bond lengths of Cu-O and Cu-(O)-Cu in the spectra of calcined Cu-ZnO catalysts become longer due to the interaction between Cu and over-coated ZnO. So, the good interaction between the metallic sites and basic sites helps to improve the selectivity of the catalyst.

2.5. CO_2 Catalytic Activation to Produce MeOH

According to the study of the relationship between the catalyst structure and activity, a theory of CO_2 catalytic activation can be proposed. During the reaction on the 5Cu-1CyZn-30s, CO_2 is chemisorbed and activated by the basic sites, ZnO related, which try to push the atom O out of the CO_2 molecule. At the same time, the H_2 is activated by the metallic site, Cu related, to form the H atom, which then diffuses to the basic site. Since more points where basic sites and metallic sites are close to each other exist, enough H atoms help to pull out the O atom and add into the CO_2 to form C-H and O-H bonds, which result in MeOH formation. Due to the synergistic effects of both the basic and metallic sites providing the forces of pushing and pulling, CO_2 can be activated and reacted with H_2 to form MeOH.

This proposed theory was proved by the performance of our catalysts; it has also been verified by others' work. The Cu-ZnO catalyst supported by reduced graphene oxide nanosheets exhibits a higher CO_2 reaction rate compared to other Cu-ZnO catalysts [5]. According to the TEM image of the catalyst, it shows greatly even-dispersed Cu and ZnO phases, which indicates excellent metallic and basic sites. Due to the contribution of these two sites, the catalyst has high CO_2 conversion. Since they randomly connect with each other, the selectivity to MeOH is not outstanding. Another Cu-ZnO catalyst [7] prepared by precipitated ZnO on the surface of Cu powders show 100% of selectivity to MeOH. The catalyst has a very small amount of ZnO precipitated on the surface of metallic Cu, which represent a number of basic sites, a large number of metallic sites, and very good interaction between them. During the reaction, there are enough H atom formation and spillover to the basic sites to reduce the activated CO_2 to MeOH. However, this catalyst has very low CO_2 conversion due to the weak basic sites. According to the analysis of our catalysts and others, the conclusion is much clearer that the catalyst with high activity and selectivity should have both excellent metallic and basic sites, and stronger interaction between them.

3. Experiment

3.1. Catalyst Preparation

Cu nanoparticles were prepared by Imp method or SEA method. Silica gel (6–12 mesh, grad 40; Fisher Chem, Canada) ground to 255 to 350 nm was used as the substrate. In the Imp method, a certain amount of copper nitrate hemi-pentahydrate ($Cu(NO_3)_2 \cdot 2.5\ H_2O$, 98%~102%; Aldrich, Canada) was dissolved in 2 mL of distilled water, and ammonia hydroxide (28 wt%; Sigma-Aldrich, Canada) was added dropwise to form a copper hydroxide solid. Then, the solid was re-dissolved to be a dark blue solution. The ammonia was added until the volume of the solution of Cu precursor was the same as the pore volume of the substrate. The dark blue solution, which is tetraamine copper(II) ion, $[Cu(NH_3)_4]^{2+}$ was added to silica gel dropwise while stirring with a glass rod for 10 min. The resultant solid was dried overnight at 110 °C, and calcined at 250 °C for 3 h. In the SEA method, 10 g of silica gel was pretreated in 50 mL of ammonia solution for 30 min to adjust the surface charge to negative. The pH of the ammonia solution was higher than 10. To prepare the $[Cu(NH_3)_4]^{2+}$-containing solution, a certain amount of copper nitrate was dissolved in 50 mL of H_2O. Then, ammonia solution was added to the nitrate solution. The pretreated silica gel mixture and $[Cu(NH_3)_4]^{2+}$ solution were mixed and stirred for 2 h. The solid was filtered, washed, dried at 110 °C overnight, and calcined at 250 °C for 3 h.

The ZnO layer on the resultant solid from SEA was prepared by the ALD method using Beneq TFS 200 equipment. Because smaller Cu particles with a good distribution formed on samples prepared by SEA method according to TEM images. Diethyl zinc (DEZ) (98.4 wt%; Sigma-Aldrich, USA) was used as the Zn precursor and H_2O was employed as the oxidant. Nitrogen (99.999%; Praxair, Canada) was used to purge the gas lines and the reactor. The temperature of the reaction was set at 150 °C to avoid the formation of metallic Zn. The ALD cycle follows the reactions [32]:

$$Zn(C_2H_5)_2 + H_2O \rightarrow ZnO + 2C_2H_6 \text{ (g) } \Delta H = -70 \text{ kcal.} \tag{2}$$

This reaction splits into two half-reactions:

$$\text{Sub-Surf-OH} + Zn(C_2H_5)_2 \rightarrow \text{Sub-Surf-O-}Zn(C_2H_5) + C_2H_6 \text{ (g)}, \tag{3}$$

$$\text{Sub-Surf-O-}Zn(C_2H_5) + H_2O \rightarrow \text{Sub-Surf-O-Zn-OH} + C_2H_6 \text{ (g)}. \tag{4}$$

The procedure of each ALD cycle is as follows: (1) DEZ vapor was introduced to the chamber and reacted with hydroxyls that are on the surface of the substrate to form monoethylzinc. The monotheylzinc replaced hydrogen and connected to the oxygen atom with the release of ethane as depicted in Equation (3); (2) the chamber was purged with N_2 to remove the unreacted precursor and gaseous products; (3) H_2O vapor was allowed into the chamber to react with the monoethylzinc, forming hydroxyls bonded with Zn, and releasing ethane (Equation (4)); and (4) the chamber was again purged with N_2 to remove the unreacted oxidant and the gaseous products. Before steps (1) and (3), the pressure of the chamber was reduced to 2.0 kPa and the valve that separated the chamber and the pump was closed until the purging procedure started. Finally, the pulse time and/or number of cycles were changed to obtain catalysts with different atomic layers and properties.

3.2. Catalyst Characterization

The elemental composition of catalysts was determined by inductively coupled plasma mass spectrometry (ICP-MS) analysis at Saskatchewan Research Council. The analysis was carried out by the equipment of PerkinElmer Optima 5300 DV/Optima 8300 DV.

The BET surface area, average pore diameter, and pore volume of the support and catalysts were measured by N_2 adsorption using Micromeritics ASAP 2020 with 0.2 g of the sample. The analysis bath temperature was maintained at -196 °C through the experiment. Degassing was carried out at 200 °C for 2 h. From room temperature to 200 °C, the heating rate was 10 °C·min^{-1}. The average pore diameter and pore volume were obtained from the result of the BJH analysis.

The H_2 temperature-programmed reduction (H_2-TPR) and CO_2 temperature-programmed desorption (CO_2-TPD) studies were performed using a Micromeritics 2950 HP Chemisorption Analyzer, equipped with a thermal conductivity detector (TCD). Before the analysis of H2-TPR, the sample was purged at 30 °C until the baseline signal was stable. The H_2-TPR analysis was conducted by heating samples from 30 to 650 °C at a ramp of 10 °C·min^{-1}. The flow rate of 10% H_2/argon was 50 mL·min^{-1}. During the CO_2-TPD analysis, the catalyst sample was first pretreated at 250 °C for 20 min in He atmosphere to remove physically adsorbed CO_2. Then, the CO_2 adsorption was carried out at 50 °C. Afterwards, CO_2 desorption was performed with a linear temperature increase up to 800 °C at a rate of 10 °C·min^{-1}.

A transmission electron microscope (TEM) and energy-dispersive X-ray spectroscopy (EDX) were used to investigate the Cu particle size and distribution of the elements of the reduced catalysts on a HITACHI HT-7700 with Bruker X-ray detector at Western College of Veterinary Medicine imaging center, University of Saskatchewan. Before the TEM and EDX tests, the samples were first reduced by 40 v/v% H_2 at atmospheric pressure (ramp from room temperature to 500 °C, rate 5 °C·min^{-1}, isothermal for 3 h).

X-ray diffraction (XRD) analysis was conducted on a Bruker Advance D8 series II, equipped with Cu Kα radiation ($\lambda = 0.1541$ nm). The crystallite sizes of the reduced catalysts were estimated using Scherer's equation:

$$\text{size} = k\lambda/\beta(2\theta)\cos\theta,$$

where k is the crystallite shape constant (0.94), λ is the X-ray wavelength, and $\beta(2\theta)$ is the full width at half maximum in radians. The lattice constants for the reduced catalysts were determined by Bragg's law:

$$n\lambda = 2d\ \sin\theta,$$

where n is the order of diffraction and d is the spacing between the different planes of atoms in the crystal lattice.

The active metal dispersion of the Cu–ZnO catalysts was obtained using CO chemisorption, which was performed on a Micrometrics ASAP 2020 instrument. After the sample was loaded and degassed, it was reduced in-situ with pure hydrogen at 500 °C for 2 h and cooled down under vacuum to 35 °C following purging in the flow of He for 30 min. Then, chemisorption analysis was conducted by passing the pulses of CO gas.

The Cu K-edge and Zn K-edge X-ray absorption spectroscopy (XAS) measurements were conducted at the Soft X-ray Micro-characterization Beamline (SXRMB) and Industry, Development, Education, and Students (IDEAS) beamline, respectively at the Canadian Light Source (CLS, Saskatoon, SK, Canada). A few milligrams of calcined or reduced samples were loaded on carbon tape. Both Cu K-edge and Zn K-edge were collected in fluorescence and transmission modes.

3.3. Catalyst Evaluation

The evaluations of the chosen catalysts with CO_2 hydrogenation reactions were conducted in an Inconel fixed bed reactor (ID: 22 mm and length: 450 mm) with a single-pass downward flow. In total, 1 g of catalyst was diluted with 90 mesh size silicon carbide particles and loaded in the constant temperature zone of the reactor. Other parts of the reactor were filled with different sizes of silicon carbide by layers. The catalyst was then reduced under H_2 (99.9% purity, Praxair, Canada) flow (120 mL·min^{-1}) at atmospheric pressure (ramp from room temperature to 500 °C, rate 5 °C·min^{-1}, isothermal for 3 h). The reduced catalyst was cooled down to 250 °C under H_2 flow. The desired reaction gas flow rates were adjusted to a total volumetric flow rate of 160 mL·min^{-1} (H_2:CO_2 = 3:1) (CO_2: 99.9% purity, Praxair, Canada). Then, the reactor was pressurized to the reaction pressure of 4.0 MPa. The gas product was cooled to 0 °C in a cold trap to separate the condensable gas in the product stream into the liquid phase at the reaction pressure.

The non-condensable gases in the stream were monitored by a bubble flow column installed at the exit of the reactor, and analyzed by an off-line Agilent Technologies 7890A gas chromatograph instrument equipped with two thermal conductivity detectors (TCDs) and a flame ionization detector (FID). The liquid products were collected during a 24-h reaction period and analyzed with an off-line Agilent Technologies 7890A gas chromatograph equipped with a DB-Wax capillary column and FID detector. The weight of liquid products was measured after the collection to ensure the mass balance.

The conversion of CO_2 was calculated as the average values for the total test duration:

$$X_{CO_2} = 100\% \times \frac{n_{CO_2,in} - n_{CO_2,out}}{n_{CO_2,in}}.$$

Methanol selectivity was calculated with respect to carbon-containing products (no other product obtained except MeOH and CO):

$$S_{MeOH} = 100\% \times \frac{n_{MeOH}}{n_{MeOH} + n_{CO}}.$$

The turnover frequency (TOF) of CO_2 or H_2 was calculated based on the number of basic or metallic sites on the catalysts:

$$TOF = \frac{reaction\ rate}{number\ of\ active\ sites}.$$

4. Conclusions

The synthesis of the catalysts as planned—the isolated Cu nanoparticles distributed in ZnO layers, using SEA and ALD over-coating—was successful. By varying the exposure time of the Zn-containing precursor and the number of ZnO deposition layers, there was a slight change in the environment of the ZnO sites to Cu sites and vice versa, and their interactions.

As expected, these catalysts facilitated a greater CO formation rate than the MeOH formation rate. The more uniformed Cu site and ZnO site distribution in 5Cu-1CyZn-30s allowed it to have relatively more MeOH formation.

This study partially supports the mechanism theory in CO_2 hydrogenation to form MeOH. Activated CO_2 on the ZnO sites needs activated hydrogen in the right position. Our following work is to synthesize the catalysts with isolated ZnO sites distributed in Cu sites, in the hope that the activated CO_2 will have enough activated hydrogen around it in the right position to form MeOH.

Author Contributions: Conceptualization, J.G. and H.W.; Data curation, J.G.; Formal analysis, J.G.; Funding acquisition, H.W. and A.D.; Investigation, J.G. and P.E.B.; Methodology, J.G.; Resources, Y.H., A.D. and H.W.; Supervision, H.W. and A.D.; Writing—original draft, J.G.; Writing—review and editing, P.E.B., Y.H., A.D. and H.W.

Funding: This research was funded by the Natural Sciences and Engineering Research Council of Canada (NSERC), Canadian Foundation for Innovation (CFI) and the China Scholarship Council (CSC).

Conflicts of Interest: The authors declare no conflict of interest.

References

1. Le Quéré, C.; Barbero, L.; Hauck, J.; Andrew, R.M.; Canadell, J.G.; Sitch, S.; Korsbakken, J.I. Global Carbon Budget 2018. *Earth Syst. Sci. Data* **2018**, *10*, 2141–2194. [CrossRef]

2. Wang, W.; Wang, S.; Ma, X.; Gong, J. Recent advances in catalytic hydrogenation of carbon dioxide. *Chem. Soc. Rev.* **2011**, *40*, 3703–3727. [CrossRef] [PubMed]

3. Ma, J.; Sun, N.; Zhang, X.; Zhao, N.; Xiao, F.; Wei, W.; Sun, Y. A short review of catalysis for CO_2 conversion. *Catal. Today* **2009**, *148*, 221–231. [CrossRef]

4. Duan, H.; Yang, C.Y.; Singh, D.R.; Chiang, B.K.; Wang, S.; Xiao, C.P.; Patel, B.J.; Danaci, C.D.; Burke, B.N.; Zhai, Y.; et al. Mesoporous Carbon-supported Cu/ZnO for Methanol Synthesis from Carbon Dioxide. *Aust. J. Chem.* **2014**, *67*, 907–914. [CrossRef]

5. Deerattrakul, V.; Dittanet, P.; Sawangphruk, M.; Kongkachuichay, P. CO_2 hydrogenation to methanol using Cu-Zn catalyst supported on reduced graphene oxide nanosheets. *J. CO_2 Util.* **2016**, *16*, 104–113. [CrossRef]

6. Dong, X.; Li, F.; Zhao, N.; Xiao, F.; Wang, J.; Tan, Y. CO_2 hydrogenation to methanol over Cu/ZnO/ZrO$_2$ catalysts prepared by precipitation-reduction method. *Appl. Catal. B Environ.* **2016**, *191*, 8–17. [CrossRef]

7. Le Valant, A.; Comminges, C.; Tisseraud, C.; Canaff, C.; Pinard, L.; Pouilloux, Y. The Cu-ZnO synergy in methanol synthesis from CO_2, Part 1: Origin of active site explained by experimental studies and a sphere contact quantification model on Cu + ZnO mechanical mixtures. *J. Catal.* **2015**, *324*, 41–49. [CrossRef]

8. Liang, Z.; Gao, P.; Tang, Z.; Lv, M.; Sun, Y. Three dimensional porous Cu-Zn/Al foam monolithic catalyst for CO_2 hydrogenation to methanol in microreactor. *J. CO_2 Util.* **2017**, *21*, 191–199. [CrossRef]

9. Karelovic, A.; Bargibant, A.; Fernández, C.; Ruiz, P. Effect of the structural and morphological properties of Cu/ZnO catalysts prepared by citrate method on their activity toward methanol synthesis from CO_2 and H_2 under mild reaction conditions. *Catal. Today* **2012**, *197*, 109–118. [CrossRef]

10. Gesmanee, S.; Koo-Amornpattana, W. Catalytic hydrogenation of CO_2 for methanol production in fixed-bed reactor using Cu-Zn supported on gamma-Al$_2$O$_3$. *Energy Procedia* **2017**, *138*, 739–744. [CrossRef]

11. Behrens, M.; Studt, F.; Kasatkin, I.; Kühl, S.; Hävecker, M.; Abild-pedersen, F.; Zander, S.; Girgsdies, F.; Kurr, P.; Kniep, B.; et al. The Active Site of Methanol Synthesis over Cu/ZnO/Al$_2$O$_3$ Industrial Catalysts. *Science* **2012**, *759*, 893–898. [CrossRef] [PubMed]

12. Behrens, M. Meso-and nano-structuring of industrial Cu/ZnO/(Al$_2$O$_3$) catalysts. *J. Catal.* **2009**, *267*, 24–29. [CrossRef]

13. Fujimoto, K.; Yu, Y. Spillover effect on the stabilization of Cu-Zn catalyst for CO_2 hydrogenation to methanol. *Stud. Surf. Sci. Catal.* **1993**, *77*, 393–396.

14. Hu, B.; Yin, Y.; Liu, G.; Chen, S.; Hong, X.; Tsang, S.C.E. Hydrogen spillover enabled active Cu sites for methanol synthesis from CO_2 hydrogenation over Pd doped CuZn catalysts. *J. Catal.* **2018**, *359*, 17–26. [CrossRef]

15. Waugh, K.C. Methanol Synthesis. *Catal. Today* **1992**, *15*, 51–75. [CrossRef]

16. Ren, H.; Xu, C.-H.; Zhao, H.-Y.; Wang, Y.-X.; Liu, J.J.-Y.; Liu, J.J.-Y. Methanol synthesis from CO_2 hydrogenation over Cu/γ-Al_2O_3 catalysts modified by ZnO, ZrO_2 and MgO. *J. Ind. Eng. Chem.* **2015**, *28*, 261–267. [CrossRef]

17. Choi, Y.; Futagami, K.; Fujitani, T.; Nakamura, J. The difference in the active sites for CO_2 and CO hydrogenations on Cu / ZnO-based methanol synthesis catalysts. *Catal. Lett.* **2001**, *73*, 27–31. [CrossRef]

18. Kanai, Y.; Watanabe, T.; Fujitani, T.; Saito, M.; Nakamura, J.; Uchijima, T. Evidence for the migration of ZnOx in a Cu/ZnO methanol synthesis catalyst. *Catal. Lett.* **1994**, *27*, 67–78. [CrossRef]

19. Tisseraud, C.; Comminges, C.; Belin, T.; Ahouari, H.; Soualah, A.; Pouilloux, Y.; Le Valant, A. The Cu-ZnO synergy in methanol synthesis from CO_2, Part 2: Origin of the methanol and CO selectivities explained by experimental studies and a sphere contact quantification model in randomly packed binary mixtures on Cu-ZnO coprecipitate catalysts. *J. Catal.* **2015**, *330*, 533–544. [CrossRef]

20. Tisseraud, C.; Comminges, C.; Pronier, S.; Pouilloux, Y.; Le Valant, A. The Cu–ZnO synergy in methanol synthesis Part 3: Impact of the composition of a selective Cu@ZnOxcore–shell catalyst on methanol rate explained by experimental studies and a concentric spheres model. *J. Catal.* **2016**, *343*, 106–114. [CrossRef]

21. Tisseraud, C.; Comminges, C.; Habrioux, A.; Pronier, S.; Pouilloux, Y.; Le Valant, A. Cu-ZnO catalysts for CO_2 hydrogenation to methanol: Morphology change induced by ZnO lixiviation and its impact on the active phase formation. *Mol. Catal.* **2018**, *446*, 98–105. [CrossRef]

22. Din, I.U.; Shaharun, M.S.; Naeem, A.; Tasleem, S.; Johan, M.R. Carbon nanofiber-based copper/zirconia catalyst for hydrogenation of CO_2 to methanol. *J. CO_2 Util.* **2017**, *21*, 145–155. [CrossRef]

23. Musić, S.; Filipović-Vinceković, N.; Sekovanić, L. Precipitation of amorphous SiO_2 particles and their properties. *Braz. J. Chem. Eng.* **2011**, *28*, 89–94. [CrossRef]

24. Gawande, M.B.; Goswami, A.; Felpin, F.X.; Asefa, T.; Huang, X.; Silva, R.; Zou, X.; Zboril, R.; Varma, R.S. Cu and Cu-Based Nanoparticles: Synthesis and Applications in Catalysis. *Chem. Rev.* **2016**, *116*, 3722–3811. [CrossRef] [PubMed]

25. Ayodele, O.B.; Tasfy, S.F.H.; Zabidi, N.A.M.; Uemura, Y. Co-synthesis of methanol and methyl formate from CO_2 hydrogenation over oxalate ligand functionalized ZSM-5 supported Cu/ZnO catalyst. *J. CO_2 Util.* **2017**, *17*, 273–283. [CrossRef]

26. Huang, C.; Chen, S.; Fei, X.; Liu, D.; Zhang, Y. Catalytic hydrogenation of CO_2 to methanol: Study of synergistic effect on adsorption properties of CO_2 and H_2 in $CuO/ZnO/ZrO_2$ system. *Catalysts* **2015**, *5*, 1846–1861. [CrossRef]

27. Gao, P.; Li, F.; Zhan, H.; Zhao, N.; Xiao, F.; Wei, W.; Zhong, L.; Sun, Y. Fluorine-modified Cu/Zn/Al/Zr catalysts via hydrotalcite-like precursors for CO_2 hydrogenation to methanol. *Catal. Commun.* **2014**, *50*, 78–82. [CrossRef]

28. Liu, Y.; Sun, K.; Ma, H.; Xu, X.; Wang, X. Cr, Zr-incorporated hydrotalcites and their application in the synthesis of isophorone. *Catal. Commun.* **2010**, *11*, 880–883. [CrossRef]

29. Bhuiyan, M.M.R.; Lin, S.D.; Hsiao, T.C. Effect of calcination on Cu-Zn-loaded hydrotalcite catalysts for C-C bond formation derived from methanol. *Catal. Today* **2014**, *226*, 150–159. [CrossRef]

30. Velu, S.; Suzuki, K.; Gopinath, C.S.; Yoshida, H.; Hattori, T. XPS, XANES and EXAFS investigations of $CuO/ZnO/Al_2O_3/ZrO_2$ mixed oxide catalysts. *Phys. Chem. Chem. Phys.* **2002**, *4*, 1990–1999. [CrossRef]

31. Dadlani, A.; Acharya, S.; Trejo, O.; Nordlund, D.; Peron, M.; Razavi, J.; Berto, F.; Prinz, F.B.; Torgersen, J. Revealing the Bonding Environment of Zn in ALD Zn(O,S) Buffer Layers through X-ray Absorption Spectroscopy. *Appl. Mater. Interfaces* **2017**, *9*, 39105–39109. [CrossRef] [PubMed]

32. Janocha, E. Electronic Properties of ALD Zinc Oxide Interfaces and its Implication for Chalcopyrite Absorber Materials. Ph.D. Thesis, Fakultät IV—Elektrotechnik und Informatik, Technische Universität Berlin, Berlin, Germany, 2011. [CrossRef]

 *catalysts* 

Article

Hydroprocessing of Oleic Acid for Production of Jet-Fuel Range Hydrocarbons over Cu and FeCu Catalysts

Afees A. Ayandiran, Philip E. Boahene, Ajay K. Dalai * and Yongfeng Hu

Catalysis and Chemical Reaction Engineering Laboratories, University of Saskatchewan, Saskatoon, SK S7N 5A9, Canada; aaa273@mail.usask.ca (A.A.A.); peb225@mail.usask.ca (P.E.B.); yongfeng.hu@lightsource.ca (Y.H.)
* Correspondence: ajay.dalai@usask.ca; Tel.: +1-306-966-4771; Fax: +1-306-966-4777

Received: 25 October 2019; Accepted: 8 December 2019; Published: 11 December 2019

Abstract: In the present study, a series of monometallic Cu/SiO_2-Al_2O_3 catalysts exhibited immense potential in the hydroprocessing of oleic acid to produce jet-fuel range hydrocarbons. The synergistic effect of Fe on the monometallic Cu/SiO_2-Al_2O_3 catalysts of variable Cu loadings (5–15 wt%) was ascertained by varying Fe contents in the range of 1–5 wt% on the optimized 13% Cu/SiO_2-Al_2O_3 catalyst. At 340 °C and 2.07 MPa H_2 pressure, the jet-fuel range hydrocarbons yield and selectivities of 51.8% and 53.8%, respectively, were recorded for the $Fe(3)$-$Cu(13)/SiO_2$-Al_2O_3 catalyst. To investigate the influence of acidity of support on the cracking of oleic acid, ZSM-5 (Zeolite Socony Mobil–5) and HZSM-5(Protonated Zeolite Socony Mobil–5)-supported 3% Fe-13% Cu were also evaluated at 300–340 °C and 2.07 MPa H_2 pressure. Extensive techniques including N_2 sorption analysis, pyridine- Fourier Transform Infrared Spectroscopy (Pyridine-FTIR), X-ray Diffraction (XRD), X-ray Photoelectron Spectroscopy (XPS), and H_2-Temperature Programmed Reduction (H_2-TPR) analyses were used to characterize the materials. XPS analysis revealed the existence of Cu^{1+} phase in the $Fe(3)$-$Cu(13)/SiO_2$-Al_2O_3 catalyst, while Cu metal was predominant in both the ZSM-5 and HZSM-5-supported FeCu catalysts. The lowest crystallite size of $Fe(3)$-$Cu(13)/SiO_2$-Al_2O_3 was confirmed by XRD, indicating high metal dispersion and corroborated by the weakest metal–support interaction revealed from the TPR profile of this catalyst. CO chemisorption also confirmed high metal dispersion (8.4%) for the $Fe(3)$-$Cu(13)/SiO_2$-Al_2O_3 catalyst. The lowest and mildest Brønsted/Lewis acid sites ratio was recorded from the pyridine–FTIR analysis for this catalyst. The highest jet-fuel range hydrocarbons yield of 59.5% and 73.6% selectivity were recorded for the $Fe(3)$-$Cu(13)/SiO_2$-Al_2O_3 catalyst evaluated at 300 °C and 2.07 MPa H_2 pressure, which can be attributed to its desirable textural properties, high oxophilic iron content, high metal dispersion and mild Brønsted acid sites present in this catalyst.

Keywords: hydroprocessing; FeCu catalysts; jet fuel; oleic acid

---

1. Introduction

The aviation sector is a large growing sector which bridges large distances within relatively short time. The total number of international air passengers worldwide in 2018 was 4.4 billion, and this is expected to increase to 7.8 billion in 2036 with Compound Annual Growth Rate (CAGR) of 3.6% according to the prediction made by the International Air Transport Association (IATA). The aviation sector facilitates 35% of world trade by value and it is responsible for transporting 54% of international tourists. In spite of the significance of this sector, it is being faced by challenges over the years. The worldwide aviation industry consumes about 1.7 billion barrels of conventional jet fuel annually [1]. The development of the aviation industry is paralleled with increase in greenhouse gas emissions [2]. According to the air transport action group (ATAG), 895 million tonnes of CO_2 was emitted by flights in 2018, which represents about 2% of human-induced emissions. As a result, aviation experts unanimously agreed on three targets that are carbon neutral growth by 2020, fifty percent greenhouse gas emission reduction by 2050 with respect to 2005 as a baseline and lastly, improvement of efficiency of fuel by 1.5% from 2009 to 2020. These are the first set of climate change targets that are streamlined to a particular sector in the world [3]. In order to achieve these targets, different carbon footprint mitigation strategies were devised. They are technological improvements, use of aviation biofuels, airline operations improvements and market-based measures. Sgouridis et al. [4] used the Global Aviation Dynamics (GAID) model to prove that the potential contribution of aviation biofuels and market-based measures is significantly higher than the contribution of the other strategies [4].

Biochemical and thermochemical processes are among the technologies utilized for production of jet fuel from these biomass-based materials [1]. The alternative fuels for air transport include hydroprocessed renewable jet fuels (HRJ), Fischer–Tropsch jet fuels (FTJ), liquid biohydrogen, biomethane and bioalcohol [1]. Hydroprocessed renewable jet fuel has been proven to have the highest jet fuel-to-feed ratio, highest overall energy efficiency, lowest total capital investment and the lowest jet-fuel selling price [4]. Hydroprocessed renewable jet fuels are produced from oil-based feedstocks using hydroprocessing technologies that encompass hydrotreating, hydrodeoxygenation, hydrocracking and isomerization reactions. They find application in conventional aviation turbine engines without modification and without blending with conventional petroleum-derived jet fuels. Hydroprocessed jet fuels are suitable for high altitude flights due to their high cold flow properties [1].

Copper-based catalysts are well known for their preferential selectivity for hydrogenation of C=O bond in carbonyl compounds due to the presence of the unfilled 3D electron levels of metallic copper. These catalysts are also known for their high H_2 adsorption and activation abilities [5]. Bykoeva et al. [6] carried out hydrodeoxygenation over a reduced NiCu bimetallic supported catalysts and reported that high catalyst selectivity and catalyst stability were obtained as a result of the inclusion of Cu in the bimetallic catalyst. In spite of all the advantages of using s copper-based catalyst for the hydroprocessing of vegetable oils, they have low affinities for oxygen. To develop a novel catalyst with higher catalyst activity and selectivity, bimetallic catalysts are used due to their often appreciable chemical and electronic behavior as compared to their corresponding monometallic catalysts. Iron has high oxophilicity and can be used to tune the activity and selectivity of copper. Iron has the highest natural abundance, lowest price and rich redox chemistry [7]. Kandel et al. [7] performed hydrodeoxygenation of microalgal oil over the reduced iron-based catalyst. It was found that the high selectivity towards liquid alkanes was due to the Fe-O bond strength; thus providing effective reduction of the microalgal oil to form paraffins. Iron can facilitate hydroprocessing of vegetable oils through binding and subsequent activation of the oxygenates. Addition of iron to copper can also improve the surface area of copper and inhibit its sintering [4]. The supports used in the hydroprocessing of vegetable oil need to complement the activity role played by the active site phase of the catalyst. This can be achieved by synthesizing catalysts' support of high surface area and pore volume. Light hydrocarbons production can be avoided by using catalyst support of moderate Brønsted acid sites [8].

Alumina, silico-alumino phosphates, zeolites, zirconium oxide and mesoporous materials have been utilized by researchers for the hydroprocessing of vegetable oils [9–11]. Kazuhisa et al. [12] worked on the hydroprocessing of jatropha oil over a catalyst supported on ZSM-5 (23 wt% silica–alumina ratio) in a 100 mL autoclave batch reactor at 300 °C, 6.5 MPa H_2 pressure, 1 g catalyst/g feed and 12 h reaction time. An 83.8% conversion of jatropha oil, 3.06% CO_2 selectivity, 16% C_1–C_4 hydrocarbon selectivity and 80.7% C_{10}–C_{20} hydrocarbons yield were obtained. This shows the high activity and selectivity strength of ZSM 5 zeolite-supported catalyst for liquid hydrocarbons. It also shows the cracking effects of ZSM 5, thus indicating it can also be suitable for the hydrodeoxygenation and hydrocracking of vegetable oils to produce jet-fuel hydrocarbon fractions. Mixed support of SiO_2 and Al_2O_3 has also been used for hydroprocessing of vegetable oil. Yanyong et al. [13] examined the hydroprocessing of vegetable oils over the $NiMo/SiO_2$, $NiMo/Al_2O_3$ and $NiMo/SiO_2$-Al_2O_3 catalysts at 350 °C and 4 MPa H_2 pressure. The catalysts supported by the mixture of silica and alumina gave the highest and the most desirable iso/normal ratio (0.26). This shows clearly that SiO_2-Al_2O_3-supported catalysts have more preference for hydroprocessing of vegetable oils for bio-jet fuel production as compared to the Al_2O_3- and SiO_2-supported catalysts. Amorphous silica alumina support is viewed as a polymer of Al_2O_3 on a backbone of SiO_2, while crystalline ZSM-5 and HZSM materials are viewed as copolymers of Al_2O_3 and SiO_2 with capacities for ion exchange [14]. These three materials have varying Brønsted acid sites concentrations. Hydroprocessing of vegetable oils for production of jet-fuel range hydrocarbon largely depends on Brønsted acid sites concentrations of the catalysts used [8].

In this work, the catalytic performance on the conversion of model compound of vegetable oils (oleic acid), yield and selectivity of jet-fuel range hydrocarbons were studied over the Cu/SiO_2-Al_2O_3, $FeCu/SiO_2$-Al_2O_3, $FeCu/ZSM$-5 and $FeCu/HZSM$-5 catalysts. These catalysts were characterized for determination of their physicochemical properties and their impacts on product selectivity. The influence of reaction temperature, contact time and catalyst acidity were also investigated.

2. Results and Discussion

2.1. N_2- Adsorption/Desorption Measurement

The N_2- adsorption/desorption isotherms of all the silica–alumina-supported catalysts and its support are shown in Figure S1. For all the profiles shown in Figure S1, at relatively low pressure, no significant adsorption was observed for the support and the catalysts showed the formation of monolayer of adsorbed molecules of nitrogen gas. Significant adsorption at high relative pressure as shown in Figure S1 indicates adsorption in mesoporous materials [15]. The profiles of the support (SiO_2-Al_2O_3) and the monometallic catalysts exhibit type IV isotherms indicating that the support and the catalysts are mesoporous. Despite different loadings of iron, the profiles of all the $FeCu/SiO_2$-Al_2O_3 catalysts exhibit type IV isotherm indicating mesoporosity [15]. The pore diameters of the catalysts as shown in Table 1 for all the silica–alumina-supported catalysts and its support confirmed the type IV isotherm in Figure S1. The pore diameters were 4.4–5.7 nm indicating mesoporous nature of material. Out of all the monometallic catalysts, $Cu(13)/SiO_2$-Al_2O_3 catalyst samples have the largest pore diameter of 5.3 nm. Pore diameter decreases with increase in iron loading (1–5 wt%) as shown in Table 1. This trend can be ascribed to pore blockage.

Table 1. Textural properties of catalyst samples and their respective supports.

Sample ID	BET Surface Area (m²/g)	Micropore Volume (cm³/g)	Mesopore Volume (cm³/g)	Total Pore Volume (cm³/g)	Pore Diameter (nm)	Crystallite Size (nm)
SiO_2-Al_2O_3	660	0.00	0.94	0.94	5.7	-
ZSM	393	0.17	0.07	0.24	2.4	-
HZSM	321	0.12	0.08	0.20	2.5	-
$Cu(5)/SiO_2$-Al_2O_3	623	-	0.94	0.94	5.2	7.3
$Cu(10)/SiO_2$-Al_2O_3	611	-	0.81	0.81	5.2	7.6
$Cu(13)/SiO_2$-Al_2O_3	455	-	0.81	0.80	5.3	8.7
$Cu(15)/SiO_2$-Al_2O_3	510	-	0.60	0.60	5.0	8.7
$Fe(1)$-$Cu(13)/SiO_2$-Al_2O_3	458	-	0.58	0.58	5.1	24.0
$Fe(2)$-$Cu(13)/SiO_2$-Al_2O_3	483	-	0.59	0.59	4.9	18.1
$Fe(3)$-$Cu(13)//SiO_2$-Al_2O_3	446	-	0.50	0.50	4.5	5.9
$Fe(5)$-$Cu(13)/SiO_2$-Al_2O_3	430	-	0.47	0.47	4.4	8.3
$Fe(3)$-$Cu(13)/ZSM$-5	266	0.10	0.06	0.16	2.4	33.8
$Fe(3)$-$Cu(13)/HZSM$-5	193	0.09	0.01	0.10	2.1	29.1

The N_2- adsorption/desorption isotherms of the three catalysts ($Fe(3)$-$Cu(13)/SiO_2$-Al_2O_3, $Fe(3)$-$Cu(13)/HZSM$-5 and $Fe(3)$-$Cu(13)/ZSM$-5) and their respective supports (SiO_2-Al_2O_3, HZSM and ZSM-5 zeolite) are shown in Figure S2. The profiles of SiO_2-Al_2O_3 support and $Fe(3)$-$Cu(13)/SiO_2$-Al_2O_3 catalyst exhibit type-IV isotherms. This indicates that the catalysts are mesoporous. Unlike $Fe(3)$-$Cu(13)/SiO_2$-Al_2O_3 catalyst and its respective support, $Fe(3)$-$Cu(13)/HZSM$-5 and

Fe(3)-Cu(13)/ZSM-5 and their respective supports, show no significant adsorption capacity within the relative pressure of 0–0.8. Low nitrogen adsorption occurs at very high relative pressure ($p/p_o > 0.8$), indicating a mixed type I–type IV isotherm. It shows the presence of both micro- and mesoporosity in the two catalyst samples and their respective supports [15].

The textural properties of the three catalysts, (Fe(3)-Cu(13)/SiO$_2$-Al$_2$O$_3$, Fe(3)-Cu(13)/HZSM-5 and Fe(3)-Cu(13)/ZSM-5), and their respective supports (SiO$_2$-Al$_2$O$_3$, HZSM-5 and ZSM-5), are also shown in Table 1. The Fe(3)-Cu(13)/SiO$_2$-Al$_2$O$_3$ catalyst and its support have only mesoporous volumes, while other samples have both mesoporous and microporous volumes. The pore volume of all the supports decreases after metals loading due to blockage of the pores. Of all the catalyst samples, Fe(3)-Cu(13)/SiO$_2$-Al$_2$O$_3$ has the largest surface area, pore volume and pore diameter.

2.2. XRD Analysis

The XRD patterns of the silica–alumina-supported catalysts and its support are shown in Figure 1. The 23° diffraction peak on the diffuse XRD pattern of silica alumina support coincides with the literature [16–20] and the broadness of the peaks shows the material is amorphous. The two peaks at 36° and 43° diffraction angles are ascribed to copper (I) oxide [21]. The intensity of the peaks of copper (I) oxide at a 36° diffraction angle increases with copper loading. Diffraction peaks of copper (II) oxide, copper metal, iron oxides and iron were not observed due to their high dispersion on the support. The 42.5° diffraction angle peak attributed to copper (I) oxide [21] in the diffractogram of Fe(3)-Cu(13)/SiO$_2$-Al$_2$O$_3$ catalyst is the most diffuse peak. The decrease in the crystallite size of Cu$_2$O in Fe(1)-Cu(13)/SiO$_2$-Al$_2$O$_3$, Fe(2)-Cu(13)/SiO$_2$-Al$_2$O$_3$ and Fe(3)-Cu(13)/SiO$_2$-Al$_2$O$_3$ with iron loading indicates the promotional effect of iron on the dispersion of copper (see Table 1). Of all the silica–alumina-supported bimetallic catalysts, Fe(3)-Cu(13)/SiO$_2$-Al$_2$O$_3$ catalyst has the lowest Cu$_2$O crystallite size of 5.9 nm. This indicates its potential of being active for hydroprocessing of oleic acid for production of jet-fuel range hydrocarbons (C$_8$–C$_{16}$).

The XRD patterns of the Fe(3)-Cu(13)/SiO$_2$-Al$_2$O$_3$, Fe(3)-Cu(13)/HZSM-5 and Fe(3)-Cu(13)/ZSM-5 catalyst samples and their respective supports are shown in Figure 2. The X-ray light incident in a periodically arranged crystalline materials scatters in a specific direction and results in high intensity narrow peaks, while the X-ray light incident in amorphous materials scatters in random directions and gives broad peaks. In Figure 2, the discrete X-ray diffraction patterns of HZSM-5 and ZSM-5 supports are sharp Bragg peaks. This shows that these two materials have high degree of crystallinity with long range order. The X-ray diffraction patterns of HZSM-5 and ZSM-5 coincide with that reported in the literature [22,23]. The broad Bragg peak at 23° diffraction angle on the diffuse XRD pattern of amorphous silica alumina shows that it is amorphous and it also coincides with that reported in the literature [16]. The X-ray diffraction patterns of Fe(3)-Cu(13)/HZSM-5 and Fe(3)-Cu(13)/ZSM-5 show clearly the phases of Cu nanoparticles with the 23° sharp peak confirming the supports. In all the XRD patterns of FeCu/HZSM-5 and Fe(3)-Cu(13)/ZSM-5 catalysts, the Bragg peaks at 43°, 51° and 74° diffraction angles, respectively, are ascribed to the presence of Cu nanoparticles [24]. These three characteristic diffraction peaks correspond to the (111), (200) and (220) planes of face-centred cubic structure of copper. The peaks at 36° and 42.5° in the diffractogram of Fe(3)-Cu(13)/SiO$_2$-Al$_2$O$_3$ catalyst are attributed to the presence of copper (I) oxide [21]. The absence of the diffraction peaks of the reduced and oxidized phases of iron in all the samples can be ascribed to the fact that iron may be either present in its noncrystalline phase or in minute quantities below XRD sensitivity.

From the N$_2$- adsorption/desorption measurement, the BET surface area of the silica–alumina support is the highest followed by the ZSM-5 support and HZSM-5 support. The catalyst, Fe(3)-Cu(13)/SiO$_2$-Al$_2$O$_3$ with the highest surface area, pore diameter and pore volume has the lowest copper phase crystallite size.

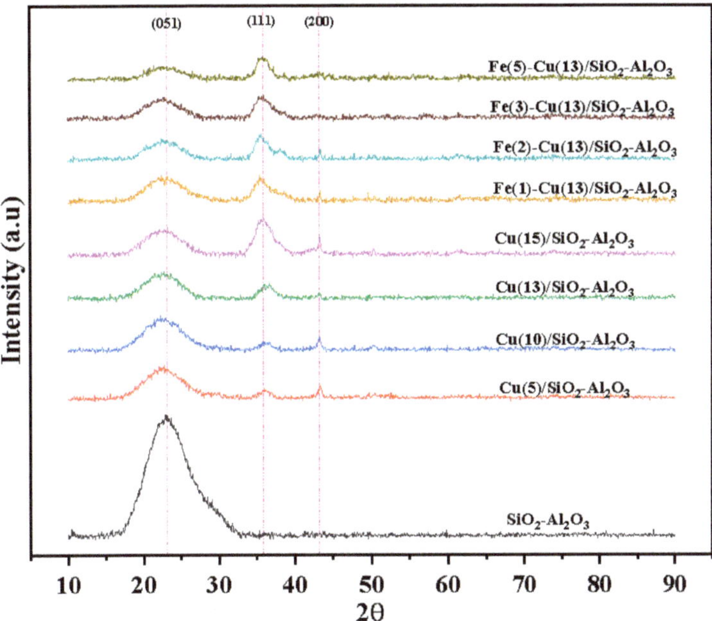

Figure 1. Wide angle X-ray Diffraction (XRD) patterns of SiO_2-Al_2O_3, Cu(5)/SiO_2-Al_2O_3, Cu(10)/SiO_2-Al_2O_3, Cu(13)/SiO_2-Al_2O_3 and Cu(15)/SiO_2-Al_2O_3, Fe(1)-Cu(13)/SiO_2-Al_2O_3, Fe(2)-Cu(13)/SiO_2-Al_2O_3, Fe(3)-Cu(13)/SiO_2-Al_2O_3 and Fe(5)-Cu(13)/SiO_2-Al_2O_3 catalysts.

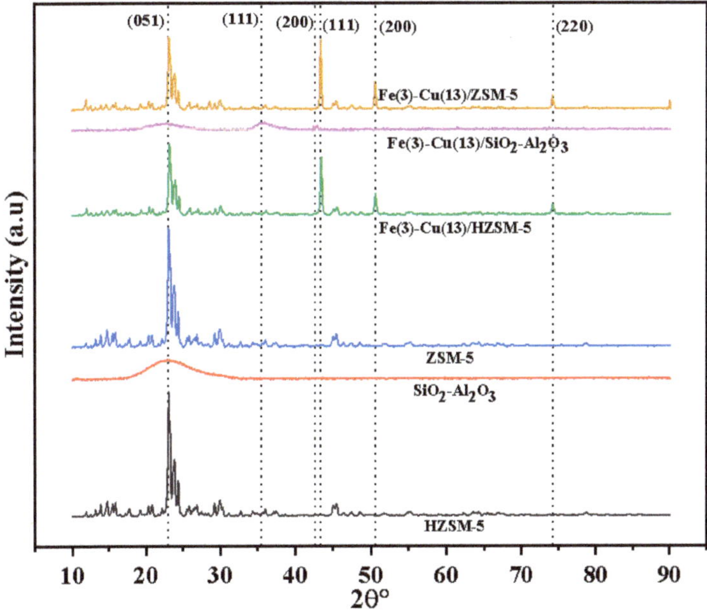

Figure 2. XRD patterns of SiO_2-Al_2O_3, ZSM-5 (Zeolite Socony Mobil–5) and HZSM-5(Protonated Zeolite Socony Mobil–5) supports; Fe(3)-Cu(13)/SiO_2-Al_2O_3, Fe(3)-Cu(13)/ZSM-5 and Fe(3)-Cu(13)/HZSM catalysts.

2.3. Inductively Coupled Plasma-Optical Emission Spectrometry (ICP-OES) and CO Chemisorption Analyses

The actual loadings of iron and copper in Fe(3)-Cu(13)/SiO$_2$-Al$_2$O$_3$, Fe(3)-Cu(13)/ZSM-5 and Fe(3)-Cu(13)/HZSM-5 catalyst samples evaluated using the ICP-OES shows that they are approximately the same as the targeted loading, if rounded up to the nearest whole number. The exact actual loadings of iron and copper of these three catalyst samples in terms of their weight percentage were used in their CO chemisorption analysis. The crystallite size and percentage dispersion surface area of Cu and Fe metals were calculated using the CO chemisorption method and tabulated in Table 2. The metals crystallite size increases in the order of Fe(3)-Cu(13)/SiO$_2$-Al$_2$O$_3$ < Fe(3)-Cu(13)/ZSM-5 < Fe(3)-Cu(13)/HZSM-5. Fe(3)-Cu(13)/SiO$_2$-Al$_2$O$_3$ catalyst has the lowest metals crystallite size as observed from both CO chemisorption analysis and XRD peaks. The Scherrer equation was used to calculate the crystallite size of the copper phase detected from X-ray diffractograms of the catalysts. Cu was detected in the Fe(3)-Cu(13)/ZSM-5 and Fe(3)-Cu(13)/HZSM-5 catalysts with metal crystallite sizes of 33.8 nm and 29.1 nm, respectively, while Cu$_2$O was detected in Fe(3)-Cu(13)/SiO$_2$-Al$_2$O$_3$ catalyst with a crystallite size of 5.9 nm. All the reduced and oxidized phases of iron were not detected in the XRD patterns of the catalysts. These undetected iron phases were detected from XPS and TPR as discussed in later sections. Metals dispersion of the catalyst decreases with increase in their crystallite size as shown in Table 2. Metal dispersion of the three catalysts increases with the surface area of their respective support shown in Table 1. This trend can be attributed to increase in the proportion of catalysts' surface atoms with respect to the bulk catalysts. Fe(3)-Cu(13)/SiO$_2$-Al$_2$O$_3$ catalyst has the highest metal dispersion and largest surface area as compared to Fe(3)-Cu(13)/ZSM-5 and Fe(3)-Cu(13)/HZSM-5 catalysts.

Table 2. Inductively Coupled Plasma-Optical Emission Spectrometry (ICP-OES) and CO chemisorption analyses of catalyst samples.

Catalyst Samples	ICP-OES		Metals Dispersion (%)	Metallic Surface Area (m^2/g Sample)	Metallic Surface Area (m^2/g of Metal)	Metals Crystallite Size (nm)
	Cu (wt%)	Fe (wt%)				
Fe(3)-Cu(13)/SiO$_2$-Al$_2$O$_3$	13.0	2.5	8.4	8.1	54.3	8.4
Fe(3)-Cu(13)/ZSM-5	13.1	3.1	7.6	7.6	46.7	9.8
Fe(3)-Cu(13)/HZSM-5	13.4	3.2	6.3	6.5	38.4	11.9

Metals Surface area $= \frac{V_m N_a}{S_f S_d}$, *Metals dispersion* $= \frac{10^4 V_m M}{S_f W_m}$, *Metals crystal size* $= \frac{10^4 F}{MSS\, D_m}$, $V_m =$ *monolayer volume* $\left(\frac{mol\ of\ CO}{g\ of\ sample}\right)$, $M =$ *molar mass of metals* $\left(\frac{g}{mol}\right)$, $F =$ *shape factor*, $W_m =$ *metals loading in* %, $D_m =$ *metals density*, $N_a =$ *Avogadro number*, $S_d =$ *surface density of metals* $\left(number\ of\ metal\ atoms/m^2\right)$, $S_f =$ *stoichiometric factor* (*mole of CO per metal atom*).

2.4. Fourier Transform Infra-Red Analysis

The molecular structure of Fe(3)-Cu(13)/SiO$_2$-Al$_2$O$_3$, Fe(3)-Cu(13)/ ZSM-5 and Fe(3)-Cu(13)/ HZSM-5 catalysts and their respective supports was analyzed by the FTIR technique in the absorption region of 400–1400 cm^{-1}, as shown in Figure 3. The absorption band detected at 439 cm^{-1} wavenumber can be ascribed to Si-O bending. The framework vibration at 537 cm^{-1} wavenumber on the ZSM-5 and HZSM-5 support are characteristics of five membered rings tetrahedron shaped MFI zeolites. The absorption detected at 791 cm^{-1}, 1065 cm^{-1} and 1210 cm^{-1} wavenumbers can be ascribed to the external symmetric stretch, internal asymmetric stretch and external asymmetric stretch, which are typical for extremely siliceous materials [18,25]. The peaks were more intense in ZSM-5 and HZSM because of their higher silica–alumina ratio as compared to that of the silica–alumina support. There was a slight peak shift to a higher wavenumber at 537 cm^{-1} after Cu and Fe impregnation on HZSM-5 in the framework vibration ascribed to five membered rings tetrahedron shaped MFI zeolites. There was also slight shift of peaks to a higher wavenumber at 1065 cm^{-1} after Cu and Fe impregnation on ZSM-5 and HZSM-5 in the absorption band ascribed to internal asymmetric stretch of extremely siliceous materials. These shifts of FTIR peaks after impregnation of Cu and Fe are due to change in bond length of the aluminosilicate frameworks in the catalyst samples [18,25].

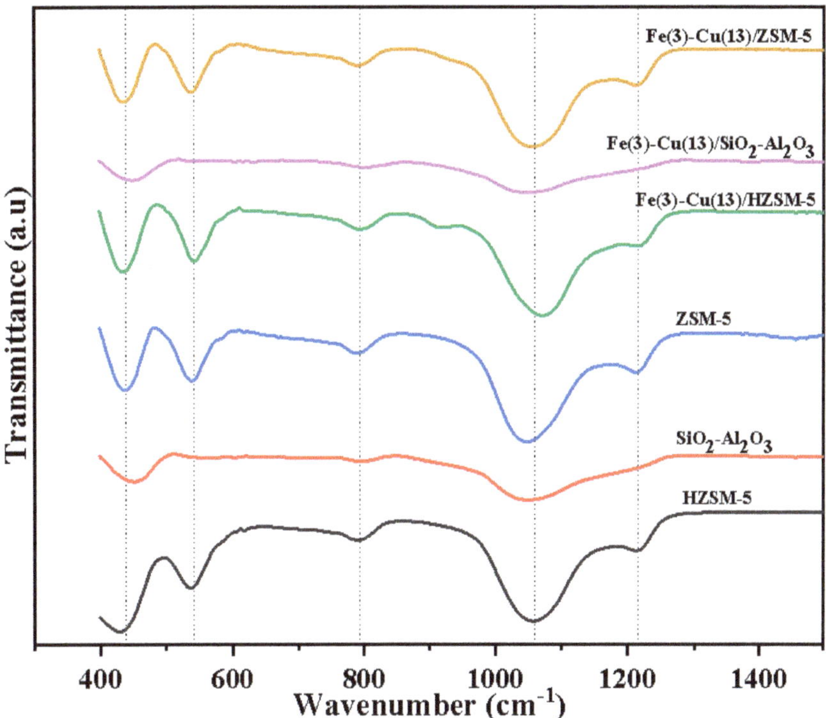

Figure 3. FTIR spectra of SiO$_2$-Al$_2$O$_3$, ZSM-5 and HZSM-5 supports; Fe(3)-Cu(13)/SiO$_2$-Al$_2$O$_3$, Fe(3)-Cu(13)/ZSM-5 and Fe(3)-Cu(13)/HZSM catalysts.

2.5. XPS

The XPS spectra of the Fe(3)-Cu(13)/SiO$_2$-Al$_2$O$_3$, Fe(3)-Cu(13)/ZSM-5 and Fe(3)-Cu(13)/HZSM-5 catalyst samples were fitted for both Cu2p and Fe2p using the CasaXPS software as shown in Figure 4. The analysis confirmed the presence of iron and copper in their oxides and reduced states in all the three catalyst samples [26–35]. Cu2p$_{1/2}$ and Cu2p$_{3/2}$ XPS peaks were observed at 935–937 eV and 954–956 eV respectively, while Fe2p$_{1/2}$ and Fe2p$_{3/2}$ XPS peaks were observed at 721–723 eV and 708–713 eV respectively. The weak satellite peak at 945 eV binding energy on the XPS spectra of Fe(3)-Cu(13)/HZSM and Fe(3)-Cu(13)/ZSM catalysts confirms the presence of Cu$_2$O, while the strong satellite peak at the similar binding energy on the XPS spectra of Fe(3)-Cu(13)/SiO$_2$-Al$_2$O$_3$ can be attributed to Cu^{2+} phase [31]. The atomic compositions of the metal oxides and reduced metals obtained from the XPS spectra fitting were tabulated in Table 3. All the three catalyst samples consist of significant atomic composition of copper and iron in their oxide state, owing to the passivation of the surface of the catalysts during their synthesis. The results also show that all the catalyst samples consist of Fe$_2$O$_3$ in larger quantities as compared to FeO at the surface. The highest surface atomic composition of oxophilic iron metal was observed in the FeCu/SiO$_2$-Al$_2$O$_3$ catalyst as revealed from XPS fitting. The varying composition of metals and their oxides in all the catalysts can be ascribed to their different metal–support interactions.

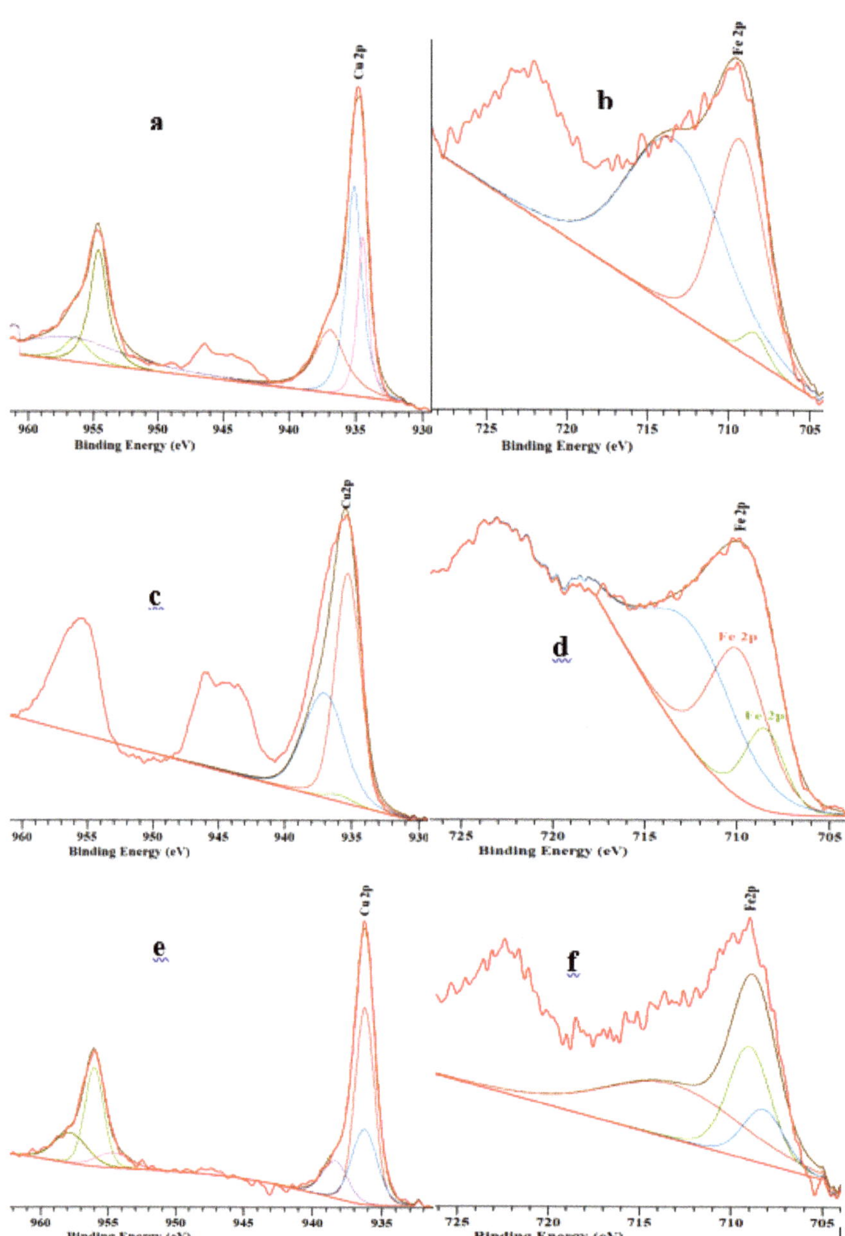

Figure 4. Cu2p and Fe2p X-ray Photoelectron Spectroscopy (XPS) spectra of (**a**,**b**) Fe(3)-Cu(13)/HZSM (**c**,**d**) Fe(3)-Cu(13)/SiO$_2$-Al$_2$O$_3$ (**e**,**f**) Fe(3)-Cu(13)/ZSM-5 catalysts.

Table 3. Cu2p$_{3/2}$ and Fe 2p$_{3/2}$ values for Fe(3)-Cu(13)/SiO$_2$-Al$_2$O$_3$, Fe(3)-Cu(13)/ZSM-5 and Fe(3)-Cu(13)/HZSM catalysts.

Catalysts	Component	Binding Energy of Cu or Fe 2p$_{3/2}$ (eV)	Atomic Composition (%)
Fe(3)-Cu(13)/HZSM	Cu$_2$O	935	24.3
	Cu	936	47.4
	CuO	937	28.3
	Fe	708	2.4
	FeO	709	38.1
	Fe$_2$O$_3$	713	58.5
Fe(3)-Cu(13)/SiO$_2$-Al$_2$O$_3$	Cu$_2$O	935	57.1
	Cu	936	1.7
	CuO	937	41.2
	Fe	708	25.9
	FeO	709	35.6
	Fe$_2$O$_3$	713	38.5
Fe(3)-Cu(13)/ZSM	Cu$_2$O	935	31.4
	Cu	936	52.9
	CuO	937	15.7
	Fe	708	15.5
	FeO	709	33.3
	Fe$_2$O$_3$	713	51.2

2.6. H$_2$-TPR Analysis

The H$_2$-TPR profiles of the Fe(3)-Cu(13)/SiO$_2$-Al$_2$O$_3$, Fe(3)-Cu(13)/HZSM-5 and Fe(3)-Cu(13)/ZSM-5 catalysts are shown in Figure 5. The profiles confirmed the findings from XRD and XPS analyses that despite reduction of the catalyst samples during synthesis, they still consist of oxides of Cu. The H$_2$ uptake from the TPR result is in the order of Fe(3)-Cu(13)/SiO$_2$-Al$_2$O$_3$ >Fe(3)-Cu(13)/ZSM-5 > Fe(3)-Cu(13)/HZSM-5 as shown in Table 4. The reduction peak temperature at 207 °C, 289 °C and 321 °C in the respective profiles of Fe(3)-Cu(13)/SiO$_2$-Al$_2$O$_3$, Fe(3)-Cu(13)/ZSM-5 and Fe(3)-Cu(13)/HZSM-5 are ascribed to the reduction of copper oxides to copper while the hump above 300 °C in all the three TPR profiles are attributed to reduction of iron oxides to iron [33]. The reduction peak temperature from the TPR profiles and the crystallite size of the catalysts obtained from CO chemisorption analysis are in the same order of Fe(3)-Cu(13)/SiO$_2$-Al$_2$O$_3$ < Fe(3)-Cu(13)/ZSM-5 < Fe(3)-Cu(13)/HZSM-5. Low reduction peak temperature and low crystallite size indicate weak metal–support interaction, therefore the Fe(3)-Cu(13)/SiO$_2$-Al$_2$O$_3$ catalyst has the weakest metal–support interaction and highest metal dispersion [34,35].

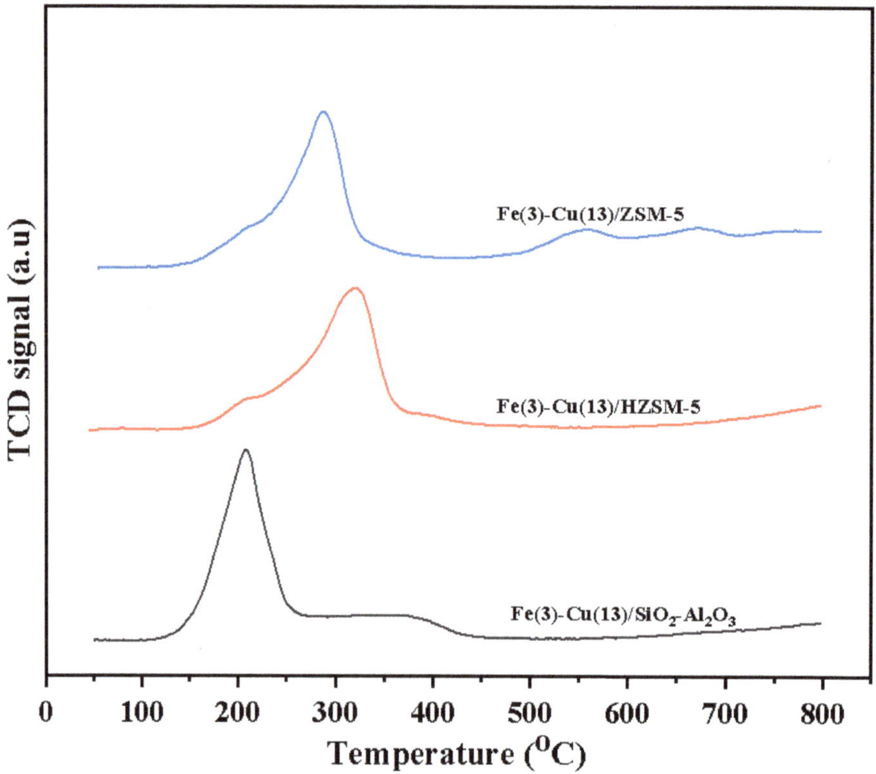

Figure 5. Temperature Programmed Reduction (TPR) profiles of Fe(3)-Cu(13)/SiO$_2$-Al$_2$O$_3$, Fe(3)-Cu(13)/ZSM and Fe(3)-Cu(13)/HZSM catalysts.

Table 4. H$_2$ uptake and reduction peak temperatures of catalyst samples.

Catalyst Samples	Reduction Peak Temperature (°C)	H$_2$ Uptake (mmol g^{-1} cat)
Fe(3)-Cu(13)/SiO$_2$-Al$_2$O$_3$	207	3.1
Fe(3)-Cu(13)/ZSM	289	2.6
Fe(3)-Cu(13)/HZSM	321	1.6

2.7. Pyridine FTIR Analysis

Pyridine, ammonia and acetonitrile can be used to determine the Brønsted and Lewis acid sites of the catalysts. In this work, pyridine was used as molecular probe and it shows a clear distinction between the Brønsted and Lewis acid sites. Moreover, the kinetic diameter of pyridine is 0.57 nm which is lower than the 2.1–5.7 nm pore size of the catalysts [36]. Catalysts of high Brønsted/Lewis acidity ratio favours cracking and is also not selective for dehydrogenation [37]. Cracking is desired for hydroprocessing of oleic acid for production of jet-fuel range hydrocarbons and dehydrogenation is undesirable. Pyridine FTIR spectra and Brønsted/Lewis acid sites ratio of Fe(3)-Cu(13)/SiO$_2$-Al$_2$O$_3$, Fe(3)-Cu(13)/ZSM-5 and Fe(3)-Cu(13)/HZSM-5 catalysts were shown in Figure S3 and Figure 6, respectively. The Brønsted/Lewis acid sites ratio increases in the order of Fe(3)-Cu(13)/SiO$_2$-Al$_2$O$_3$ < Fe(3)-Cu(13)/ZSM-5 < Fe(3)-Cu(13)/HZSM-5 as shown in Figure 6. Fe(3)-Cu(13)/HZSM-5 and Fe(3)-Cu(13)/ZSM-5 catalysts have higher Brønsted/Lewis acid sites ratio as compared to the Fe(3)-Cu(13)/SiO$_2$-Al$_2$O$_3$ catalyst. This indicates that the MFI zeolite catalysts are richer in Brønsted acid sites. Fe(3)-Cu(13)/HZSM-5 catalyst have higher Brønsted/Lewis acid sites ratio in comparison with Fe(3)-Cu(13)/ZSM-5 due to the protonation of the ZSM-5 support. Brønsted/Lewis sites ratio of the catalyst as shown in Figure 6 has significant influence on selectivity of jet-fuel range hydrocarbons as discussed in the later section.

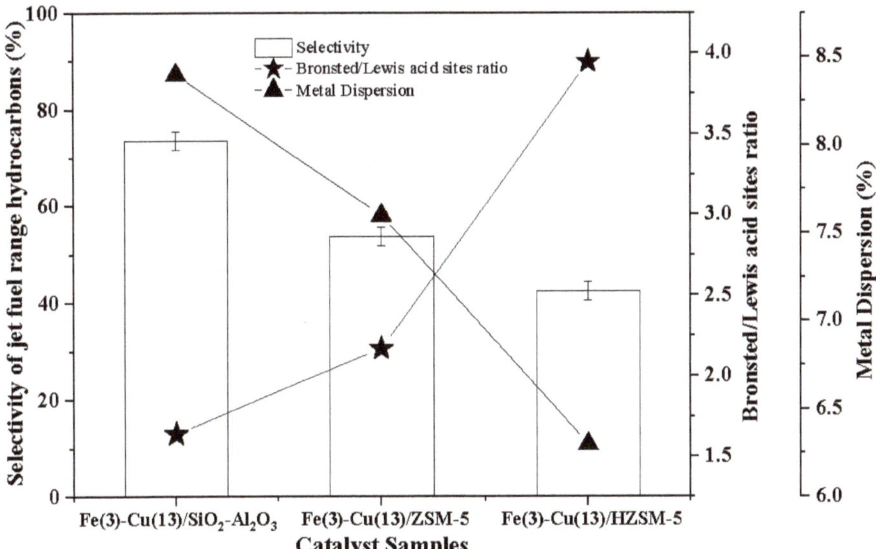

Figure 6. Selectivity of jet fuel hydrocarbons at 300 °C and 2.07 MPa H$_2$ pressure over the Brønsted/Lewis acid sites ratio and metal dispersion of Fe(3)-Cu(13)/SiO$_2$-Al$_2$O$_3$, Fe(3)-Cu(13)/ZSM-5 and Fe(3)-Cu(13)/HZSM-5 catalysts.

2.8. Catalyst Evaluation

The products obtained from the hydroprocessing reaction were analyzed based on oleic acid conversion, yield and selectivity of jet-fuel range hydrocarbons. The results obtained from the gas chromatography (GC) analysis show that reaction time, temperature and catalysts have significant effects on the conversion of oleic acid, yield and selectivity of the jet-fuel range hydrocarbons. The error % for all the result was within ± 5%. Figure 7 shows the results obtained from the evaluation of Cu(5)/SiO$_2$-Al$_2$O$_3$, Cu(10)/SiO$_2$-Al$_2$O$_3$, Cu(13)/SiO$_2$-Al$_2$O$_3$ and Cu(15)/SiO$_2$-Al$_2$O$_3$ catalysts at 340 °C, 2.07 MPa hydrogen pressure and 5% catalyst/feed ratio. The effects of reaction times (2–10 hours) on the conversion of oleic acid is shown in Figure 7. The conversion of oleic acid increases with time for all the four catalysts. The highest oleic acid conversion obtained was 78.4% at 10 hours from the evaluation of Cu(13)/SiO$_2$-Al$_2$O$_3$ catalyst. The effects of reaction time (2–10 hours) on the selectivity of jet-fuel range hydrocarbons are shown in Figure 7. The highest selectivity of jet-fuel range hydrocarbons obtained was 41.9% respectively at 6 hours. Unlike conversion of oleic acid, which increases with time, the selectivity of jet-fuel range hydrocarbons initially increases and later decreases with reaction time due to subsequent cracking of the jet-fuel range hydrocarbons to lighter hydrocarbons shown in Table S1. Oleic acid consists of macromolecules requiring catalyst of high pore size for easy internal diffusion. Cu(13)/SiO$_2$-Al$_2$O$_3$ performs better than the other three monometallic catalysts due to its high pore diameter (5.3 nm), which implies high accessibility of oleic acid and hydrogen to the active site (copper) of the catalyst.

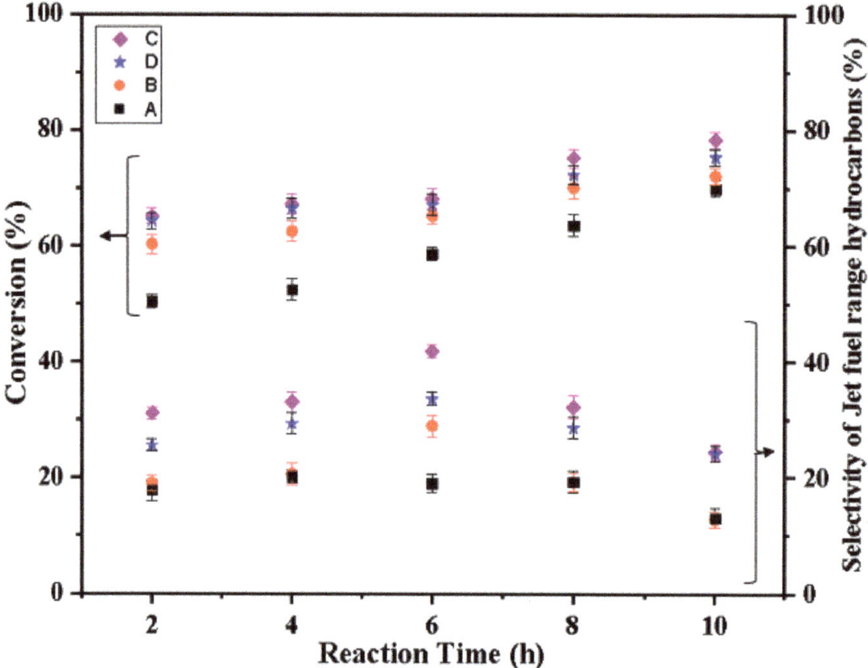

Figure 7. Conversion of oleic acid and selectivity of jet fuel hydrocarbons at different temperatures and 2.07 MPa H_2 pressure over the (A) Cu(5)/SiO_2-Al_2O_3, (B) Cu(10)/SiO_2-Al_2O_3, (C) Cu(13)/SiO_2-Al_2O_3 and (D) Cu(15)/SiO_2-Al_2O_3 catalysts.

The effects of reaction time (2–10 hours) on oleic acid conversion and selectivity of jet-fuel range hydrocarbons over the FeCu/SiO_2-Al_2O_3 catalysts are shown in Figure 8. The conversion of oleic acid increases with time for all the catalysts. The highest oleic acid conversion obtained was 98% at 10 hours from the evaluation of Fe(3)-Cu(13)/SiO_2-Al_2O_3 catalyst. The highest selectivity of jet-fuel range hydrocarbons obtained was 53.8% at 10 hours. The better performance of Fe(3)-Cu(13)/SiO_2-Al_2O_3 catalyst is due to its smaller crystallite size and high dispersion of copper and iron metals and high reducibility of metals as compared to the other three catalysts.

The effects of reaction times (2–10 h) and temperature (300–340 °C) on oleic acid conversion, yield and selectivity of jet-fuel range hydrocarbons over the Fe(3)-Cu(13)/SiO_2-Al_2O_3, Fe(3)-Cu(13)/HZSM-5 and Fe(3)-Cu(13)/ZSM-5 catalysts are shown in Figures 9–11 and Table 5. The product distribution of C_8–C_{16} is also shown in Table S2. The conversion of oleic acid increases with reaction time and temperature for all the three catalyst samples. The conversion approximately stabilizes between 8 to 10 h for Fe(3)-Cu(13)/SiO_2-Al_2O_3 catalyst at 320–340 °C, Fe(3)-Cu(13)/ZSM catalyst at 300–340 °C and, Fe(3)-Cu(13)/HZSM catalyst at 320–340 °C.

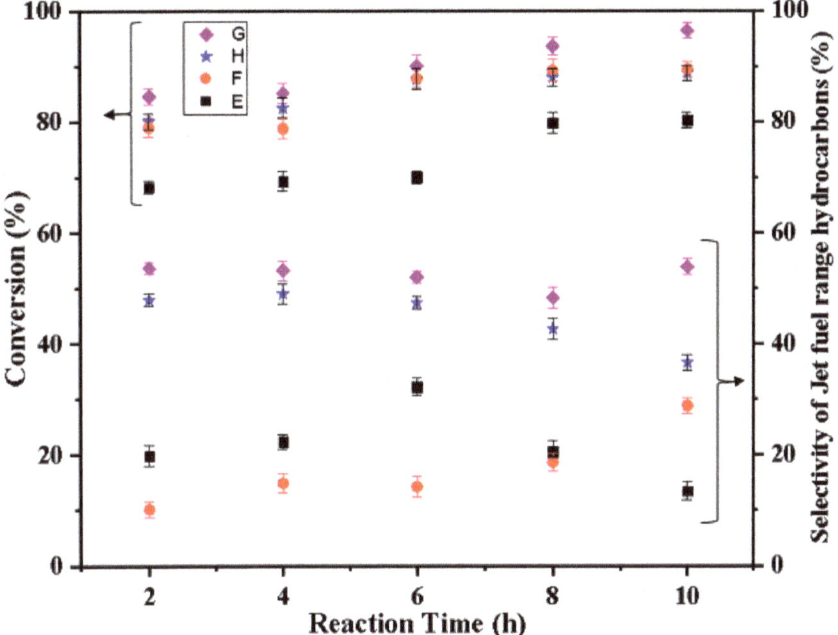

Figure 8. Conversion of oleic acid and selectivity of jet fuel hydrocarbons at different temperatures and 2.07 MPa H_2 pressure over the (E) Fe(1)-Cu(13)/SiO$_2$-Al$_2$O$_3$, (F) Fe(2)-Cu(13)/SiO$_2$-Al$_2$O$_3$, (G) Fe(3)-Cu(13)/SiO$_2$-Al$_2$O$_3$ and (H) Fe(5)-Cu(13)/SiO$_2$-Al$_2$O$_3$ catalysts.

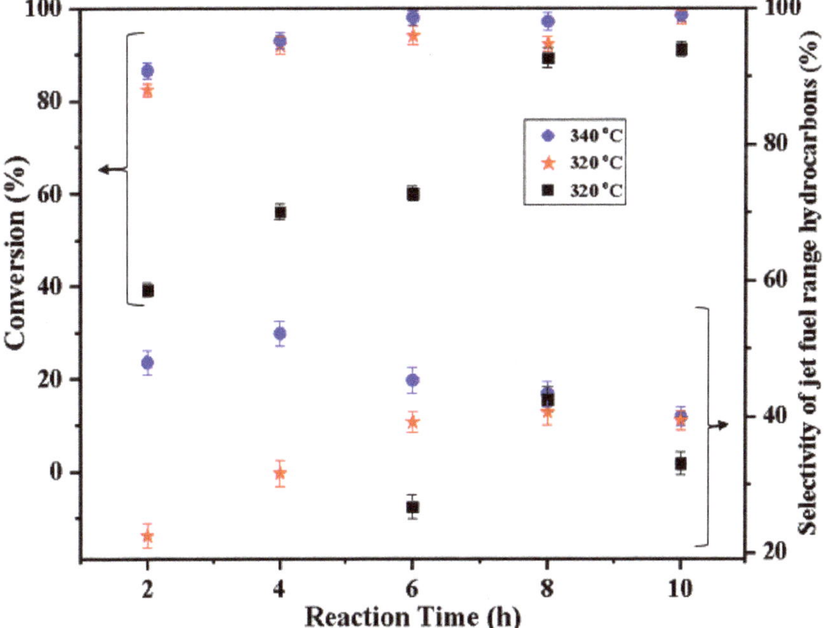

Figure 9. Conversion of oleic acid and selectivity of jet fuel hydrocarbons at different temperatures and 2.07 MPa H_2 pressure over the Fe(3)-Cu(13)/HZSM-5 catalyst.

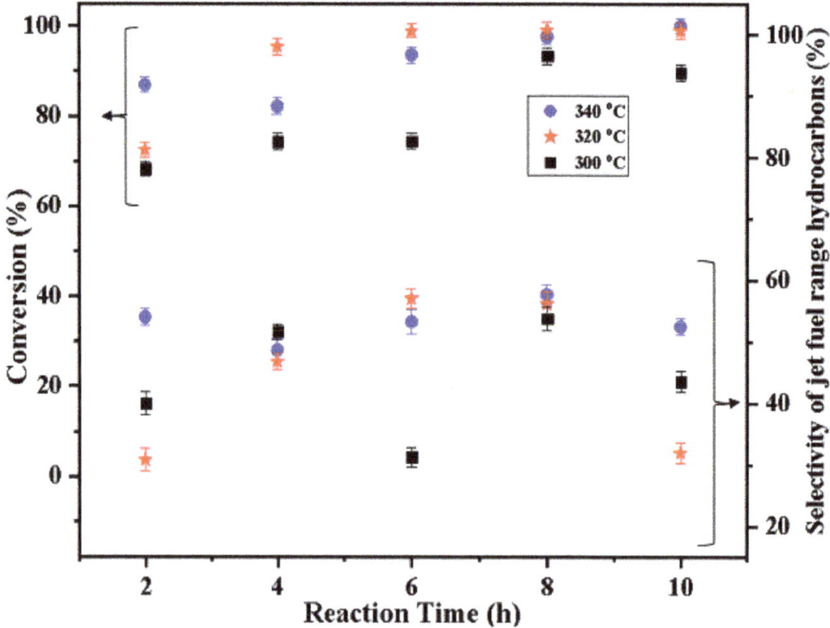

Figure 10. Conversion of oleic acid and selectivity of jet fuel hydrocarbons at different temperatures and 2.07 MPa H$_2$ pressure over the Fe(3)-Cu(13)/ZSM-5 catalyst.

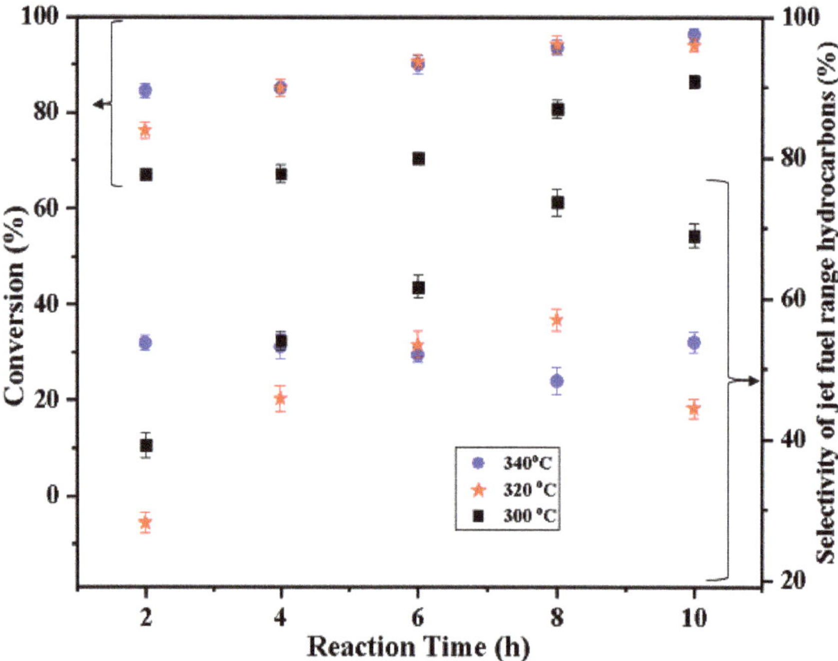

Figure 11. Conversion of oleic acid and selectivity of jet fuel hydrocarbons at different temperatures and P$_{H2}$: 2.07 MPa H$_2$ pressure over the Fe(3)-Cu(13)/SiO$_2$-Al$_2$O$_3$ catalyst.

Table 5. Yield of C_8–C_{16} hydrocarbons at t: 2–10 hours; T: 300–340 °C, and P_{H2}: 2.07 MPa H_2 pressure.

Catalyst	Temperature (°C)	Residence Time (h)				
		2	**4**	**6**	**8**	**10**
Fe(3)-Cu(13)/SiO$_2$-Al$_2$O$_3$	300	26.3	36.3	43.5	59.5	59.7
	320	21.5	39.0	48.4	53.7	41.8
	340	45.4	45.3	45.2	45.2	51.8
Fe(3)-Cu(13)/ ZSM	300	27.3	38.4	23.3	50.2	39.1
	320	22.3	44.6	56.4	55.7	32.4
	340	47.0	40.0	49.9	56.4	52.5
Fe(3)-Cu(13)/ HZSM-5	300	4.6	8.5	14.7	38.0	30.1
	320	19.2	29.5	38.3	39.5	38.9
	340	39.6	48.0	42.64	40.0	39.2

The highest yield and selectivity of jet fuel hydrocarbons were recorded for Fe(3)-Cu(13)/ SiO$_2$-Al$_2$O$_3$ catalyst at 300 °C as shown in Figure 11 and Table 5. The total number of surface atoms of Cu and Fe per the total number of atoms present in the catalyst increases with surface area. Catalysts of high Brønsted/Lewis acid site ratio favours cracking, while oligomerization and dehydrogenation are favoured by catalysts of low Brønsted/Lewis acid sites ratio [37]. The internal diffusion calculation was carried out using the Weisz–Prater criterion [38]. The Weisz–Prater parameter for this reaction was 3.6×10^{-7}, which is much less than 1 indicating that the internal diffusion in catalyst particles is absent. The experimental data show that with reaction time, more lighter hydrocarbons (C_5–C_7) were produced from jet-fuel range hydrocarbons (C_8–C_{16}) due to deep and mild cracking. This may be due to acidic nature of the catalysts used in this study. Hydroprocessing reactions of fatty acids, triglycerides and vegetable oils require catalysts of mild Brønsted acid sites to produce jet-fuel range hydrocarbons. Catalysts of high Brønsted sites have higher tendency of producing hydrocarbons that are lighter than the jet-fuel range hydrocarbons [8]. Catalysts of low Brønsted acid sites favor only the hydrodeoxygenation reaction with no sufficient cracking strength. The Fe(3)-Cu(13)/SiO$_2$-Al$_2$O$_3$ catalyst has mild Brønsted/Lewis acid sites ratio as proven from Figure 6. The relationship between the metal dispersion, Brønsted/Lewis acid sites ratio and selectivity of jet-fuel range hydrocarbons at 300 °C and 2.07 MPa H_2 pressure over the Fe(3)-Cu(13)/SiO$_2$-Al$_2$O$_3$, Fe(3)-Cu(13)/ZSM-5 and Fe(3)-Cu(13)/HZSM-5 catalysts are shown in Figure 6. Fe(3)-Cu(13)/HZSM-5 catalyst has the lowest metal dispersion and Brønsted/Lewis acid sites ratio and the lowest selectivity of jet-fuel range hydrocarbons, while the Fe(3)-Cu(13)/SiO$_2$-Al$_2$O$_3$ catalyst has the highest metal dispersion and Brønsted/Lewis acid sites ratio and the highest selectivity of jet-fuel range hydrocarbons. The productivity of Fe(3)-Cu(13)/SiO$_2$-Al$_2$O$_3$, Fe(3)-Cu(13)/ZSM-5 and Fe(3)-Cu(13)/HZSM-5 catalysts in terms of g jet fuel/g catalyst/h and g jet fuel/m^2 metals surface area/g catalyst are shown in Table S3. The Fe(3)-Cu(13)/SiO$_2$-Al$_2$O$_3$ catalyst is the most productive catalyst, with 1.0 g jet fuel/g catalyst/h and 2.6 g jet fuel/m^2 metals surface area/g catalyst. The Fe(3)-Cu(13)/SiO$_2$-Al$_2$O$_3$ catalyst is the most promising when the entire temperature range(300–340 °C) is considered. This can be attributed to the relatively low Brønsted/Lewis acid sites ratio [8], high Cu and Fe metals dispersion, high pore volume, specific surface area of the Fe(3)-Cu(13)/SiO$_2$-Al$_2$O$_3$ catalyst and high surface composition of oxophilic iron metal observed from XPS fitting.

3. Experimental Methods

3.1. Catalyst Synthesis

Reduced Cu/SiO$_2$-Al$_2$O$_3$, FeCu/SiO$_2$-Al$_2$O$_3$, FeCu/ZSM-5 and FeCu/HZSM-5 bimetallic catalysts were synthesized using diffusional impregnation and co-impregnation techniques.

3.1.1. Chemical and Materials

Oleic acid (code number: 364525) and amorphous silica–alumina catalyst support (SiO_2/Al_2O_3: 6.25, code number: 343358) were procured from Sigma-Aldrich (Missouri, U.S.A). Copper (II) nitrate hemipentahydrate (code number: 467855) and iron nitrate nonahydrate (code number: 254223) were purchased from ES company (New Jersey, U.S.A) and Millipore (Canada) Ltd (Ontario, Canada) respectively. ZSM-5 zeolite (code number: CB2314, SiO_2/Al_2O_3: 23) was supplied by zeolyst international (Kansas, U.S.A).

3.1.2. Support Preparation

HZSM support (SiO_2/Al_2O_3: 23) was prepared by exchanging ZSM-5 with a 1.0 M ammonium nitrate solution at 100 °C for 3 h followed by calcination in air at 550 °C for 4 h [39]. Two commercial supports (amorphous silica alumina and ZSM-5) were used alongside with the synthesized HZSM.

3.1.3. Catalyst Preparation by Impregnation

Monometallic 5, 10, 13 and 15 wt% Cu/SiO_2-Al_2O_3 catalysts were synthesized using copper (II) nitrate hemipentahydrate as precursor for copper. The solution of the precursor was impregnated on amorphous silica alumina using diffusional impregnation technique and then the mixture was dried overnight at 100 °C in a drying oven for each of the four catalyst samples. For instance, 15 g of 13 wt% Cu/SiO_2-Al_2O_3 was synthesized by impregnating 7.1 g of copper (II) nitrate hemipentahydrate on 13.1 g of amorphous silica alumina. The dried mixture was calcined in air in a muffle furnace at 450 °C for 6 hours at 1 °C /min heating rate, reduced under 50 mL/min H_2 flow at 400 °C for 4 hours at 1 °C/min heating rate and passivated with 1% O_2/N_2 flowing at 50 mL/min for 1 hour ambient temperature. Thirteen wt% Cu/SiO_2-Al_2O_3 was selected from the four catalyst samples for iron promotional effect studies.

Bimetallic Fe(1)-Cu(13)/SiO_2-Al_2O_3, Fe(2)-Cu(13)/SiO_2-Al_2O_3, Fe(3)-Cu(13)/SiO_2-Al_2O_3, Fe(1)-Cu(13)/ SiO_2-Al_2O_3 catalysts were synthesized using copper (II) nitrate hemipentahydrate and iron nitrate nonahydrate as precursors for copper and iron respectively. Copper (II) nitrate hemipentahydrate solution and iron nitrate nonahydrate solution was first impregnated on amorphous silica alumina using diffusional co-impregnation method and then the mixture was dried overnight at 100 °C. For instance, 15 g of Fe(3)-Cu(13)/SiO_2-Al_2O_3 was synthesized by co-impregnating 7.1 g of copper (II) nitrate hemipentahydrate and 3.3 g of iron nitrate nonahydrate solution on 12.6 g of amorphous silica alumina. The dried co-impregnated mixture was calcined in air in a muffle furnace at 450 °C for 6 hours by heating at 1 °C/min, reduced under 50 mL/min H_2 flow at 400 °C for 4 hours at 1 °C/min heating rate and passivated with 1% O_2/N_2 flowing at 50 mL/min for 1 hour ambient temperature. Fe(3)-Cu(13)/SiO_2-Al_2O_3 catalyst was selected as the catalyst from the four bimetallic catalysts for supports optimization studies.

For effective support optimization, Fe(3)-Cu(13)/ZSM-5 and Fe(3)-Cu(13)/HZSM-5 catalysts were also synthesized using the same procedure as Fe(3)-Cu(13)/SiO_2-Al_2O_3 catalyst.

3.2. Catalyst Characterization

3.2.1. ICP-OES

The mass compositions of copper and iron in Fe(3)-Cu(13)/SiO_2-Al_2O_3, Fe(3)-Cu(13)/ZSM-5 and Fe(3)-Cu(13)/HZSM-5 catalysts were evaluated using a concentrated mixture of $HF/HNO_3/HClO_3$ to digest 0.125 g of each catalyst samples to dryness in a Teflon tube to analyze copper and also utilizing Ox automated fusion instrument to fuse the combination of Lithium metaborate and Lithium tetraborate mixture with 0.1 g of each catalyst samples in a graphite crucible for analyze iron. Dilute HNO_3 was then used to dissolve the dry residue and fused product obtained respectively from the copper and iron analyses and they were analyzed with Perkin Elmer ICP-OES (Optima 5300 DV) in the geoanalytical laboratory of the Saskatchewan Research Council.

3.2.2. N_2-Adsorption/Desorption Measurement

Micrometrics ASAP 2020 instrument was used to characterize all the catalyst samples and their respective supports with the BET method. Each catalyst sample was degassed in a sealed tube in vacuum conditions at 250 °C for 5 h and evacuated until a static pressure of less than 1.33 Pa was obtained. Physisorption analysis was then carried out with N_2 at −196 °C.

3.2.3. XRD

X-ray diffraction patterns of all the catalyst samples and their respective supports were obtained with a monochromatic Cu Kα radiation source of 0.15406 nm wavelength using an Ultima IV instrument from Rigaku Instruments. The normal scan rate of the equipment is 2° scan/min within 10–80° diffraction angle and 0.02 step size. The average crystallite size of the intense XRD peaks obtained for all the catalyst samples was calculated using Equation (4).

$$L = \frac{k\lambda}{\beta \cos \theta} \tag{1}$$

k = constant; λ = wavelength of the source of X-ray; β = Full width at half maxima; θ = half of the Bragg angle.

3.2.4. XPS

XPS was used to study the phases of the active sites (Copper and Iron) at the surface of $Fe(3)$-$Cu(13)/SiO_2$-Al_2O_3, $Fe(3)$-$Cu(13)/ZSM$-5 and $Fe(3)$-$Cu(13)/HZSM$-5 catalyst samples. Kratos (Manchester city, United Kingdom) AXIS Supra system equipped Rowland circle monochromated source of 1486.6 eV Al K-α was used for XPS measurements for all three synthesized catalyst samples. Three high-resolution regional scans were carried out using 0.05 eV steps with 20 eV pass energy. An accelerating voltage of 15,000 eV and an emission current of 0.015 A were used for the analysis.

3.2.5. FTIR Spectroscopy.

A JASCO FT-IR 4100 instrument was used to identify the functional groups of the all the synthesized catalyst samples and their respective supports. For analysis, 3 mg of sample was uniformly mixed with 0.4 g of KBr pellets. Qualitative analysis of the functional groups of the catalyst samples were obtained with 32 scans of 4 cm^{-1} nominal resolution. The IR spectra of pelletized samples were later recorded in transmission mode in the wavenumber range of 400–1400 cm^{-1}.

3.2.6. CO Chemisorption

The metal dispersion and crystallite size of $Fe(3)$-$Cu(13)/SiO_2$-Al_2O_3, $Fe(3)$-$Cu(13)/ZSM$-5 and $Fe(3)$-$Cu(13)/HZSM$-5 catalyst samples were measured using the Micrometrics ASAP 2020 chemisorption system. The catalyst samples were heated to 350 °C at 10 °C/min ramp rate in the presence of H_2. They were then held for 2 h and later cooled down to 35 °C and evaluated to a static pressure below 1.3×10^{-5} N/m^2. Pulses of CO were passed over the evacuated sample and the total CO uptake was measured at 35 °C. Stoichiometric factor of 0.5 mole of CO per metal atom was used for copper and iron.

3.2.7. Temperature Programmed Reduction (TPR)

H_2-TPR of $Fe(3)$-$Cu(13)/SiO_2$-Al_2O_3, $Fe(3)$-$Cu(13)/ZSM$-5 and $Fe(3)$-$Cu(13)/HZSM$-5 catalyst samples were conducted at 101.3 KPa using Micrometrics Auto Chem II 2920 analyzer. 10% H_2/Ar was circulated to 0.05 g of each sample in a steel tube at 50 cm^3/min and the temperature was increased to 850 °C from ambient temperature at 10 °C/min. As the temperature increases, the reaction of the catalyst and hydrogen gas proceeds to produce water vapour, which was trapped through a cold trap by outlet stream circulation. The exit gas stream was channeled via a calibrated thermal conductivity detector (TCD) for the detection of varying H_2 concentrations due to catalyst reduction.

3.2.8. Pyridine FTIR

The FTIR technique was utilized to study the nature of acid sites of Fe(3)-Cu(13)/SiO_2-Al_2O_3, Fe(3)-Cu(13)/ZSM-5 and Fe(3)-Cu(13)/HZSM-5 catalyst samples and their respective supports using a wavenumber region of pyridine (1400–1700 cm^{-1}). A sample cylindrical cup in a Spectrotech diffuse reflectance in situ cell equipped with a thermocouple and zinc selenide windows was loaded with 0.01 g of each catalyst sample. These three catalyst samples and their respective supports were pretreated at 350 °C in order to remove any adsorbed water on the catalyst surface. Pyridine vapor was then passed over each catalyst sample at 100 °C for 1 hour to obtain pyridine chemisorbed samples. After adsorption of pyridine, nitrogen gas was used for the stabilization of the catalyst samples at 100 °C for 30 min with a ramping rate of 5 °C/min, then allowed to cool to ambient temperature. The samples were analyzed with a JASCO FT-IR 4100 instrument in the wavenumber range of 1400–1700 cm^{-1} and their respective IR spectra were recorded. Brønsted/Lewis acid sites ratio of all the three catalyst samples were calculated using Equation (5):

$$\frac{C_B}{C_L} = \frac{IMEC(B)^{-1}}{IMEC(L)^{-1}} \cdot \frac{IT(B)}{IT(L)} \tag{2}$$

where, $\frac{C_B}{C_L}$ = ratio of concentration of Bronsted and Lewis acid sites. IMEC(B) and IMEC(L) are integrated molar extinction coefficients (cm/μmol) of Brønsted and Lewis acid sites, respectively. IT(B) and IT(L) are integrated transmittances of Brønsted and Lewis acid sites, respectively [40,41].

3.3. Catalyst Evaluation

The catalytic reactions were carried out in a Parr stirred batch reactor. It is made in a bench top with moveable vessel mounting style. The capacity of the reactor vessel is 300 mL, with dimension of 2.5 inches diameter and 4 inches depth and the reactor is connected to a Parr 4848 reactor controller. The catalyst samples were evaluated in this Parr batch reactor. Two g of each of the catalyst samples and 40 g of oleic acid were placed in the reactor and hydrogen gas was used to pressurize the reactor to 2.07 MPa. Cu/SiO_2-Al_2O_3 and FeCu/SiO_2-Al_2O_3 catalysts were evaluated at 340 °C. The temperatures of the reaction involving support optimization studies were set at 300, 320 and 340 °C. The impeller speed and reaction time were 500 rpm and 10 h, respectively. The liquid product samples were collected at 2 h interval, filtered and diluted using chloroform as the diluent. The solution of liquid product samples was analyzed with GC (Agilent 7890A) equipped with a flame ionization detector (FID). A 30 m long DB-5 capillary column with 0.25 mm inner diameter was used. The temperature of the oven was programmed to start from 40 °C for 2 min and increased to 280 °C at 10 °C/min ramping rate with a 5 min final hold time. One μL of each product sample was injected with a split ratio of 10:1 into the column. C_6–C_{20} aliphatic hydrocarbons were used as external standard to quantify the liquid hydrocarbons produced. The gaseous products were analyzed using an online GC equipped with FID and catalyst performance was evaluated based on oleic acid conversion and jet-fuel range hydrocarbons (C_8–C_{16}) selectivity as given below.

$$\text{Oleic acid conversion (\%)} = \left(\frac{\text{amount of oleic acid reacted}}{\text{amount of oleic acid initially taken}}\right) \times 100 \tag{3}$$

$$\text{Selectivity of jet} - \text{fuel range hydrocarbons (\%)} = \left(\frac{\text{amount of jet} - \text{fuel range hydrocarbons}}{\text{amount of products formed}}\right) \times 100 \tag{4}$$

$$\text{Yield of jet} - \text{fuel range hydrocarbons } (\%) = \left(\frac{\text{amount of jet} - \text{fuel range hydrocarbons}}{\text{amount of oleic acid initially taken}} \right) \times 100 \quad (5)$$

4. Conclusions

In summary, copper metal with optimized loading on silica alumina support was suitable for hydroprocessing of oleic acid for production of jet-fuel range hydrocarbons. The best monometallic catalyst performance with 41.9% selectivity of jet-fuel range hydrocarbons (C_8–C_{16}) was achieved at 340 °C, 2.07 MPa H_2 pressure and 6 hours reaction time over the catalyst with the largest pore size of 5.3 nm ($Cu(13)/SiO_2$-Al_2O_3).

Optimization studies of iron promotional effects for hydroprocessing of oleic acid for production of jet-fuel range hydrocarbons on the $Cu(13)/SiO_2$-Al_2O_3 catalyst showed more promising result in comparison with the monometallic copper catalysts due to effect of iron loading in lowering metal crystallite size indicating increase in metal dispersion. The best catalyst performance with 51.8% yield and 53.8% selectivity of jet-fuel range hydrocarbons was achieved over the iron-promoted copper catalyst with the lowest crystallite size ($Fe(3)$-$Cu(13)/ SiO_2$-Al_2O_3) at 340 °C, 2.07 MPa and 10 h of reaction time.

Support optimization studies on HZSM-5, ZSM-5 and SiO_2-Al_2O_3 supports reveal that $Fe(3)$-$Cu(13)/ SiO_2$-Al_2O_3 catalyst gives the best catalyst performance with 59.5% yield and 73.6% selectivity of jet-fuel range hydrocarbons. This promising performance was attributed to its large pore diameter, large pore volume and large surface area; low crystallite size and weak metal–support interaction from H_2-TPR analysis, indicating, high metal dispersion from CO chemisorption analysis, high oxophilic iron content from XPS fitting and mild Brønsted acid sites from pyridine FTIR analysis.

Supplementary Materials: The following are available online at http://www.mdpi.com/2073-4344/9/12/1051/s1, Figure S1. N_2-adsorption isotherms of SiO_2-Al_2O_3, $Cu(5)/SiO_2$-Al_2O_3, $Cu(10)/SiO_2$-Al_2O_3, $Cu(13)/SiO_2$-Al_2O_3 and $Cu(15)/SiO_2$-Al_2O_3, $Fe(1)$-$Cu(13)/SiO_2$-Al_2O_3, $Fe(2)$-$Cu(13)/SiO_2$-Al_2O_3, $Fe(3)$-$Cu(13)/SiO_2$-Al_2O_3 and $Fe(5)$-$Cu(13)/SiO_2$-Al_2O_3 catalysts. Figure S2. N_2-adsorption isotherms of SiO_2-Al_2O_3, ZSM-5 and HZSM-5 supports; $Fe(3)$-$Cu(13)/SiO_2$-Al_2O_3, $Fe(3)$-$Cu(13)/ZSM$-5 and $Fe(3)$-$Cu(13)/HZSM$ catalysts. Figure S3. Py-FTIR spectra of SiO_2-Al_2O_3, ZSM-5 and HZSM-5 supports; $Fe(3)$-$Cu(13)/SiO_2$-Al_2O_3, $Fe(3)$-$Cu(13)/ZSM$-5 and $Fe(3)$-$Cu(13)/HZSM$ catalysts. Table S1: Selectivity of lighter hydrocarbons at t: 8 h; T: 300 °C, and P_{H2}: 2.07 MPa H_2 pressure. Table S2: Selectivity of C_8–C_{16} hydrocarbons at t: 8 h; T: 300 °C, and P_{H2}: 2.07 MPa H_2 pressure. Table S3: Catalysts productivity towards C_8–C_{16} hydrocarbons at t: 8 hours; T: 300–340 °C, and P_{H2}: 2.07 MPa H_2 pressure.

Author Contributions: Conceptualization, A.A.A.; writing—original draft preparation, A.A.A.; writing—review and editing, A.A.A., P.E.B. and A.K.D.; methodology, A.A.A., A.K.D., Y.H. and P.E.B.; formal analysis, A.A.A.; investigation, A.A.A. and P.E.B.; resources, A.K.D.; data curation, A.A.A.; supervision, A.K.D. and Y.K.D.; project administration, A.K.D.; funding acquisition, A.K.D.

Funding: This Research was funded by Canada Research Chairs (CRC) and Natural Sciences and Engineering Research Council of Canada (NSERC).

Acknowledgments: The authors acknowledge Canada Research Chair (CRC) Program and Natural Sciences and Engineering Research Council of Canada (NSERC) for funding this research.

Conflicts of Interest: The authors declare no conflicts of interest.

References

1. Wang, W.; Tao, L. Bio-jet fuel conversion technologies. Renew. *Sust. Energ. Rev.* **2016**, *53*, 801–822. [CrossRef]

2. Kieckhäfer, K.; Quante, G.; Müller, C.; Spengler, T.S.; Lossau, M.; Jonas, W. Simulation-Based Analysis of the Potential of Alternative Fuels towards Reducing CO_2 Emissions from Aviation. Simulation-Based Analysis of the Potential of Alternative Fuels towards Reducing CO_2 Emissions from Aviation. *Energies* **2018**, *11*, 186. [CrossRef]

3. Bwapwaa, J.K.; Anandraj, A.; Trois, C. Possibilities for conversion of microalgae oil into aviation fuel: A review. *Renew. Sust. Energ. Rev.* **2017**, *80*, 1345–1354. [CrossRef]

4. Sgouridis, S.; Bonnefoy, P.A.; Hansman, R.J. Air transportation in a carbon constrained world: Long-term dynamics of policies and strategies for mitigating the carbon footprint of commercial aviation. *Transp. Res. Part A* **2011**, *45*, 1077–1091. [CrossRef]

5. Diederichs, G.W.; Mandegari, M.A.; Farzad, S.; Görgens, J.F. Techno-economic comparison of biojet fuel production from lignocellulose, vegetable oil and sugar cane juice. *Bioresour. Technol.* **2016**, *216*, 331–339. [CrossRef] [PubMed]

6. Bykova, M.V.; Yu, D.; Ermakov, V.V.; Kaichev, O.A.; Bulavchenko, A.A.; Saraev, M.; Yu, L.; Yakovlev, V.A. Ni-based sol–gel catalysts as promising systems for crude bio-oil upgrading: Guaiacol hydrodeoxygenation study. *Appl. Catal. B Environ.* **2012**, *113*, 296–307. [CrossRef]

7. Kandel, K.; Anderegg, J.W.; Nelson, N.C.; Chaudhary, U.; Slowing, I.I. Supported iron nanoparticles for the hydrodeoxygenation of microalgal oil to green diesel. *J. Catal.* **2014**, *314*, 142–148. [CrossRef]

8. Rabaev, M.; Landau, M.V.; Vidruk-Nehemya, R.; Koukouliev, V.; Zarchin, R.; Herskowitz, M. Conversion of vegetable oils on Pt/Al$_2$O$_3$/SAPO-11 to diesel and jet fuels containing aromatics. *Fuel* **2015**, *161*, 287–294. [CrossRef]

9. Deepak, V.; Rohit, K.; Bharat, S.R.; Anil, K.S. Aviation fuel production from lipids by a single-step route using hierarchical mesoporous zeolites. *Energ. Environ. Sci.* **2011**, *4*, 1667. [CrossRef]

10. Verma, D.; Rana, B.S.; Sibi, M.G.; Sinha, A.K. Diesel and aviation kerosene with desired aromatics from hydroprocessing of jatropha oil over hydrogenation catalysts supported on hierarchical mesoporous SAPO-11. *Appl. Catal. A Gen.* **2015**, *490*, 108–116. [CrossRef]

11. Bennici, S.; Gervasini, A.; Ravasio, N.; Zaccheria, F. Optimization of Tailoring of CuO*x* Species of Silica Alumina Supported Catalysts for the Selective Catalytic Reduction of NO*x*. *J. Phys. Chem. B* **2003**, *107*, 5168–5176. [CrossRef]

12. Kazuhisa, M.; Yanyong, L.; Megumu, I.; Isao, T. Production of Synthetic Diesel by Hydrotreatment of Jatropha Oils Using Pt-Re/H-ZSM-5 Catalyst. *Energy Fuel* **2010**, *24*, 2404–2409.

13. Yanyong, L.; Rogelio, S.; Kazuhisa, M.; Tomoaki, M.; Kinya, S. Hydrotreatment of Vegetable Oils to Produce Bio-Hydrogenated Diesel and Liquefied Petroleum Gas Fuel over Catalysts Containing Sulfided NiMo and Solid Acids. *Energy Fuel* **2011**, *25*, 4675–4685.

14. Ali, M.A.; Tatsumi, T.; Masuda, T. Development of heavy oil hydrocracking catalysts using amorphous silica-alumina and zeolites as catalyst supports. *Appl. Catal. A Gen.* **2002**, *233*, 77–90. [CrossRef]

15. Leofanti, G.; Padovan, M.; Tozzola, G.; Venturelli, B. Surface area and pore texture of catalysts. *Cat. Today* **1998**, *41*, 207–219. [CrossRef]

16. Toyama, N.; Inoue, N.; Ohki, S.; Tansho, M.; Shimizu, T.; Umegaki, T.; Kojima, Y. Influence of hollow silica-alumina composite spheres prepared using various amount of L(+)-arginine on their activity for hydrolytic dehydrogenation of ammonia borane. *Adv. Mater. Lett.* **2016**, *7*, 339–343. [CrossRef]

17. Saber, O.; Gobara, H.M. Optimization of silica content in alumina-silica nanocomposites to achieve high catalytic dehydrogenation activity of supported Pt catalyst. *Egypt. J. Petrol.* **2014**, *23*, 445–454. [CrossRef]

18. Shalaby, N.H.; Elsalamony, R.A.; El Naggar, A.M.A. Mesoporous waste-extracted SiO$_2$–Al$_2$O$_3$-supported Ni and Ni–H$_3$PW$_{12}$O$_{40}$ nano-catalysts for photo-degradation of methyl orange dye under UV irradiation. *New J. Chem.* **2018**, *42*, 9177–9186. [CrossRef]

19. Morettia, G.; Dossib, C.; Fusib, A.; Recchiab, S.; Psarob, R. A comparison between Cu-ZSM-5, Cu/S-1 and Cu/mesoporous silica alumina as catalysts for NO decomposition. *Appl. Catal. B Environ.* **1999**, *20*, 67–73. [CrossRef]

20. Ishihara, A.; Negura, H.; Hashimoto, T.; Nasu, H. Catalytic properties of amorphous silica-alumina prepared using malic acid as a matrix in catalytic cracking of n-dodecane. *Appl. Catal. A Gen.* **2010**, *388*, 68–76. [CrossRef]

21. Chen, L.; Chen, C.; Liang, K.; Chang, S.H.; Tseng, Z.; Yeh, S.; Chen, C.; Wu, W.; Wu, C. Nano-structured CuO-Cu$_2$O Complex Thin Film for Application in CH$_3$NH$_3$PbI$_3$ Perovskite Solar Cells. *Nanoscale Res. Lett.* **2016**, *11*, 402. [CrossRef] [PubMed]

22. Tao, Y.; Kanoh, H.; Kaneko, K. ZSM-5 Monolith of Uniform Mesoporous Channels. *J. Am. Chem. Soc.* **2003**, *125*, 6044–6045. [CrossRef] [PubMed]

23. Sanchez, M.Z.; Mauricio, J.E.; Paredes, A.R.; Gamero, P.; Cortés, D. Antimicrobial properties of ZSM-5 type zeolite functionalized with silver. *Mater. Lett.* **2017**, *191*, 65–68. [CrossRef]

24. Betancourt-Galindo, R.; Reyes-Rodriguez, P.Y.; Puente-Urbina, B.A.; Avila-Orta, C.A.; Rodríguez-Fernández, O.S.; Cadenas-Pliego, G.; García-Cerda, L.A. Synthesis of Copper Nanoparticles by Thermal Decomposition and Their Antimicrobial Properties. *J. Nanomater.* **2014**, *2014*, 1–5. [CrossRef]

25. Hosseinpour, M.; Golzary, A.; Saber, M.; Yoshikawa, K. Denitrogenation of biocrude oil from algal biomass in high temperature water and formic acid mixture over HZSM-5 nanocatalyst. *Fuel* **2017**, *206*, 628–637. [CrossRef]

26. Biesinger, M.C.; Payne, B.P.; Grosvenor, A.P.; Laua, L.W.M.; Gerson, A.R.; Smart, R.S.C. Resolving surface chemical states in XPS analysis of first row transition metals, oxides and hydroxides: Cr, Mn, Fe, Co and Ni. *Appl. Surf. Sci.* **2011**, *257*, 2717–2730. [CrossRef]

27. Biesingera, M.C.; Laua, L.W.M.; Gerson, A.R.; Smart, R.S.C. Resolving surface chemical states in XPS analysis of first row transition metals, oxides and hydroxides: Sc, Ti, V., Cu and Zn. *Appl. Surf. Sci.* **2010**, *257*, 887–898. [CrossRef]

28. Feng, X.; Cao, Y.; Lan, L.; Lin, C.; Li, Y.; Xu, H.; Gong, M.; Chen, Y. The promotional effect of Ce on CuFe/beta monolith catalyst for selective catalytic reduction of NOx by ammonia. *Chem. Eng. J.* **2016**, *302*, 697–706. [CrossRef]

29. Gao, W.; Zhao, Y.; Liu, J.; Huang, Q.; He, S.; Li, C.; Zhao, J.; Wei, M. Catalytic conversion of syngas to mixed alcohols over CuFe-based catalysts derived from layered double hydroxides. *Catal. Sci. Technol.* **2013**, *3*, 1324. [CrossRef]

30. He, Q.; Yang, X.; He, R.; Bueno-López, A.; Miller, H.; Ren, X.; Yang, W.; Koel, B.E. Electrochemical and spectroscopic study of novel Cu and Fe-based catalysts for oxygen reduction in alkaline media. *J. Power Sources* **2012**, *213*, 169–179. [CrossRef]

31. He, Q.; Yang, X.; Ren, X.; Bruce, E.; Koel, B.E.; Ramaswamy, N.; Mukerjee, S.; Kostecki, R. A novel CuFe-based catalyst for the oxygen reduction reaction in alkaline media. *J. Power Sources* **2011**, *196*, 7404–7410. [CrossRef]

32. Wang, M.; Shu, Z.; Zhang, L.; Fan, X.; Tao, G.; Wang, Y.; Chen, L.; Wu, M.; Shi, J. Amorphous Fe^{2+}-rich FeOx loaded in mesoporous silica as a highly efficient heterogeneous Fenton catalyst. *Dalton Trans.* **2014**, *43*, 9234. [CrossRef]

33. Xiao, Z.; Jin, S.; Wang, X.; Li, W.; Wang, J.; Liang, C. Preparation, structure and catalytic properties of magnetically separable Cu–Fe catalysts for glycerol hydrogenolysis. *J. Mater. Chem.* **2012**, *22*, 16598. [CrossRef]

34. Manikandan, M.; Venugopal, A.K.; Nagpure, A.S.; Chilukuri, S.; Raja, T. Promotional effect of Fe on the performance of supported Cu catalyst for ambient pressure hydrogenation of furfural. *RSC Adv.* **2016**, *6*, 3888. [CrossRef]

35. Sheng, H.; Lobo, R.F. Iron-Promotion of Silica-Supported Copper Catalysts for Furfural Hydrodeoxygenation. *ChemCatChem* **2016**, *8*, 3402–3408. [CrossRef]

36. Barzetti, T.; Selli, E.; Moscotti, D.; Forn, L. Pyridine and ammonia as probes for FTIR analysis of solid acid catalysts. *J. Chem. Soc.* **1996**, *92*, 1401–1407. [CrossRef]

37. Ma, T.; Zhang, L.; Song, Y.; Shang, Y.; Zhai, Y.; Gong, Y. A comparative synthesis of ZSM-5 with ethanol or TPABr template: Distinction of Brønsted/Lewis acidity ratio and its impact on n-hexane cracking. *Catal. Sci. Technol.* **2018**, *8*, 1923. [CrossRef]

38. Fogler, H.S. *Elements of Chemical Reaction Engineering*, 5th ed.; Pearson Education Incorporation: Upper Saddle River, NJ, USA, 2016; pp. 734–735.

39. Jothimurugesan, K.; Gangwal, S.K. Titania-Supported Bimetallic Catalysts Combined with HZSM-5 for Fischer-Tropsch Synthesis. *Ind. Eng. Chem. Res.* **1998**, *37*, 1181–1188. [CrossRef]

40. Platon, A.; Thomson, W.J. Quantitative Lewis/Bro1nsted Ratios Using DRIFTS. *Ind. Eng. Chem. Res.* **2003**, *24*, 24. [CrossRef]

41. Li, D.; Bui, P.; Zhao, H.Y.; Oyama, S.T.; Dou, T.; Shen, Z.H. Rake mechanism for the deoxygenation of ethanol over a supported Ni_2P/SiO_2 catalyst. *J. Catal.* **2012**, *290*, 1–12. [CrossRef]

 *catalysts*

Article

Catalytic Decomposition of Oleic Acid to Fuels and Chemicals: Roles of Catalyst Acidity and Basicity on Product Distribution and Reaction Pathways

Wanpeng Hu [1,†], **Hui Wang** [1,2,*,†], **Hongfei Lin** [2], **Ying Zheng** [1,3,*], **Siauw Ng** [4], **Manlin Shi** [1], **Ying Zhao** [1] **and Ruoqian Xu** [5]

1   College of Biological, Chemical Science and Engineering, Jiaxing University, 118 Jiahang road, Jiaxing 314001, China; huwanpeng2002@163.com (W.H.); s1160298455@163.com (M.S.); zy782914598@163.com (Y.Z.)
2   Department of Chemical Engineering, University of New Brunswick, 15 Dineen Drive, Fredericton, NB E3B 5A3, Canada; linhongfei@hotmail.com
3   Department of Chemical and Biochemical Engineering, Western University, 1151 Richmond Street, London, ON N6A 3K7, Canada
4   National Centre for Upgrading Technology, Canmet ENERGY-Devon, 1 Oil Patch Drive, Edmonton, AB T9G 1A8, Canada; siauw.ng@canada.ca
5   Xi'an modern Chemistry Research Institute, Xi'an 710065, China; nutar@163.com
*   Correspondence: huiwang@zjxu.edu.cn (H.W.); ying.zheng@uwo.ca (Y.Z.); Tel.: +86-573-8364-3264 (H.W.); +86-151-9661-2138 (Y.Z.)
†   As the co-first author.

Received: 28 October 2019; Accepted: 6 December 2019; Published: 13 December 2019

Abstract: The roles of catalyst acidity and basicity playing in catalytic conversion of oleic acid were studied in a fixed-bed micro-reactor at atmospheric pressure. The chemical compositions of the petroleum-like products were obtained and the reaction pathways of different catalysts are discussed. The metal oxides are suitable for upgrading oleic acid into organic liquid products (OLPs). Over 98% oxygen was removed when CaO, MgO, and TiO_2 were implemented, whereas a minimum oxygen removal lower than 20% was obtained by using quartz. The oxygen removal was 73% by alumina; however, the light oil yield (to feed) and the valuable product yield received were the highest in all investigated catalysts. The hydrocarbons in OLPs, overwhelmingly presenting in the product, were found to be alkenes and cycloalkenes, followed by saturated hydrocarbons, and then aromatics lower than 4%. For Lewis acidic catalysts, higher acidity of the catalyst is beneficial to deoxygenation but also secondary cracking. CaO has higher dehydrogenation capability than MgO does.

Keywords: catalytic conversion; oleic acid; catalyst acidity and basicity; product distribution; reaction pathways

1. Introduction

With the development of the economy, the consumption of crude oil and its products keeps increasing. On the other hand, petroleum is non-renewable energy; therefore, oil products are becoming scarce. Furthermore, the ongoing strengthen environmental policies make the emission limitation continuously strict. Therefore, researchers have transferred their focus onto biofuels [1,2]. However, biofuels cannot be directly utilized because of the existence of oxygenates and unsaturated carbon–carbon double bonds [3]. Catalytic cracking is one of effective processes for biofuel upgrading [4–7].

As the structure, the active (acidic or basic) sites, and the bond between the metal and oxygen enhance the adsorption of reactants on the active sites, metal oxides are often used as biofuel upgrading catalysts [6]. The early acidic and basic cracking catalysts for biofuels upgrading dated back to the

1980s [8]. At 300~500 °C, Both acid and base metal oxide catalysts were found to be effective catalysts in the deoxygenation of fatty acids or triglycerides to a mixture of oxygen containing products and hydrocarbons [4,9–12]. The various impacts of oxygenates upgraded by both acid and base catalysts were also compared [13–16]. The Al_2O_3 was able to remove 94 wt.% of oxygen from waste cooking oil at the temperature of 470 °C [4]. The basic sites in a catalyst were found to be capable of strongly inhibiting secondary cracking, which resulted in high residual oil yields and low gas yields. Xu et al. found that the amounts of carboxylic acids and aldehydes, as well as the high acid value of liquid products, were significantly decreased by using base catalysts, and that the catalysts were modified by using base catalysts such as CaO rather than Al_2O_3. Furthermore, at low temperature, the product of the formers showed good solubility in diesel, good cold flow properties, and high heat value [14–16]. This is contributed to by the neutralization reactions taking place between the basic sites on catalysts and acidic intermediates in liquid products. The produced neutral hydrocarbons had more similar molecular structure to diesel fuels and better properties than acidic oxygenate intermediates.

Some literature reported the reaction processes of the catalytic cracking oxygenates to petroleum-like products over metal oxide catalysts; however, most of the catalytic cracking work has been performed using acid catalysts, and the effect of the basic sites on catalyst performance and reaction mechanisms have not been sufficiently discussed. In this study, the specific roles that the acidity and basicity, and the types of acidic sites of the catalyst, played in the product distribution of oleic acid were investigated.

2. Results and Discussion

2.1. Catalyst Characterization

2.1.1. Thermal Treatment

Catalyst thermal treatment results are shown in Figure 1a,b. The weight loss of the titania and alumina are only due to the absorbed water, and there was no weight loss during quartz thermal treatment. There was obvious weight loss in CaO and MgO thermal treatment process, which corresponded to the released water and CO_2. Therefore, CaO and MgO were undergone in-situ drying with nitrogen at the temperature of 800 °C for 2 h before activity performance evaluation.

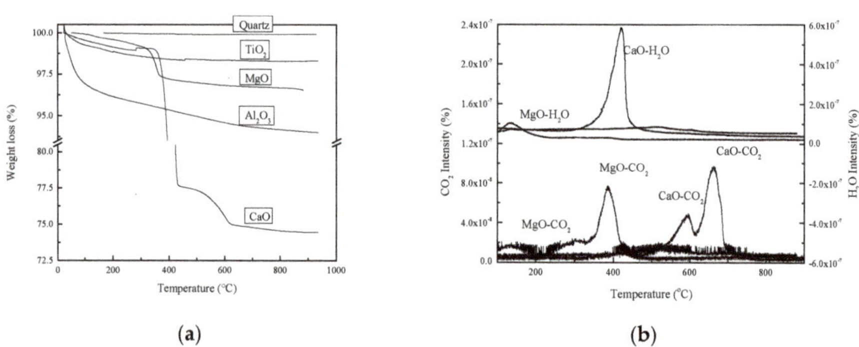

Figure 1. (a) Weight loss and derived weight loss of catalysts; (b) released substrates.

2.1.2. Physical Properties

The properties of all catalysts are given in Table 1.

Table 1. Physical and chemical properties of catalysts.

Properties	Surface Area (m^2/g)	Pore Volume (cm^3/g)	Amount of Active Sites $(\times 10^{-2}$ mmol/g$)$ *		
			Weak	Medium	Strong
Quartz [1]	0.1	4.8×10^{-4}	0.0	0.0	0.0
MgO [1]	11.6	0.2	20.7	0.0	0.0
CaO [1]	13.7	0.1	0.0	0.0	153.8
TiO_2 [2]	56.9	0.9	39.8	10.8	8.0
Al_2O_3 [2]	152.9	0.5	14.0	3.9	7.5

* the specific testing procedures are shown in Section 3.4 Catalyst Characterization. [1] was tested by NH_3-TPD. [2] was tested by CO_2-TPD.

Compared to other catalysts, quartz, the neutral catalyst, is bare of pore structures, and has extremely low surface area, and rather small pore volume. Generally speaking, acidic catalysts have higher surface areas and lager pore volume than alkaline catalysts do, which are in accordance with pore size distribution in Figure S1.

2.1.3. Amount, Types, and Strength of Active Sites

The amount and strength of active sites on acidic and alkaline catalysts are listed in Table 1. Compared to Al_2O_3, TiO_2 has a higher amount of acids with higher strength. Both the amount and the strength of active sites on CaO are higher than those on MgO. There is no active site on quartz.

Figure S2a,b shows the Py-FTIR spectra of Al_2O_3 and TiO_2, respectively. The peaks around 1450 cm^{-1}, 1540 cm^{-1}, and 1490 cm^{-1} are indicative of Lewis acids, Bronsted acid, and both [17]. Lewis acids are the only one appearing on these two catalysts.

2.2. Products

2.2.1. Catalytic Upgrading by TGA

Figure S3 shows the oleic acid decomposition process catalyzed by different catalysts by TGA tests. The corresponded MS spectra are shown in Figure S4. From Figure S3, the decomposition temperature of oleic acid ranged from 200 to 500 °C. As shown in Figure S3, 450–500 °C is an appropriate temperature range for oleic acid deoxygenation by alumina under experimental conditions and therefore then temperature of 475 °C was selected in the current study.

2.2.2. Product Distribution

The operational parameters and the product yield distributions are shown in Table S1 and Table 2, respectively. The total product yields reached 100 ± 5.0% during all investigated experiments. As expected, the conversion of quartz upgraded organic liquids was as low as 35%. Around 65% of the oleic acid was left in organic liquid products, and catalytic cracking resulted in lower liquid, higher gas, and higher coke yields.

Table 2. Product distributions and oxygen removal.

Yields (%)	Oil	Coke	Gas Hydrocarbons	Water	CO/CO_2	H_2	Light Oil * Yields	OR
Quartz	93.4	0.1	0.9	0.5	2.6	0.1	20.8	18.5
Al_2O_3	82.5	5.1	3.0	1.6	10.6	0.4	40.8	73.0
TiO_2	75.3	2.0	3.6	0.9	16.3	1.4	29.1	100.0
MgO	67.5	5.3	5.5	2.3	14.9	0.1	22.0	99.4
CaO	29.0	51.2	3.8	0.0	19.8	1.0	16.6	97.7

* light oil represents that the oil product fractions with the boiling point lower than 350 °C.

All liquids spontaneously stratified into two phases: Oil and water. According to the type of catalysts, the yields of oil products are in the following order: Silica (93.4 wt.%) > alumina (82.5 wt.%) > titania (75.3 wt.%) > MgO (67.5 wt.%) > CaO (29.0 wt.%). In contrast, the total amount of water, carbon oxide, and hydrogen were found to increase during the catalytic process, indicating that dehydration, decarbonoxide reaction, and dehydrogenation were enhanced by catalyst. It has been reported that activated alumina was an effective catalyst in the decarboxylation of fatty acids under atmospheric pressure and at 450 °C, and the liquid product yields vary in the range between 65% and 79% with n-alkanes and n-alkenes [11].

Deoxygenation of oleic acid under the current condition showed variations when upgrading with different types of catalysts. All these catalysts underwent more deoxygenation by the removal of CO and CO_2 than by water. The liquid yields, the gas hydrocarbon yields, and the coke yields appear to be closely associated with the types of active sites on the catalysts. Higher liquid but lower gas hydrocarbon and coke yields are achieved first by quartz, then by acidic catalysts, and finally by alkaline catalysts. The stronger the active sites on catalyst were, the lower liquid yields and the higher gas hydrocarbon yields reached. The more active sites on catalyst, the lower coke yields. However, CaO is an exception, which has lower gas hydrocarbon yield and higher coke yield than MgO.

Moreover, certain amount of hydrogen was detected by GC-RGA. Serrano et al. studied the feasibility to produce hydrogen, and received 0.12% hydrogen through the decomposition of vegetable and microalgal oils [18].

Oxygen removal and light oil yields in oil products are also shown in Table 2. The amount of active sites strongly affects the oxygen removal. Even though the oxygen removal of Al_2O_3 in oil products is lower than those of TiO_2 and alkaline catalysts, the yield of light oils is the highest, which is supported by the previous publication [12].

2.2.3. Product Compositions

Table S2 shows the contents of gas hydrocarbons. C1–C3 olefin and paraffin are the principal organic gas products. Generally, the contents of olefins are higher than those of paraffin with the same carbon number. Only CaO caused different results that the content of methane is relatively higher than that of other organic gases, which might be because the high concentration of hydrogen reacting with CO forms methane. It is worth of noting that the contents of paraffin upgraded by CaO are higher than those of olefin with the same carbon number. The high concentration of hydrogen may make the olefins in CaO upgrading gases saturated [19–21].

The group and elemental compositions, and the carbon number distribution of hydrocarbons of oil products, are shown in Figures 2 and 3, respectively. The types and carbon number distributions of oxygenates in organic liquid products (OLPs) are shown in Figures 4 and 5. In general, the chemical compounds does not differ greatly between acidic and alkaline upgrading oils. The OLPs contained hydrocarbons of C6 to C18, and C22 to C24, including of alkanes, alkenes, cycloalkanes, cycloalkenes, aromatics, and oxygenates. Most oxygenates were removed by acidic and alkaline catalysts. It is easy to find the low aromatics content, which is in accordance with the molar ratio of hydrogen to carbon, which was 1.81 ± 0.06 in both acidic and alkaline catalyst upgrading oil products. The oxygen content well resembled oxygenate contents in all catalytic upgrading oil products. Demirbas et al. [22] reported the same trend with low concentration of aromatics, while some researchers obtained different results.

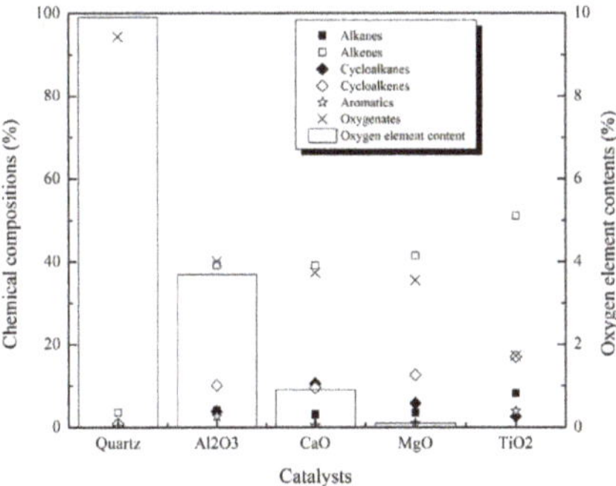

Figure 2. Chemical compositions of organic liquid products (OLPs).

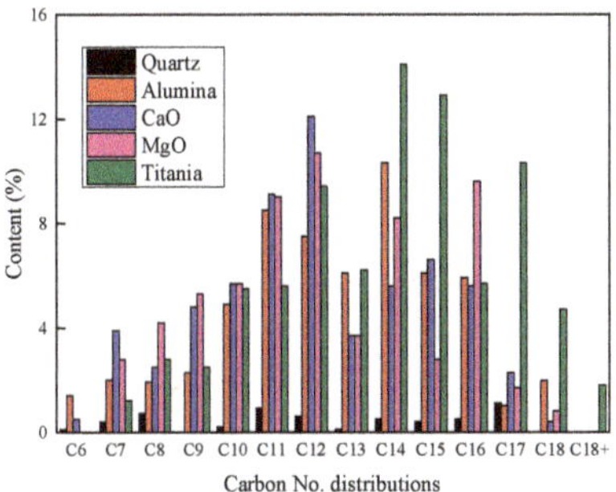

Figure 3. Carbon number distribution of hydrocarbons.

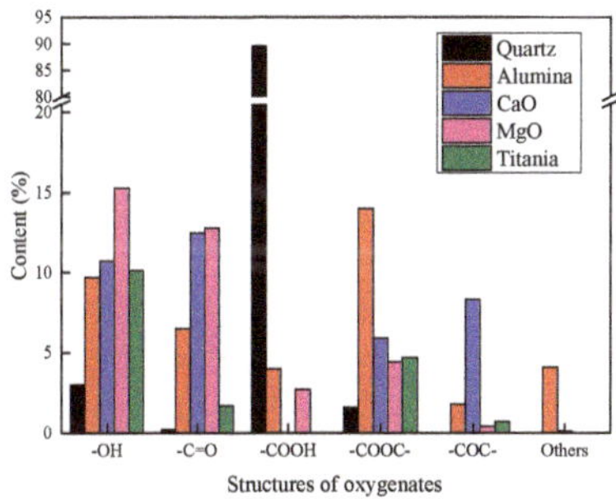

Figure 4. Types of oxygenates.

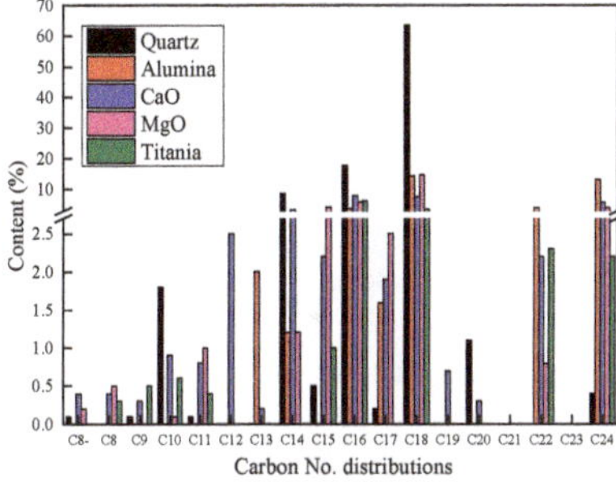

Figure 5. Carbon number distribution of oxygenates.

Aldehydes, esters, ketones, and alcohols were found in the OLPs. Certain amount of di-epoxide, aldehyde, epoxide, acid, and ester were presented in some samples. Low conversion with the high concentration of fatty acids was achieved when working with quartz. Ketones and aldehydes were found in the catalytic reactions, but were not detected during thermal cracking. Esters were found in both catalytic and non-catalytic upgrading oil products.

Figure S5 shows the TPO analysis results of all investigated catalysts except for the quartz. The reason is neither oxygen was consumed, nor water and CO_2 were released (same to oxygen TGA results in Figure 6). As shown in Figure 6 and Figure S5, different catalysts have significantly different TPO results. For Alumina, the released water below 250 °C should be the produced water that was absorbed by the alumina; and the weight loss between 250 °C and 700 °C is from the coke because the oxygen was consumed within this range. For titania, TPO result is very simple and clear, and only coke is burnt out in the range of 300–600 °C. For CaO, the released water and CO_2 were from coke when the temperature was lower than 500 °C, while the CO_2 should be decomposed from $CaCO_3$ when the temperature was higher than 500 °C for no oxygen was consumed. The MgO TPO result is similar to the CaO one, and only the temperature and the amount of released CO_2 from metal carbonate are not the same. The coke was burnt out when the temperature was higher than 400 °C, while the released CO_2 ranging from 150 to 400 °C is from $MgCO_3$. All the relevant calculations, such as the produced water and CO_2, as well as the coke content, were based on those results.

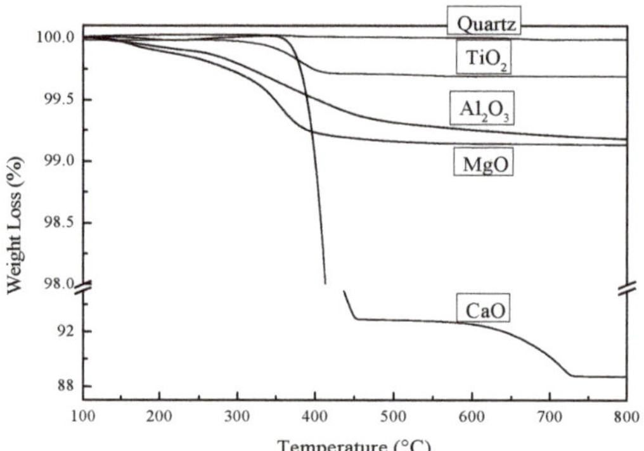

Figure 6. Weight loss of spent catalysts.

2.3. Reaction Pathways

The possible reaction mechanisms for the deoxygenation of oleic acid illustrated in Scheme 1 are included the followings [5,23]: (1) The production of anhydrides from the dehydration of acids, the decarboxylation of anhydrides to aldehyde/ketones, and the decarbonylation (or reduction) of aldehyde/ketones to alkenes (or alcohols); (2) the breaking of the C–H bond was promoted by the presence of Lewis acids and the generation of active hydrogen; the reduction of oleic acids to aldehydes, alcohols, alkenes, and eventually alkanes in turn by the produced active hydrogen; (3) the neutralization of acid and alkali to form salt, and the decarboxylation of the produced salt to form alkenes. The molarity compositions of deoxygenation products are shown in Table 3. The deoxygenation of oleic acid with nitrogen at 470 °C mainly proceed via DCO, yielding high concentration of carbon monoxide, and suggesting decarbonylation is the major deoxygenation pathway. Decarboxylation also occurred, demonstrated by the presence of carbon dioxide in Table 3. As discussed previously, CaO is still an exception, which favors decarboxylation over decarbonylation.

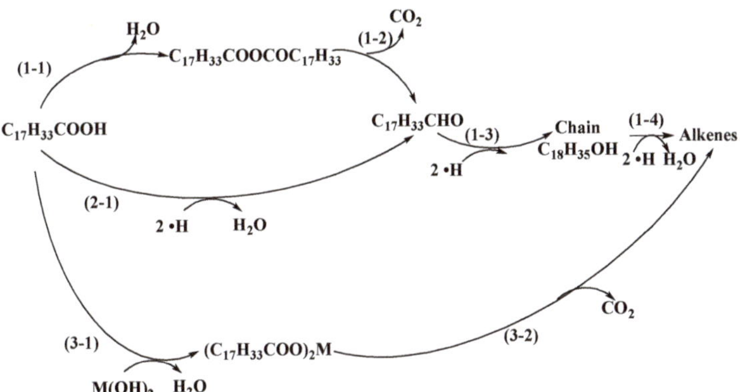

Scheme 1. Possible reaction mechanisms in the current study. M stands for the metal elements.

Table 3. Molarity compositions of deoxygenation products.

Molarity (mol/mol Oleic Acid)	CO	CO_2	H_2O
Quartz	0.13	0.05	0.06
Al_2O_3	0.52	0.21	0.19
TiO_2	1.11	0.39	0.14
MgO	0.82	0.27	0.30
CaO	0.02	0.89	0.00

Deoxygenation of oleic acid under the current condition favors decarbonylation over decarboxylation. Tani et al. had the same conclusion when upgrading palm oil by silica and MgO-silica catalysts [24].

CaO is a good dehydrogenation catalyst, however, with high coke content. For other catalysts, more active sites result in higher concentrations of hydrocarbon gases and coke. When the reaction temperature is at 470 °C, CaO reacts with CO_2, producing $CaCO_3$ (see Figure 6), accelerating the water gas shift reaction (Equation (3)), resulting in a phenomenon in which carbon monoxide and water are consumed to release CO_2 and H_2, consequently leading to the high hydrogen concentration in CaO upgrading gas products. Gosselink gave the same explanation [19].

$$CO + H_2O \leftrightarrow CO_2 + H_2 \tag{1}$$

2.3.1. Acidic Catalysts

Alumina and titania were used to evaluate the effect of Lewis acids on the product distribution. Compared to other catalysts, yields of various products and compositions of gas products obtained with titania were rather close to those obtained with alumina. This is because both of them have relative high surface area, large pore volume, and rich of pore structures; proper amount and strength of acidic sites; and same acid type (Lewis acids only).

On the other hand, compositions of hydrogen and OLP obtained with titania (both higher acidity and strength) were significantly different from those with alumina. A comparison of the compositions of hydrogen obtained with various catalysts (Table 2 and Table S2) shows that the hydrogen content in the gas product with alumina catalysts was lower than those with titania. This result is in accordance with the chemical composition of OLPs shown in Figure 2, where the unsaturated hydrocarbon (alkene, cycloalkane, and aromatics) contents of OLP upgraded by titania are much higher than those by alumina. Moreover, the oxygenate amount of titania upgrading OLP is much lower than that of alumina. These results show that there was a larger involvement of dehydrogenation and deoxygenation reactions with titania than with alumina, which may be caused by the higher amount of acidic site on titania. In addition, Table 2 shows that the gas hydrocarbon yield with titania was higher than that with alumina. The high gas hydrocarbon yield with titania may be attributed to the accelerated secondary cracking due to the stronger acidity and the higher density of acidity (the acidity on per surface area of catalyst).

The reaction mechanisms for the deoxygenation of oleic acid over Lewis acidic catalysts include the followings [5,23]: The breaking of the C–H bond and the generation of active hydrogen were promoted by Lewis acids, and therefore, the reduction of oleic acids to aldehydes, alcohols, alkenes, and finally alkanes by the produced active hydrogen were enhanced. As shown in Figure S4 and Table 3, the contents of alcohols, ketones, and ethers in the acidic catalyst upgrading products were lower than those in the basic catalyst upgrading products.

2.3.2. Neutral Catalyst

Analysis of the gas products upgraded by quartz showed the existence of CO and CO_2, indicating that deoxygenation reaction proceeded through decarbonylation or decarboxylation mechanisms [25]. Analysis results of the OLP revealed that alkenes and fatty acids were the primary products, especially for fatty acid, which is as high as 94.4%, and predominance of C18 (65%), C16 (18%), and C14 (8%) fatty acids. Analysis of oxygenates in OLP shows the presence of a mixture of alcohols, esters, and acids. The results indicated that at 470 °C with quartz, the conversion of oleic acid to the anhydrides was low, which resulted in the lack of ketones in the OLP [26]. Bare aromatic compounds and only a small number of alkane and cycloalkane species were detected, which implies that less secondary thermal cracking occurred. All these results agree with other researchers' [27]. Compared to the yields and compositions obtained by all investigated catalysts, the presence of the catalyst reduced the amount of the liquid product and caused further removal of the acid groups [9].

2.3.3. Alkaline Catalysts

Tests running over calcium oxide and magnesium oxide catalysts were used to assess the effect of basic sites on the product distribution. The yields of various products and the compositions of gas and OLPs are shown in Figures 2–5, Figure S3, and Figure S4 and Tables 2 and 3, and Table S2. Compared to other catalysts, the yields of OLPs, coke, and gas hydrocarbons, the oxygen removal, as well as the compositions and carbon number distributions of OLPs and oxygenates in OLPs obtained with CaO, were rather close to those obtained with MgO. This is probably due to their similar surface area, pore volume, and basic sites. It was an interesting observation that there was a difference between the yield of water and that of carbon oxide and hydrogen for calcium oxide, as well as the concentration of those for magnesium oxide. As discussed in Section 2.2.3, at a reaction temperature of 470 °C, CaO reacts with CO_2, producing $CaCO_3$ (see Figure 6), accelerating the water gas shift reaction, resulting in a phenomenon where carbon monoxide and water are consumed to produce CO_2 and H_2. This leads to the higher yields and concentration of carbon oxide and hydrogen obtained by CaO than by MgO. In addition, CaO has a relative high dehydrogenation capability, which is confirmed by the higher aromatic, cycloalkane, and alkene concentration in CaO upgrading OLP. Furthermore, the water yield obtained by MgO was rather higher than by CaO. Different from the literature [8], the alkaline catalysts (CaO and MgO) in the current study did not inhibit the secondary cracking, which was deduced by the low oil yields and the high gas yields.

The amount of fatty acid left in the liquid products decreased most drastically and reached a rather lower level than that in other catalyst upgrading OLPs, except with titania. This was probably attributed to the saponification of the fatty acid by the base [9]. However, the removal of the acid functions in the liquid products did not necessarily lead to oxygen groups-free products. The liquid product consisted of an certain amount of ketones and aldehydes with –C=O, alcohols with –OH, esters (especially di-esters) with –COOC-, and epoxides (especially di-epoxides) with –C-O-C-, as shown in Figure 4.

MgO/CaO reacts with the produced CO_2 to form $MgCO_3/CaCO_3$, which promote the decarboxylation of oleic acid. It should be noted that $CaCO_3$ is so stable under reaction condition (470 °C) that a high temperature (>700 °C) is needed to decompose it to CaO [28]. However, $MgCO_3$ is easy to be decomposed to CO_2 and MgO under the investigated reaction conditions. In practice, the TGA-MS analysis results (Figure 1) showed that both $MgCO_3$ and $Mg(OH)_2$ were decomposed to MgO at 450 °C.

The reaction mechanisms for the deoxygenation of oleic acid over basic catalysts include the following: (1) The production of anhydrides from the dehydration of acids, the decarboxylation of anhydrides to aldehyde/ketones, and the decarbonylation of aldehyde/ketones to alkenes were enhanced by the basic catalysts; (2) the neutralization of acid and alkali to form salt, and the decarboxylation of the produced salt to form alkenes; and (3) the reduction of oleic acids to aldehydes, alcohols, alkenes, and eventually alkanes in succession by the produced active hydrogen. As shown in Figure S4 and Table 3, the acid and ester contents in the acidic catalyst upgrading products were much higher than those in the basic catalyst upgrading products, which further illustrates the neutralization of oleic acids with basic catalysts.

2.3.4. Aromatization

As shown in Figure 2, the aromatic contents upgraded by all investigated catalysts (neutral, base, Lewis acids only) are lower than 3.9 wt.%. Thus, the similarity in product distribution between acidic catalysts (titania and alumina) and other catalysts shows clearly that Lewis acid is not the dominative factor in aromatization involving oleic acid as feed, although Cerny et al. has reported over 40% and around 85% aromatics in the <150 °C and >150 °C fractions when a ultra-stable zeolite Y (USY, with the same amount Bronsted and Lewis acid) was used as the catalyst [29]. In addition, Omar et al. found that adding ZSM-5 (with both Bronsted and Lewis acids) catalyst produced an appreciable quantity of aromatic compounds [27]. Therefore, a key property, the existence of Bronsted acids on catalyst surface, was important to the formation of aromatics. Frety et al. got the same conclusion [30]. Part of the acidic sites (Lewis and Brønsted acids) will change to each other during the reaction process with the presence of water, which is confirmed by the higher yields of aromatic hydrocarbons when running with titania and alumina (Lewis acid only catalysts) as compared to quartz (non-acid catalyst). Other reasons, such as pore structures of catalysts [12] and high-power treatment (by microwave [27]), were found to affect the formation of aromatics.

3. Materials and Methods

3.1. Catalyst Reaction Performance Evaluation

Oleic acid (CAS No.: 112-80-1, purchased from Fisher Scientific, Canada) was used as the feed in the current study. For obtaining the information between the catalytic decomposition of oleic acid and reaction conditions, a mixture of oleic acid and catalysts was tested first by the TGA combined with an Residue gas analyzer (mass spectrometer), which was used to record the decomposed volatile gases. An amount of 10 mg catalyst was loaded in the TGA testing pan and was pretreated at 700 °C for 2 h. Then, around 10 mg oleic acid was added into the testing pan and tested with a heating rate of 10 °C/min to 800 °C. A quartz fixed-bed reactor was used as the reaction system under atmospheric pressure. The schematic of the reaction system is shown in Figure S6. This system consists of a quartz tube reactor (inner diameter: 10 mm; length: 100 mm), a nitrogen gas feed system, a liquid feeding syringe, a heater and temperature control system, a condenser, and a mixture of iced ethanol-saturated salt cooling system; in addition, liquid and gas products collection systems were also introduced. The reaction conditions are shown in Table S1 in the Supplemental Files.

3.2. Catalyst Preparation

Alumina was prepared from Boehmite. An amount of 100 g Boehmite was dissolved in about 150 mL deionized water and was acidified by HCl to the PH value at 3–4 under a temperature of 60–65 °C until the colloidal substances were produced. The solution was dried in an oven at 105 °C for 5 h and then was calcined in a muffle oven under 700 °C for 2 h [31]. TiO_2 and quartz were used to maintain the state when they were purchased. CaO and MgO were undergone in-situ drying with nitrogen at 800 °C in order to remove the metal hydroxide and metal carbonate, and to remain in the state of metal oxide only during the reaction process (see Section 2.1).

3.3. Products Analysis

Compositions of inorganic gases, such as CO, CO_2, and H_2, were tested by GC (Varian GC-3400, Varian, Agilent Technologies, Inc., Santa Clara, CA, USA), which is equipped with an Agilent HP-PLOT Q column and combined with the residue gas analyzer (RGA-200, Stanford Research Systems, Sunnyvale, CA, USA). Compositions of hydrocarbons in gas products were determined by GC/FID (GC-17A, Shimadzu Scientific Instruments, Kyoto, Japan).

The contents of carbon, hydrogen, and oxygen were tested by LECO CHNS-932. Samples were combusted in an oxygen atmosphere at 1150 °C. Boiling point distribution of liquid products was tested in accordance with ASTM D2887, using a gas chromatography (Shimadzu GC-2010, Shimadzu, Shimadzu Scientific Instruments, Kyoto, Japan), which is equipped with Agilent Technologies DB-2887 column. The compositions of liquid products were analyzed by a gas chromatography (SHIMADZU GC-17A, Shimadzu Scientific Instruments, Kyoto, Japan) coupled with mass spectrometry (SHIMADZU MS-QP5000, Shimadzu Scientific Instruments, Kyoto, Japan) (GC/MS), which was equipped with an Agilent HP-5MS column.

The coke content of spent catalysts was determined by TGA-Q500 (TA Instruments, New Castle, DE, USA). The sample treatment procedures and analyzing program were the same to the TPO testing indicated in Section 3.4.

3.4. Catalysts Characterization

Surface area, pore volume, and pore size distribution were measured at 77K on an Autosorb-1 (ANTON PAAR USA INC., Ashland, VA, USA). The specific surface area of the catalyst powder was calculated by employing the Brunauer-Emmet-Teller (BET) method. The total pore volume of nitrogen was adsorbed under relative pressure p/p0 0.995. Pore size distribution was calculated by the Barrett-Joyner-Halenda (BJH) method.

The strength and amount of acidic/alkaline sites on the surface of catalysts was determined by NH_3/CO_2 temperature program desorption (TPD) using a Quantachrome Autosorb 1-C and a Residue Gas Analyzer (RGA-200, Stanford Research Systems, Sunnyvale, CA, USA). About 100–1000 mg catalyst sample was loaded into a U-shaped quartz tube, which was then installed in a furnace. The catalyst was treated in-situ at 120 °C for 1 h in helium and then was vacuumed (lower than 3000 micron) for 30 min. The catalyst underwent NH_3 or CO_2 adsorption at 120 °C for 1 h, and then physi-desorption at 120 °C by helium purging. Finally, the catalyst was heated up to 900 °C (at a rate of 10 °C per minute) for chemi-desorption. The weak, medium, and strong active sites of the catalyst were assigned to the peak areas of the NH_3/CO_2-TPD curves lower than 350 °C, between 350 and 500 °C, and above 500 °C.

The types and strengths of acids on the surface of catalysts were tested by Pyridine-FTIR. A catalyst sample of 2–15 mg was pressed to form wafers (1 cm in diameter) and installed on the support located in a cell. The cell included two KBr windows through which IR could be transmitted. The IR cell was degassed at 450 °C by an N_2 flow for 1 h. When the cell's temperature cooled down to 150 °C, the pressure of the system was evacuated to 100 micron by a high vacuum pump. Pyridine was then introduced for adsorption. The excess or physisorbed pyridine was purged by N_2. Infrared spectra were recorded on a Nicolet FTIR-6700 (Thermo Fisher Scientific, Waltham, MA USA) at temperatures of 150, 250, 350, and 450 °C. Data obtained from the spectrometer were analyzed by Omnic 8 software. The characteristic bands at 1450 cm^{-1} and 1550 cm^{-1} were assigned to Lewis acid sites and Bronsted acid sites [17], respectively.

The type and the amount of coke on the spent catalysts, was determined by TPO, using a Quantachrome Autosorb 1-C and an RGA-200. About 100–200 mg sample of catalyst was loaded in a U-shaped quartz tube and installed in a furnace. The catalyst was exposed to a flow of helium (50 mL/min), and heated to 120 °C at a rate of 10 °C/min, remaining at 120 °C for 1 h. The physisorbed water in the catalyst sample was removed. Subsequently, the catalyst was heated up to 900 °C at a rate of 10 °C/min and under a flowing gas mixture of 2% vol oxygen (1 mL/min) in helium (49 mL/min).

3.5. Calculations

Oxygen removal (OR)

$$OR = (M_{OF} - M_{OO})/M_{OF} * 100\%, \tag{2}$$

where

OR: Oxygen removal, %

M_{OF}: Mass of Oxygen in the Feed, g. "M" means mass, "$_O$" and "$_F$" respectively stands for "oxygen" and "feed".

M_{OO}: Mass of Oxygen in the Oil products, g. "M" means mass, "$_O$" and "$_O$" respectively stands for "oxygen" and "oil products".

4. Conclusions

Under inert N_2 atmosphere, liquid product yields fall into the range of 29.0 and 93.4 wt.% of the feed but are lower with active sites. The higher oxygen removal (97.7–100.0 wt.%) are obtained with the presence of CaO, MgO, and Titania, whereas a minimum oxygen removal (18.5 wt.%) is obtained with quartz. Even though the oxygen removal of alumina is as high as 73.0 wt.%, light oil yield (to feed) and the valuable product yield are still the highest in all investigated catalysts. Liquid products were composed mainly of hydrocarbons ranging from 6 to 18 in the terms of carbon number, while oxygenates from 10 to 18, 20, 22, and 24. The hydrocarbons in OLPs majorly were found to be alkenes and cycloalkenes, followed by cycloalkanes and alkanes, and aromatics content was lower than 3.9 wt.%. For Lewis acidic catalysts, more acidity of the catalyst is beneficial for deoxygenation and secondary cracking as well. For basic catalysts, the saponification of oleic acid occurred. CaO has higher dehydrogenation capability than MgO does. The existence of Bronsted acids is important to the formation of aromatics.

Supplementary Materials: The following are available online at http://www.mdpi.com/2073-4344/9/12/1063/s1, Figure S1: Pore size distributions of the studied catalysts; Figure S2: Py-FTIR spectra of the acidic catalysts; Figure S3: Oleic acid decomposition process with different catalysts by TGA; Figure S4: the corresponding MS spectra of the Figure S3; Figure S5: TPO testing of the spent catalysts; Figure S6: Reaction flow chart of this study. Table S1: Operational parameters of catalyst activity performance evaluation; Table S2: Organic gas products compositions.

Author Contributions: Conceptualization, H.W., Y.Z. (Ying Zheng), and S.N.; data curation, H.W., M.S., and Y.Z. (Ying Zhao); formal analysis, H.W.; funding acquisition, H.W. and Y.Z. (Ying Zheng); investigation, H.W. and H.L.; methodology, H.W., H.L., and Y.Z. (Ying Zheng) project administration, H.W. and Y.Z. (Ying Zheng); resources, H.W. and Y.Z. (Ying Zheng); software, H.W. and H.L.; supervision, H.W. and Y.Z. (Ying Zheng); validation, H.W.; visualization, H.W. and Y.Z. (Ying Zheng); writing—original draft, H.W.; writing—review and editing, W.H., H.W., Y.Z. (Ying Zheng), S.N., and R.X.

Funding: This research was funded by the Natural Sciences and Engineering Research Council of Canada, grant number NSERC strategic: 463140-2014STPGP and NSERC DISCOVERY: RGPIN-2015-0386; the Canada Foundation for Innovation, grant number #31983; Canada Research Chairs program, grant number 950-228053; and Jiaxing University, grant number #70518034.

Conflicts of Interest: The authors declare no conflict of interest.

References

1. Wang, H.; Li, G.; Rogers, K.; Lin, H.; Zheng, Y.; Ng, S. Hydrotreating of waste cooking oil over supported CoMoS catalyst—Catalyst deactivation mechanism study. *Mol. Catal.* **2017**, *443*, 228–240. [CrossRef]
2. Wang, H.; Zhang, L.; Li, G.; Rogers, K.; Lin, H.; Seers, P.; Ledan, T.; Ng, S.; Zheng, Y. Application of uniform design experimental method in waste cooking oil (WCO) co-hydroprocessing parameter optimization and reaction route investigation. *Fuel* **2017**, *210*, 390–397. [CrossRef]
3. Lu, Q.; Li, W.Z.; Zhu, X.F. Overview of fuel properties of biomass fast pyrolysis oils. *Energy Convers. Manag.* **2009**, *50*, 1376–1383. [CrossRef]
4. Wang, H.; Lin, H.; Feng, P.; Han, X.; Zheng, Y. Integration of catalytic cracking and hydrotreating technology for triglyceride deoxygenation. *Catal. Today* **2017**, *291*, 172–179. [CrossRef]
5. Wang, H.; Lin, H.; Zheng, Y.; Ng, S.; Brown, H.; Xia, Y. Kaolin-based catalyst as a triglyceride FCC upgrading catalyst with high deoxygenation, mild cracking, and low dehydrogenation performances. *Catal. Today* **2019**, *319*, 164–171. [CrossRef]
6. Yigezu, Z.D.; Muthukumar, K. Catalytic cracking of vegetable oil with metal oxides for biofuel production. *Energy Convers. Manag.* **2014**, *84*, 326–333. [CrossRef]
7. Refaat, A.A. Biodiesel production using solid metal oxide catalysts. *Int. J. Environ. Sci. Technol.* **2011**, *8*, 203–221. [CrossRef]
8. Zhao, C.; Bruck, T.; Lercher, J.A. Catalytic deoxygenation of microalgae oil to green hydrocarbons. *Green Chem.* **2013**, *15*, 1720–1739. [CrossRef]
9. Dosanjos, J.R.S.; Gonzalez, W.D.; Lam, Y.L.; Frety, R. Catalytic decomposition of vegetable oil. *Appl. Catal.* **1983**, *5*, 299–308. [CrossRef]
10. Kirszensztejn, P.; Przekop, R.; Tolinska, A.; Mackowska, E. Pyrolytic and catalytic conversion of rape oil into aromatic and aliphatic fractions in a fixed bed reactor on Al_2O_3 and Al_2O_3/B_2O_3 catalysts. *Chem. Pap.* **2009**, *63*, 226–232. [CrossRef]
11. Vonghia, E.; Boocock, D.G.B.; Konar, S.K.; Leung, A. Pathways for the deoxygenation of triglycerides to alphatic-hydrocarbons over activated alumina. *Energy Fuels* **1995**, *9*, 1090–1096. [CrossRef]
12. Idem, R.O.; Katikaneni, S.P.R.; Bakhshi, N.N. Catalytic conversion of canola oil to fuels and chemicals: Roles of catalyst acidity, basicity and shape selectivity on product distribution. *Fuel Proc. Technol.* **1997**, *51*, 101–125. [CrossRef]
13. Smets, K.; Roukaerts, A.; Czech, J.; Reggers, G.; Schreurs, S.; Carleer, R.; Yperman, J. Slow catalytic pyrolysis of rapeseed cake: Product yield and characterization of the pyrolysis liquid. *Biomass Bioenergy* **2013**, *57*, 180–190. [CrossRef]
14. Xu, J.M.; Jiang, J.C.; Lu, Y.J.; Chen, J. Liquid hydrocarbon fuels obtained by the pyrolysis of soybean oils. *Bioresour. Technol.* **2009**, *100*, 4867–4870.
15. Xu, J.; Jiang, J.; Chen, J.; Sun, Y. Biofuel production from catalytic cracking of woody oils. *Bioresour. Technol.* **2010**, *101*, 5586–5591. [CrossRef]
16. Xu, J.; Jiang, J.; Sun, Y.; Chen, J. Production of hydrocarbon fuels from pyrolysis of soybean oils using a basic catalyst. *Bioresour. Technol.* **2010**, *101*, 9803–9806. [CrossRef]
17. Ward, J.W. Nature of active sites on zeolite. Rare earth Y zeolite. *J. Catal.* **1969**, *13*, 321–327. [CrossRef]
18. Serrano, D.P.; Dufour, J.; Iribarren, D. On the feasibility of producing hydrogen with net carbon fixation by the decomposition of vegetable and microalgal oils. *Energy Environ. Sci.* **2012**, *5*, 6126–6135. [CrossRef]
19. Gosselink, R.W.; Hollak, S.A.W.; Chang, S.W.; van Haveren, J.; de Jong, K.P.; Bitter, J.H.; van Es, D.S. Reaction Pathways for the Deoxygenation of Vegetable Oils and Related Model Compounds. *ChemSusChem* **2013**, *6*, 1576–1594. [CrossRef]
20. Rozmyslowicz, B.; Maki-Arvela, P.; Tokarev, A.; Leino, A.R.; Eranen, K.; Murzin, D.Y. Influence of Hydrogen in Catalytic Deoxygenation of Fatty Acids and Their Derivatives over Pd/C. *Ind. Eng. Chem. Res.* **2012**, *51*, 8922–8927. [CrossRef]
21. Immer, J.G.; Kelly, M.J.; Lamb, H.H. Catalytic reaction pathways in liquid-phase deoxygenation of C18 free fatty acids. *Appl. Catal. A Gen.* **2010**, *375*, 134–139. [CrossRef]
22. Demirbas, A.; Kara, H. New options for conversion of vegetable oils to alternative fuels. *Energy Sour. Part A Recovery Util. Environ. Eff.* **2006**, *28*, 619–626. [CrossRef]

23. Benson, T.J.; Daggolu, P.R.; Hernandez, R.A.; Liu, S.T.; White, M.G. Catalytic deoxygenation chemistry: Upgrading of liquids derived from biomass processing. In *Advances in Catalysis*; Gates, B.C., Jentoft, F.C., Eds.; Academic Press: Cambridge, MA, USA, 2013; Volume 56, pp. 187–353.
24. Tani, H.; Hasegawa, T.; Shimouchi, M.; Asami, K.; Fujimoto, K. Selective catalytic decarboxy-cracking of triglyceride to middle-distillate hydrocarbon. *Catal. Today* **2011**, *164*, 410–414. [CrossRef]
25. Asomaning, J.; Mussone, P.; Bressler, D.C. Thermal deoxygenation and pyrolysis of oleic acid. *J. Anal. Appl. Pyrolysis* **2014**, *105*, 1–7. [CrossRef]
26. Billaud, F.; Minh, A.K.T.; Lozano, P.; Pioch, D. Catalytic cracking of octanoic acid, Journal of Analytical and Applied Pyrolysis. *J. Anal. Appl. Pyrolysis* **2001**, *58*, 605–616. [CrossRef]
27. Omar, R.; Robinson, J.P. Conventional and microwave-assisted pyrolysis of rapeseed oil for bio-fuel production-reaction pathways analysis by GCMS results. *J. Anal. Appl. Pyrolysis* **2014**, *105*, 131–142. [CrossRef]
28. Yoshioka, T.; Handa, T.; Grause, G.; Lei, Z.G.; Inomata, H.; Mizoguchi, T. Effects of metal oxides on the pyrolysis of poly(ethylene terephthalate). *J. Anal. Appl. Pyrolysis* **2005**, *73*, 139–144. [CrossRef]
29. Cerny, R.; Kubu, M.; Kubicka, D. The effect of oxygenates structure on their deoxygenation over USY zeolite. *Catal. Today* **2013**, *204*, 46–53. [CrossRef]
30. Frety, R.; da Rocha, M.D.G.C.; Brandao, S.T.; Pontes, L.A.M.; Padilha, J.F.; Borges, L.E.P.; Gonzalez, W.A. Cracking and hydrocracking of triglycerides for renewable liquid fuels: Alternative processes to transesterification. *J. Braz. Chem. Soc.* **2011**, *22*, 1206–1220. [CrossRef]
31. Hu, Y. Study on Direct Catalytic Cracking of High TAN Crude Oil. Ph.D. Thesis, China University of Petroleum (East China), Qingdao, China, 2011.

Article

Enhancement of HDO Activity of MoP/SiO$_2$ Catalyst in Physical Mixture with Alumina or Zeolites

Ivan V. Shamanaev *, Irina V. Deliy, Evgeny Yu. Gerasimov, Vera P. Pakharukova and Galina A. Bukhtiyarova

Boreskov Institute of Catalysis, Lavrentiev Ave. 5, 630090 Novosibirsk, Russia; delij@catalysis.ru (I.V.D.); gerasimov@catalysis.ru (E.Y.G.); verapakharukova@yandex.ru (V.P.P.); gab@catalysis.ru (G.A.B.)
* Correspondence: i.v.shamanaev@catalysis.ru; Tel.: +7-383-326-9410; Fax: +7-383-330-8056

Received: 21 November 2019; Accepted: 24 December 2019; Published: 31 December 2019

Abstract: Catalytic properties of physical mixture of MoP/SiO$_2$ catalyst with SiC, γ-Al$_2$O$_3$, SAPO-11 and zeolite β have been compared in hydrodeoxygenation of methyl palmitate (MP). MoP/SiO$_2$ catalyst (11.5 wt% of Mo, Mo/P = 1) was synthesized using TPR method and characterized with N$_2$ physisorption, elemental analysis, H$_2$-TPR, XRD and TEM. Trickle-bed reactor was used for catalytic properties investigation at hydrogen pressure of 3 MPa, and 290 °C. The conversions of MP and overall oxygen-containing compounds have been increased significantly (from 59 to about 100%) when γ-Al$_2$O$_3$ or zeolite materials were used instead of inert SiC. MP can be converted to palmitic acid through acid-catalyzed hydrolysis along with metal-catalyzed hydrogenolysis, and as a consequence the addition of material possessing acid sites to MoP/SiO$_2$ catalyst could lead to acceleration of MP hydrodeoxygenation through acid-catalyzed reactions. Isomerization and cracking of alkane were observed over the physical mixture of MoP/SiO$_2$ with zeolites, but the selectivity of MP conversion trough the HDO reaction route is remained on the high level exceeding 90%.

Keywords: molybdenum phosphide; hydrodeoxygenation; methyl palmitate; isomerization

---

1. Introduction

In the last decade, the research efforts in the transportation fuels production from renewable sources have grown due to environmental issues and limitation of oil reserves [1–4]. One of the versatile and already commercial routes to produce diesel and aviation fuels is hydrodeoxygenation of triglyceride-containing feedstock, including animal fats, waste frying oils, non-edible vegetable oils and so on [5,6]. Oxygen removal from triglyceride molecules includes several alternative routes [7–9]: hydrodeoxygenation (HDO) or decarboxylation/decarbonylation (HDeCO$_x$). The HDO pathway results in alkanes with the same numbers of carbon atoms in the chain and molecules of water, while HDeCO$_x$ reactions give CO$_x$ molecules and alkanes with the reduced carbon chains [7]. The HDO reaction route is more attractive in terms of atom economy, greenhouse gases production and hydrogen recycle [10,11]. Thus, the current research issue is the design of the catalysts with high activity and high selectivity towards HDO pathway.

Sulfided Ni(Co)Mo/Al$_2$O$_3$ catalysts are widely used and studied in triglycerides and esters hydrodeoxygenation [5–7,12–15], but the necessity of the sulfiding agent feeding to save the catalyst in sulfide state stimulate the development of new catalytic systems containing supported metal Ni [16–21], base metal carbides [22–24], nitrides [25] or phosphides [26–35]. Among them, transition metal phosphide catalysts demonstrate promising catalytic properties in hydrodeoxygenation of fatty acids or alkyl esters [26,28,30,32,36], vegetable oils [33], co-processing of renewable oils with petroleum products [37]. The activity of silica-supported catalysts in the HDO of methyl laurate is decreased in the following order [26]: Ni$_2$P > MoP > CoP–Co$_2$P > WP > Fe$_2$P–FeP.

Nevertheless, the high selectivity of MoP-based catalysts in the conversion of aliphatic oxygenates through the HDO route [26] makes MoP/SiO_2 the system of choice from the consideration of carbon atom economy and effluent gases composition, because carbon oxides make difficult purification of circulating hydrogen.

It was reported that acid sites can affect appreciably the HDO of aliphatic esters [26,29,38–46]. The authors of [26,29,39,40] have proposed, that Brønsted acidic sites of silica-supported nickel phosphide catalysts participate in the reaction of carboxylic ester to carboxylic acid. Also, the rate of ethyl stearate HDO over Ru/TiO_2 catalyst was increased when the catalyst was mixed with the grains of γ-Al_2O_3 [41]. The synergetic effect of Ni_2P/SiO_2 and γ-Al_2O_3 physical mixture was observed in methyl palmitate HDO resulting in the enhancement of methyl palmitate conversion over Ni_2P/SiO_2 catalyst after mixing it with alumina grains [42]. It was observed that Ni_2P catalyst supported on γ-Al_2O_3 exhibits higher conversion of methyl palmitate HDO in comparison with silica-supported one, though both catalysts contain Ni_2P nanoparticles with close mean particle size and nearly the same nickel content [43]. The authors [42,43] proposed that the acid sites of the alumina increase the reaction rate of methyl palmitate hydrolysis, that enhance the methyl palmitate conversion. The assumption was made earlier [44,45], that the Lewis acid sites of the support enhance the rate of hydrolysis of aliphatic esters over alumina-supported sulfide catalysts.

The above examples of activity enhancement in the presence of acidic material allowed us to suggest that the replacement of silica support with the alumina may increase the catalytic activity of MoP catalyst in methyl palmitate HDO without reducing selectivity. The synthesis of supported MoP/γ-Al_2O_3 catalysts is sophisticated task due to interaction between alumina surface and phosphate groups that makes the reduction difficult and prevents the formation of MoP nanoparticles on alumina support [47–49]. Up to now, the silica-supported MoP catalysts have been studied in the HDO reactions, due to relatively easy reduction of phosphate groups that makes possible the formation of highly dispersed MoP phase on silica support [26,50].

The goal of the current study is the investigation of the catalytic properties of mechanical mixtures of MoP/SiO_2 catalyst and materials with different acidity (SiC, γ-Al_2O_3, SAPO-11 and zeolite β) in methyl palmitate hydrodeoxygenation as the representative model component of triglyceride-based feedstock.

2. Results and Discussion

2.1. Catalyst Characterization

Physicochemical properties of MoP/SiO_2, prepared using phosphate and phosphite precursors by TPR are shown in Table 1. MoP_A was prepared using ammonium paramolybdate and ammonium phosphate, MoP_I was prepared using ammonium paramolybdate and phosphorous acid. MoP is a bulk sample, prepared for comparison and XRD analysis. Initial Mo/P molar ratio in impregnating solution was 1. This ratio is usually used to prepare MoP catalysts [26,51] and the samples with this ratio were shown to be more active in hydrotreating reactions [52]. After reduction Mo/P ratio is slightly decreased to 0.91 for MoP_A and MoP and to 0.83 for MoP_I (Table 1). This can be due to Mo losses at impregnation step. Mo/P ratio determined from EDX data differ slightly from the chemical analysis results, but still remains close to Mo/P = 1.

Table 1. Physicochemical properties of SiO_2, MoP/SiO_2 and MoP samples.

Sample	Mo, wt%	P, wt%	Mo/P Molar Ratio in Reduced Catalyst	Mo/P EDX	S_{BET}, m^2/g	D_{pore}, nm	D_{TEM}, nm	XRD Phase	D_{XRD}, nm
SiO_2	–	–	–	–	300	10.6	–	–	–
MoP_A	11.5	4.0	0.91	1.02	178	10.4	1.4	amorphous phase	–
MoP_I	11.5	4.3	0.83	0.97	235	7.9	2.0	amorphous phase	–
MoP	51.7	18.4	0.91	1.17	211	2.0	–	MoP	12

Initial SiO_2 support has S_{BET} = 300 m^2/g and D_{pore} = 10.6 nm. MoP_A and MoP_I have decreased value of S_{BET} and D_{pore} in comparison with the silica support. Both active component and unreduced phosphate groups can block pores and decrease S_{BET}. Bulk MoP has quite high area due to presence of large quantities of citric acid in the precursor, which are transformed to carbon after reduction.

The precursors and PO_x/SiO_2 samples were investigated by H_2-TPR (Figure 1). There are two peaks for the bulk MoP precursor: at 490 and at 720 °C. The first corresponds to $Mo^{+6} \rightarrow Mo^{+4}$ reduction, the second corresponds to $Mo^{+4} \rightarrow Mo^{\delta+}$ and $P^{+5} \rightarrow P^{\delta-}$ phosphide formation. Reduction of phosphates in PO_x/SiO_2 sample occurs at higher temperatures (> 750 °C, Figure 1). H_2-TPR of MoP_A and MoP_I precursors have the maximums at the temperatures similar to the reduction temperature of bulk MoP. H_2-TPR shows that precursor nature does not influence the temperatures of reduction of silica-supported MoP precursors. It was shown that in the case of Ni-based phosphide catalysts during H_2-TPR of phosphite precursors the additional amount of hydrogen is produced, which is detected in H_2-TPR curves [40,43]. For the Mo-phosphide precursors there is only slight difference between H_2-TPR curves of MoP_A and MoP_I precursors (Figure 1). This can be attributed to oxidation of HPO_3^{2-} by paramolybdate after impregnation:

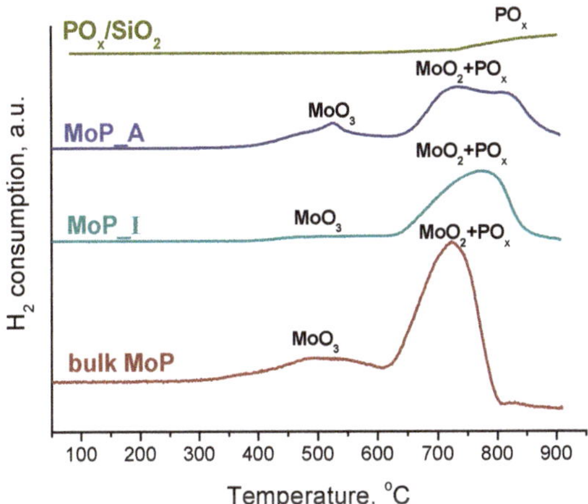

Figure 1. H_2-TPR curves of PO_x/SiO_2, supported MoP_A, MoP_I and bulk MoP oxide precursors.

$$2Mo_7O_{24}^{6-} + 7HPO_3^{2-} + H_2O \rightarrow 14MoO_3^- + 7PO_4^{3-} + 9H^+$$
$$Mo_7O_{24}^{6-} + 7HPO_3^{2-} + 4H_2O \rightarrow 7MoO_3^{2-} + 7PO_4^{3-} + 15H^+$$

These reactions are visible, because presence of Mo(IV/V) species gives the characteristic blue color of the catalysts. Moreover, the reduction peak of MoO_3 in H_2-TPR of MoP_I sample is almost invisible (Figure 1), which corresponds to the presence of Mo, preferably in Mo (IV/V) states.

The calculated consumption of H_2 during MoP_A reduction is 1.1 mmol/g, while the MoP_I consumes 0.80 mmol/g. This difference can be explained by the different states of Mo species.

XRD analysis of unreduced bulk MoP precursor did not show presence of any crystal phases (Figure 2), only increased intensity at low 2θ angles, which corresponds to amorphous carbon residues. XRD pattern of reduced bulk MoP sample include well defied maximums, corresponding to MoP phase (PDF № 24-0771). This pattern also has some amorphous phase corresponding to an increased base line in a 2θ range of 35–40°. D_{XRD} of MoP particles is 12 nm (Table 1). In supported MoP_A and MoP_I samples, there were broad signals of SiO_2 at ~20°; low-intensive broadening is also observed in the range of 35–45° that can be due to highly dispersed MoP (<3 nm) nanoparticles.

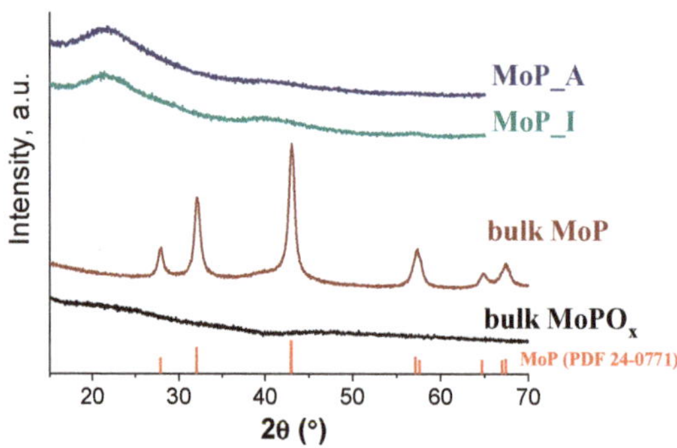

Figure 2. XRD of reduced MoP_A, MoP_I and bulk MoP samples and unreduced MoP precursor.

Figure 3 shows images of TEM analysis of MoP_A, MoP_I and bulk MoP samples. Size distributions of MoP particles are shown for MoP_A and MoP_I samples. Supported MoP_A and MoP_I samples have quite small particles (1.4 and 2.0 nm correspondingly, Table 1) and narrow particle size distributions. MoP sample has particles of 10–50 nm; therefore, there is no particle size distribution on Figure 3 for this sample.

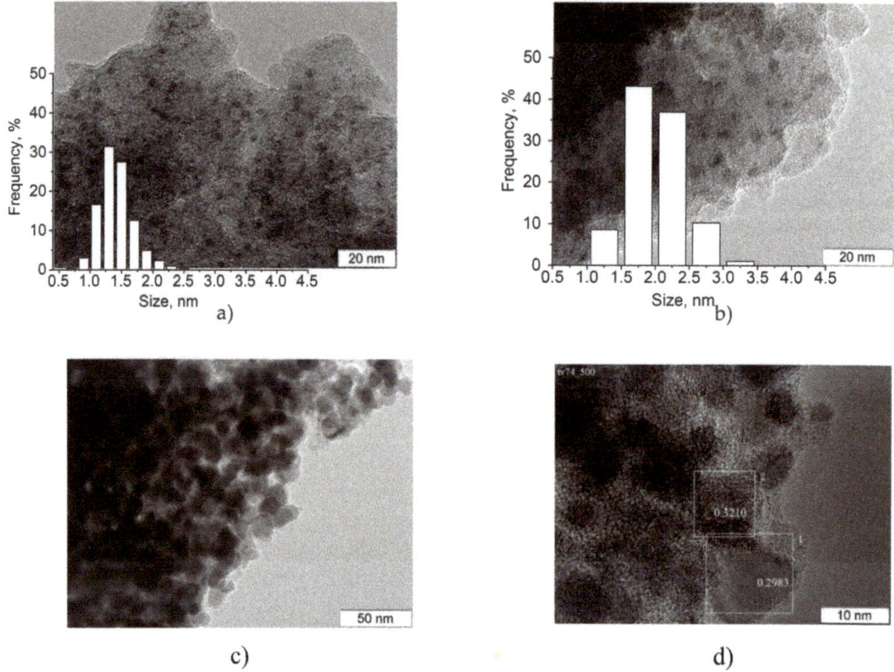

Figure 3. Images of TEM MoP_A (**a**), MoP_I (**b**), bulk MoP (**c,d**).

In preliminary MP HDO experiments (LHSV = 60 h^{-1}, T = 290 °C, P_{H2} = 3 MPa, H$_2$/feed = 600 cm^3/cm^3) 59 and 60% MP conversion were obtained over MoP_A and MoP_I, respectively. MoP_A sample was used for catalytic experiments with different diluters.

The diluters used for catalytic experiments in this work were SiC, γ-Al$_2$O$_3$, SAPO-11 and zeolite β (Table 2). XRD analysis proved corresponding crystal structures for these materials (Figure S1 in Supporting Information).

Table 2. Physicochemical properties of the diluters.

Diluter	S_{BET}, m^2/g	V_{pore}, cm^3/g	D_{pore}, nm	NH$_3$-TPD, 10^{-3} mol/g
SiC	1	–	–	–
γ-Al$_2$O$_3$	235	0.79	13.4	0.421
SAPO-11	295	0.26	3.5	1.11
zeolite β	609	0.49	3.2	1.92

Unlike SiC, an inert material with extremely low S_{BET} (1 m^2/g), γ-Al$_2$O$_3$, SAPO-11 and zeolite β have developed textural structures (Table 2). The acidity of these materials, in accordance with NH$_3$-TPD data, is decreased in the following order: zeolite β > SAPO-11 > γ-Al$_2$O$_3$ (Table 2).

Zeolite β is characterized by Brønsted acid sites and three-dimensional system of channels, with sizes of 0.56×0.56 and 0.66×0.67 nm [53]. SAPO-11 has one-dimensional system of channels with size of 0.40×0.65 nm. SAPO-11 is a mesoporous structure, which has Brønsted acid sites of middle strength [54].

2.2. MP HDO Results

Catalytic properties of physical mixtures of MoP/SiO$_2$ with SiC, γ-Al$_2$O$_3$, SAPO-11 or zeolite β were investigated in MP HDO at T = 290 °C, P_{H2} = 3 MPa, H$_2$/feed = 600 cm^3/cm^3, LHSV = 60 or 10 h^{-1}, using 0.4 or 1.4 g of catalyst and volume ratio of V_{cat}:$V_{diluter}$ = 1:10.5 or 1:1.5. Stable conversion as a function of time on stream was obtained for all tested systems (Figure S2 in Supporting Information). The results of comparative study of these systems in the methyl palmitate HDO are presented on Figure 4, which displays the values of MP conversion (X_{MP}) and oxygen conversion (X_O). MoP/SiO$_2$ sample diluted with inert SiC grains displays the lowest X_{MP} (59%) and X_O (55%) values, whereas X_{MP} and X_O data for systems containing γ-Al$_2$O$_3$, zeolite β and SAPO-11 are higher than 90%. It is well known that HDO of fatty acid esters includes consecutive and parallel reactions: hydrogenolysis of C–O (HDO) and C–C (HDeCO$_x$) bonds, hydrolysis of ester, hydrogenation of double bounds, dehydration of alcohols and esterification of acids, which occur over metal and acid active sites of catalysts, supports and diluters [26,42,55]. In particular, MP can be transformed at metal centers through hydrogenolysis to acid or aldehyde, or via acid-catalyzed hydrolysis to acid [56]. Thus, the higher conversions of MP over MoP/SiO$_2$ catalyst in the presence of Al$_2$O$_3$, SAPO-11 or zeolite β can be explained by the higher rate of methyl palmitate hydrolysis, which is favored by the acid sites of these materials.

The main HDO product over MoP/SiO$_2$–SiC and MoP/SiO$_2$–γ-Al$_2$O$_3$ is hexadecane at 100% conversion of MP; the selectivity of MP conversion through the HDO pathway exceeds 96% (Figure 5). Thus, the use of γ-Al$_2$O$_3$ instead of SiC in the mixture with MoP/SiO$_2$ increases the MP conversion, but has minor impact on the product distribution. The dilution of MoP/SiO$_2$ with the same amount of SAPO-11 or zeolite β instead of SiC decreases the selectivity of C$_{16}$ alkane formation in the course of MP HDO and increases the quantity of iso-C$_{16}$, iso-C$_{15}$ and C$_6$–C$_{14}$ hydrocarbons among the liquid products. Especially increased formation of hydrocarbons with ≤14 carbon atoms in the hydrocarbon chains is observed over MoP/SiO$_2$–zeolite β catalytic system. Apparently, Brønsted acid sites in MoP/SiO$_2$–SAPO-11 and MoP/SiO$_2$–zeolite β result in partial cracking and structural isomerization of linear C$_{15}$ and C$_{16}$ hydrocarbons, which are formed in the HDO reaction of the ester. In these cases, the secondary reactions of the alkanes that are formed in MP HDO make it unattainable to calculate selectivities of HDO and HDeCO$_x$ pathways using the results of liquid phase analysis. To estimate the effect of SAPO-11 and zeolite β on the selectivity of MP conversion the amounts of carbon oxides were compared for MoP/SiO$_2$–SiC, MoP/SiO$_2$–SAPO-11 and MoP/SiO$_2$–zeolite β (Figure 6). The results obtained let us see that the use of MoP/SiO$_2$–SAPO-11 and MoP/SiO$_2$–zeolite β induces only negligible increase in the amounts of carbon oxides in the gas phase in comparison with MoP/SiO$_2$–SiC system. Appreciable amounts of light hydrocarbons with the carbon number ≤5 were detected in the gas-phase products of MP HDO over MoP/SiO$_2$–zeolite β that is the evidence of alkane overcracking in the presence of zeolite β.

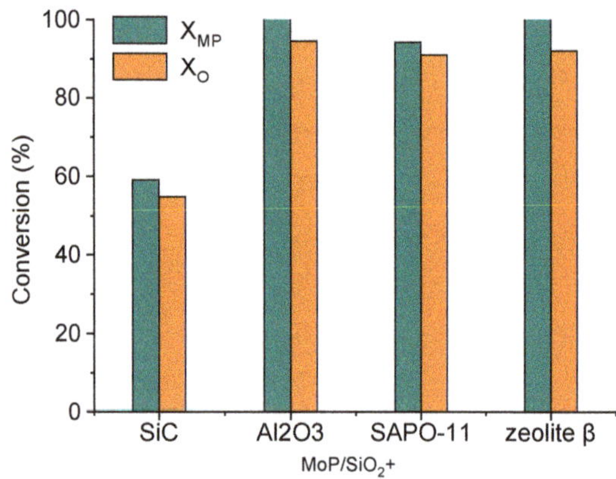

Figure 4. MP and O conversion in MP HDO over MoP/SiO$_2$ catalyst diluted with different materials. V_{cat} = 0.4 cm^3, LHSV = 60 h^{-1}, T = 290 °C, P_{H2} = 3 MPa, H$_2$/feed = 600 cm^3/cm^3.

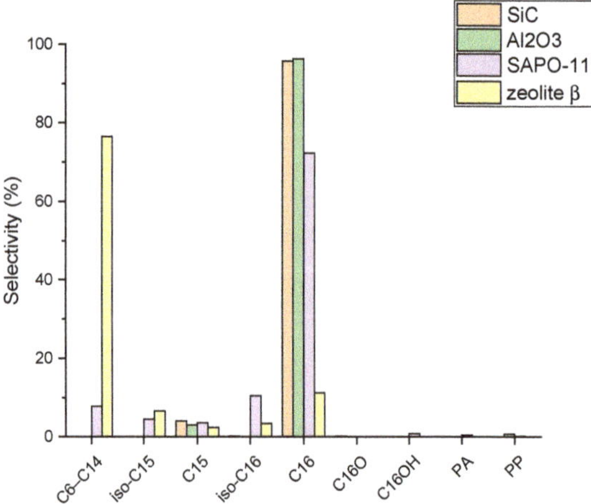

Figure 5. Selectivities of MP HDO products over MoP/SiO$_2$ catalyst diluted with different materials (X_{MP} = 100%). C16O—hexadecanal, C16OH—hexadecanol, PA—palmitic acid, PP—palmityl palmitate. V_{cat} = 1.8 cm^3, LHSV = 10 h^{-1}, T = 290 °C, P_{H2} = 3 MPa, H$_2$/feed = 600 cm^3/cm^3.

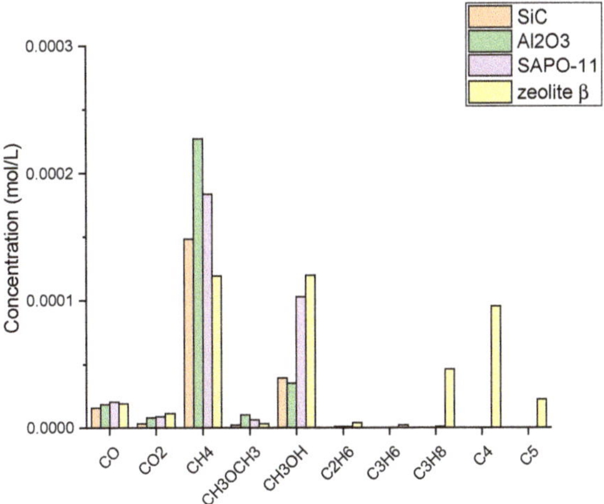

Figure 6. Concentration of gas-phase products of MP HDO over MoP/SiO$_2$ catalyst diluted with different materials (X_{MP} = 100%). V_{cat} = 1.8 cm^3, LHSV = 10 h^{-1}, T = 290 °C, P_{H2} = 3 MPa, H$_2$/feed = 600 cm^3/cm^3.

The above results show that the use of the MoP/SiO$_2$ catalyst in the mixtures with acidic materials, like γ-Al$_2$O$_3$, SAPO-11 and zeolite β improves the catalytic activity of MoP catalyst in the hydrodeoxygenation of MP without reducing selectivity through HDO pathway. MoP/SiO$_2$–γ-Al$_2$O$_3$ system gives hexadecane with the selectivity up to 96%, while MoP/SiO$_2$–SAPO-11 produces iso-alkane and some amount of C$_6$–C$_{14}$ hydrocarbons. Therefore, MoP catalysts on SAPO-11 are perspective in one-pot HDO/hydroisomerization of aliphatic esters.

3. Materials and Methods

3.1. Catalysts Preparation

Commercial silica "KSKG" (S_{BET} = 300 m^2/g, V_{pore} = 0.80 cm^3/g, D_{pore} = 10.6 nm) was obtained from ChromAnalit (Moscow, Russia), it was used as support material. Commercial SiC (Chelyabinsk Plant of Abrasive Materials, Chelyabinsk, Russia), alumina "IKGO-1" (γ-Al$_2$O$_3$, Promkataliz, Ryazan, Russia), zeolite β in H form (Angarsk catalyst and organic synthesis plant, Angarsk, Russia), SAPO-11 (Zeolyst International, Conshohocken, PA, USA) were utilized as diluters. Before catalysts preparation and experiments silica and alumina were dried (at 110 °C, 7 h) and calcined (at 500 °C, 4 h). Afterwards the materials were crushed using mortar and pestle, and sieved to obtain granules with size of 0.25–0.50 mm.

Preparation method of MoP/SiO$_2$ samples

Incipient wetness impregnation method was used to prepare precursors of MoP/SiO$_2$ samples. Mo/P molar ratio in impregnation water solution was taken equal 1 (Mo content~11.5 wt%). The label MoP_A was chosen for the MoP/SiO$_2$ sample, which was prepared from phosphate solution, the label MoP_I—for the MoP/SiO$_2$ sample, which was prepared from phosphite solution.

Phosphate method (MoP_A). To synthesis samples using this technique 2.357 g of (NH$_4$)$_2$HPO$_4$ (Alfa Aesar, Kandel, Germany) and 3.151 g of (NH$_4$)$_6$Mo$_7$O$_{24}$·4H$_2$O (Reakhim, Moscow, Russia) were dissolved under stirring in distilled water. Subsequently the support (8.5 g SiO$_2$) was impregnated by this solution. After drying on air overnight (at room temperature) the precursors were dried for 4 h at 110 °C. Calcination of the dried samples was carried out at 500 °C for 4 h, and reduction of the calcined samples (TPR) was performed in H$_2$ flow (100 mL/(min·g)). TPR has the following program: heating to 370 °C (with the rate of 3 °C/min), heating to 650 °C (with the rate of 1 °C/min), keeping the samples at 650 °C (for 1 h). Passivation of the reduced samples (to prevent oxidation on air atmosphere) was carried out at room temperature in 1 vol% O_2/He (at 40 mL/(min·g) for 2 h).

Phosphite method (MoP_I). First, silica support (8.5 g) was impregnated with aqueous solution containing 3.151 g of $(NH_4)_6Mo_7O_{24} \cdot 4H_2O$ and 1.464 g of H_3PO_3. Then, after drying overnight and subsequently at 80 °C (for 24 h), the samples were reduced (TPR) in H_2 flow (100 mL/(min·g)). TPR was carried as follows: heating to 650 °C (with the rate of 1 °C/min), keeping at 650 (for 1 h). The same passivation technique was carried out for MoP_I samples as in case of MoP_A samples.

PO_x/SiO_2 sample. SiO_2 (5 g) was impregnated by the solution of 2.357 g of $(NH_4)_2HPO_4$ in distilled water. Drying was performed at room temperature overnight and at 110 °C (for 4 h). After calcination at 500 °C (for 4 h), the precursor was reduced and passivated according the same TPR and passivation programs as for MoP_A and MoP_I precursors.

Bulk MoP reference sample preparation method. The bulk molybdenum phosphide sample (bulk MoP) was prepared as follows: to the water solution of $(NH_4)_2HPO_4$ was added water solution of $(NH_4)_6Mo_7O_{24}$. The Mo/P initial molar ratio was equal to 1. After evaporation of the water, the obtained sample was dried at 87 °C (for 48 h) and then at 124 °C (for 24 h). Calcination of the sample was performed at 500 °C (for 5 h). Afterwards, TPR was carried out (at H_2 flow of 150 mL/(min·g)): heating to 650 °C (with the rate of 1 °C/min), keeping at 650 °C (for 1 h). The obtained bulk sample was passivated in flow of 1 vol.% O_2/He (80 mL/(min·g), for 3 h).

3.2. Catalysts Characterization

Nitrogen physisorption was used to determine textural characteristics of the MoP, MoP/SiO_2 and diluters. It was carried out at 77 K on ASAP 2400 instrument (USA).

Inductively coupled plasma atomic emission spectroscopy (ICP-AES) was used to perform elemental analysis of the MoP and MoP/SiO_2 samples. ICP-AES was carried out on Optima 4300 DV (Perkin Elmer, Villebon-sur-Yvette, France).

Bruker D8 Advance X-ray diffractometer (Bruker, Bremen, Germany) was used to carry out XRD analysis. CuK_α radiation (with $\lambda = 1.5418$ Å) was used, and 2θ scanning range was from 10 to 70. To determine crystal phases JCPDS database was used [57]. Scherrer equation was used to calculate average crystallite size (D_{XRD}).

Transmission electron microscope JEM-2010 (JEOL, Tokyo, Japan) was used to obtain high resolution transmission electron microscopy (TEM) images. The accelerating voltage and resolution were 200 kV and 0.14 nm correspondingly. EDX spectrometer was used to analyze elemental composition.

H_2 temperature-programmed reduction (H_2-TPR) was performed to determine reduction temperatures of the prepared samples. The sample mass was about 0.10 g, it was heated in 10 vol.% H_2/Ar (60 mL/min, with rate of 10 °C/min) from 80 °C to 900 °C. Thermal conductivity detector (TCD) was used to determine H_2 consumption.

3.3. Experimental Setup and Procedure

Methyl palmitate (hexadecanoic acid methyl ester, ≥97%), *n*-dodecane (≥99%) and *n*-octane (≥99.5%) were obtained from Sigma-Aldrich (St. Louis, MO, USA), Acros Organics (New Jersey, NJ, USA) and Kriokhrom (Saint Petersburg, Russia), correspondingly.

A trickle-bed reactor was used to evaluate catalytic activity of the prepared MoP catalysts in methyl palmitate HDO (Figure S3 in Supporting Information). The reaction conditions: T = 290 °C, $P_{H2} = 3.0$ MPa, volume ratio H_2/feed = 600, LHSV = 60 or 10 h^{-1}. The liquid feed consisted of: 10 wt% of methyl palmitate, 0.5 wt% of *n*-octane (internal standard), the rest is *n*-dodecane. The catalysts particles were diluted with SiC, Al_2O_3, SAPO-11 or zeolite β (0.2–0.3 mm) in a ratio 1:10.5 or 1:1.5, placed in the catalytic reactor between. Above and under the catalytic layer there were SiC layers. Before methyl palmitate HDO experiments reduction of the catalyst precursors was performed in situ by TPR technique described earlier in catalysts preparation section. Collecting of liquid product effluent was carried out hourly until the reaction reaches steady-state. For all experiments time on stream = 8 h (Figure S2 in Supporting Information).

X_{MP}—methyl palmitate conversion was calculated as follows:

$$X_{MP} = \left(1 - \frac{C_{MP}}{C_{MP}^o}\right) \cdot 100\% \tag{1}$$

$c^o{}_{MP}$ and c_{MP} (mol/L)—methyl palmitate concentration in the initial feed and in the liquid product correspondingly.

X_O—oxygen-containing compounds conversion was calculated as follows:

$$X_{OCC} = \left(1 - \frac{C_O}{C_O^o}\right) \cdot 100\% \tag{2}$$

$c^o{}_O$ and c_O (mol/L)—oxygen concentration in the initial feed and in the liquid product correspondingly.

S_i—selectivity was calculated as follows:

$$S_i = \left(\frac{c_i}{c_{MP}^o - c_{MP}}\right) \cdot 100\% \tag{3}$$

c_i (mol/L)—concentration of the ith compound in the liquid product.

3.4. Analysis of the Reaction Products

7000 GC/MS Triple QQQ GC System 7890A (Agilent Technologies, Santa Clara, CA, USA) was used to identify HDO products. The GC was equipped with VF-5MS column (length 30 m, inner diameter 0.25 mm, film thickness 0.25 µm). Quantitative analysis was performed using GC with flame ionization detector (FID, Agilent 6890N, USA) and an HP-1MS column (length 30 m, inner diameter 0.32 mm, film thickness 1.0 µm). Carbon balance for all experiments was not lower than 97%.

Elemental analyzer Vario EL Cube (Elementar Analysensysteme GmbH, Langenselbold, Germany) was used to determine oxygen content in the liquid reaction mixture.

The concentration of gas-phase products was determined using GC Khromos-1000 with FID (Khromos, Dzerzhinsk, Russia). CO and CO_2 were analyzed after methanation over Pd catalyst.

4. Conclusions

In this work catalytic properties of MoP/SiO_2 catalyst in MP HDO were studied in mechanical mixture with materials having different acidity: SiC, γ-Al_2O_3, SAPO-11 and zeolite β.

A method was chosen for the preparation of SiO_2-supported MoP catalyst. Two samples were prepared using $(NH_4)_6Mo_7O_{24}$ and different P precursors: $(NH_4)_2HPO_4$ and H_3PO_3 (MoP_A and MoP_I samples, correspondingly). It was shown that H_2-TPR curves for MoP_A and MoP_I samples are coinciding regardless of P precursor that can be explained by the oxidation of H_3PO_3 in the reaction with ammonium paramolybdate during impregnation and drying of support. The silica-supported MoP catalysts samples were synthesized by reduction (TPR) of the precursors at 650 °C, and according to TEM data, the particle sizes of MoP nanoparticles were 1.4 and 2.0 nm for MoP_A and MoP_I samples, respectively. Both catalysts (MoP_A and MoP_I) display similar activity in the MP hydrodeoxygenation; MoP_A sample (designated as MoP/SiO_2) was chosen for the experiments with different diluters.

The comparative study of MoP/SiO_2–SiC, MoP/SiO_2–γ-Al_2O_3, MoP/SiO_2–SAPO-11 and MoP/SiO_2–zeolite β systems has demonstrated the enhancement of MoP/SiO_2 catalyst activity in MP HDO in mechanical mixtures with alumina, SAPO-11 and zeolite β. The higher conversions of MP over MoP/SiO_2 catalyst in the presence of Al_2O_3, SAPO-11 and zeolite β can be explained by the higher rate of methyl palmitate hydrolysis, which is favored by the acid sites of these materials. The use of γ-Al$_2$O$_3$, SAPO-11 and zeolite β instead of SiC in the mixture with MoP/SiO$_2$ increases the MP conversion, but has minor impact on the HDO selectivity. It was shown that selectivities of methyl palmitate conversion through the HDO pathways are nearly the same for the different systems, because the amounts of CO and CO$_2$ in the gas-phase products differ only slightly. The main HDO product observed in the liquid products of MP HDO over MoP/SiO$_2$–SiC and MoP/SiO$_2$–γ-Al$_2$O$_3$ is hexadecane; the selectivity of HDO pathway exceeds 96% at 100% of MP conversion. The use of MoP/SiO$_2$ in the mixture with the SAPO-11 or zeolite β instead of SiC changes the product distribution due to partial cracking and structural isomerization of linear C$_{15}$ and C$_{16}$ hydrocarbons, which are formed in HDO reaction of ester. Thus, MoP/SiO$_2$–SAPO-11 system produces iso-alkane and some amount of C$_6$–C$_{14}$ hydrocarbons, while increased formation of hydrocarbons with $\leq$ 14 carbon atoms in the hydrocarbon chains is observed over MoP/SiO$_2$–zeolite β catalytic system.

The obtained results show, that the use of the MoP/SiO$_2$ catalyst in the mixtures with γ-Al$_2$O$_3$, SAPO-11 and zeolite β enhances the catalytic activity of MoP catalyst in the hydrodeoxygenation of MP without reducing selectivity through HDO pathway. The product distribution in the liquid phase depends on the dilutor nature. Linear alkanes are produced in the presence of γ-Al$_2$O$_3$. SAPO-11 can be considered as the perspective material for the one-pot HDO/hydroisomerization of aliphatic esters.

Supplementary Materials: The following are available online at http://www.mdpi.com/2073-4344/10/1/45/s1, Figure S1: XRD patterns of materials used as catalyst diluters, Figure S2: MP conversion vs time on stream for MoP/SiO$_2$ catalysts diluted with different materials, Figure S3: The scheme of the experimental setup.

Author Contributions: Formal analysis, I.V.S. and I.V.D.; TEM analysis, E.Y.G.; XRD analysis, V.P.P.; writing and the draft preparation, I.V.D., I.V.S. and G.A.B.; reviewing and editing, I.V.D., G.A.B. and I.V.S.; supervision, G.A.B. All authors have read and agreed to the published version of the manuscript.

Funding: This work was supported by Ministry of Science and Higher Education of the Russian Federation (project AAAA-A17-117041710075-0). The authors of the paper are grateful to M.V. Shashkov and V.A. Utkin for analysis and identification of HDO reaction products by GC/MS.

Conflicts of Interest: The authors declare no conflict of interest.

References

1. Choudhary, T.V.; Phillips, C.B. Renewable fuels via catalytic hydrodeoxygenation. *Appl. Catal. A Gen.* **2011**, *397*, 1–12. [CrossRef]

2. Serrano-Ruiz, J.C.; Ramos-Fernández, E.V.; Sepúlveda-Escribano, A. From biodiesel and bioethanol to liquid hydrocarbonfuels: New hydrotreating and advanced microbial technologies. *Energy Environ. Sci.* **2012**, *5*, 5638–5652. [CrossRef]

3. Huber, G.W.; Iborra, S.; Corma, A. Synthesis of transportation fuels from biomass: Chemistry, catalysts, and engineering. *Chem. Rev.* **2006**, *106*, 4044–4098. [CrossRef] [PubMed]

4. Melero, J.A.; Iglesias, J.; Garcia, A. Biomass as renewable feedstock in standard refinery units. Feasibility, opportunities and challenges. *Energy Environ. Sci.* **2012**, *5*, 7393–7420. [CrossRef]

5. Khan, S.; Kay Lup, A.N.; Qureshi, K.M.; Abnisa, F.; Wan Daud, W.M.A.; Patah, M.F.A. A review on deoxygenation of triglycerides for jet fuel range hydrocarbons. *J. Anal. Appl. Pyrolysis* **2019**, *140*, 1–24. [CrossRef]

6. Why, E.S.K.; Ong, H.C.; Lee, H.V.; Gan, Y.Y.; Chen, W.-H.; Chong, C.T. Renewable aviation fuel by advanced hydroprocessing of biomass: Challenges and perspective. *Energy Convers. Manag.* **2019**, *199*, 112015. [CrossRef]

7. Kubička, D.; Tukač, V. Chemical Engineering for Renewables Conversion. Advances in Chemical Engineering. In *Advances in Chemical Engineering*; Murzin, D.Y., Ed.; Elsevier: Amsterdam, The Netherlands, 2013; Volume 42, pp. 141–194. ISBN 9780123865052.

8. Zharova, P.A.; Chistyakov, A.V.; Shapovalov, S.S.; Pasynskii, A.A.; Tsodikov, M.V. Original Pt-Sn/Al$_2$O$_3$ catalyst for selective hydrodeoxygenation of vegetable oils. *Energy* **2019**, *172*, 18–25. [CrossRef]

9. Kukushkin, R.G.; Reshetnikov, S.I.; Zavarukhin, S.G.; Eletskii, P.M.; Yakovlev, V.A. Kinetics of the Hydrodeoxygenation of Ethyl Ester of Decanoic Acid over the Ni–Cu–Mo/ Al$_2$O$_3$ Catalyst. *Catal. Ind.* **2019**, *11*, 191–197. [CrossRef]

10. Janampelli, S.; Darbha, S. Selective and reusable Pt-WOx/ Al$_2$O$_3$ catalyst for deoxygenation of fatty acids and their esters to diesel-range hydrocarbons. *Catal. Today* **2018**, *309*, 219–226. [CrossRef]

11. Gosselink, R.W.; Stellwagen, D.R.; Bitter, J.H. Tungsten-Based Catalysts for Selective Deoxygenation. *Angew. Chem. Int. Ed.* **2013**, *52*, 5089–5092. [CrossRef]

12. Wang, H.; Rogers, K.; Zhang, H.; Li, G.; Pu, J.; Zheng, H.; Lin, H.; Zheng, Y.; Ng, S. The Effects of Catalyst Support and Temperature on the Hydrotreating of Waste Cooking Oil (WCO) over CoMo Sulfided Catalysts. *Catalysts* **2019**, *9*, 689. [CrossRef]

13. Zhao, X.; Wei, L.; Cheng, S.; Julson, J. Review of Heterogeneous Catalysts for Catalytically Upgrading Vegetable Oils into Hydrocarbon Biofuels. *Catalysts* **2017**, *7*, 83. [CrossRef]

14. Pinheiro, A.; Dupassieux, N.; Hudebine, D.; Geantet, C. Impact of the Presence of Carbon Monoxide and Carbon Dioxide on Gas Oil Hydrotreatment: Investigation on Liquids from Biomass Cotreatment with Petroleum Cuts. *Energy Fuels* **2011**, *25*, 804–812. [CrossRef]

15. Philippe, M.; Richard, F.; Hudebine, D.; Brunet, S. Transformation of dibenzothiophenes model molecules over CoMoP/ Al$_2$O$_3$ catalyst in the presence of oxygenated compounds. *Appl. Catal. B Environ.* **2013**, *132*, 493–498. [CrossRef]

16. Liu, S.; Zhu, Q.; Guan, Q.; He, L.; Li, W. Bio-aviation fuel production from hydroprocessing castor oil promoted by the nickel-based bifunctional catalysts. *Bioresour. Technol.* **2015**, *183*, 93–100. [CrossRef]

17. Li, X.; Luo, X. Preparation of mesoporous activated carbon supported Ni catalyst for deoxygenation of stearic acid into hydrocarbons. *Environ. Prog. Sustain. Energy* **2015**, *34*, 607–612. [CrossRef]

18. Kukushkin, R.G.; Bulavchenko, O.A.; Kaichev, V.V.; Yakovlev, V.A. Influence of Mo on catalytic activity of Ni-based catalysts in hydrodeoxygenation of esters. *Appl. Catal. B Environ.* **2015**, *163*, 531–538. [CrossRef]

19. Yakovlev, V.A.; Khromova, S.A.; Sherstyuk, O.V.; Dundich, V.O.; Ermakov, D.Y.; Novopashina, V.M.; Lebedev, M.Y.; Bulavchenko, O.; Parmon, V.N. Development of new catalytic systems for upgraded bio-fuels production from bio-crude-oil and biodiesel. *Catal. Today* **2009**, *144*, 362–366. [CrossRef]

20. Zuo, H.; Liu, Q.; Wang, T.; Ma, L.; Zhang, Q.; Zhang, Q. Hydrodeoxygenation of Methyl Palmitate over Supported Ni Catalysts for Diesel-like Fuel Production. *Energy Fuels* **2012**, *26*, 3747–3755. [CrossRef]

21. Ochoa-Hernández, C.; Yang, Y.; Pizarro, P.; de la Peña O'Shea, V.A.; Coronado, J.M.; Serrano, D.P. Hydrocarbons production through hydrotreating of methyl esters over Ni and Co supported on SBA-15 and Al-SBA-15. *Catal. Today* **2013**, *210*, 81–88. [CrossRef]

22. Ki Kim, S.; Yoon, D.; Lee, S.-C.; Kim, J. Mo2C/Graphene Nanocomposite as a Hydrodeoxygenation Catalyst for the Production of Diesel Range Hydrocarbons. *ACS Catal.* **2015**, *5*, 3292–3303. [CrossRef]

23. Hollak, S.A.W.; Gosselink, R.W.; van Es, D.S.; Bitter, J.H. Comparison of tungsten and molybdenum carbide catalysts for the hydrodeoxygenation of oleic acid. *ACS Catal.* **2013**, *3*, 2837–2844. [CrossRef]

24. Wang, H.; Yan, S.; Salley, S.O.; Simon Ng, K.Y. Support effects on hydrotreating of soybean oil over NiMo carbide catalyst. *Fuel* **2013**, *111*, 81–87. [CrossRef]

25. Monnier, J.; Sulimma, H.; Dalai, A.; Caravaggio, G. Hydrodeoxygenation of oleic acid and canola oil over alumina-supported metal nitrides. *Appl. Catal. A Gen.* **2010**, *382*, 176–180. [CrossRef]

26. Chen, J.; Shi, H.; Li, L.; Li, K. Deoxygenation of methyl laurate as a model compound to hydrocarbons on transition metal phosphide catalysts. *Appl. Catal. B Environ.* **2014**, *144*, 870–884. [CrossRef]

27. Chen, J.; Yang, Y.; Shi, H.; Li, M.; Chu, Y.; Pan, Z.; Yu, X. Regulating product distribution in deoxygenation of methyl laurate on silica-supported Ni–Mo phosphides: Effect of Ni/Mo ratio. *Fuel* **2014**, *129*, 1–10. [CrossRef]

28. Shi, H.; Chen, J.; Yang, Y.; Tian, S. Catalytic deoxygenation of methyl laurate as a model compound to hydrocarbons on nickel phosphide catalysts: Remarkable support effect. *Fuel Process. Technol.* **2013**, *116*, 161–170. [CrossRef]

29. Yang, Y.; Chen, J.; Shi, H. Deoxygenation of Methyl Laurate as a Model Compound to Hydrocarbons on Ni$_2$P/SiO$_2$, Ni$_2$P/MCM-41, and Ni$_2$P /SBA-15 catalysts with different dispersions. *Energy Fuels* **2013**, *27*, 3400–3409. [CrossRef]

30. Yang, Y.; Ochoa-Hernández, C.; de la Peña O'Shea, V.A.; Coronado, J.M.; Serrano, D.P. Ni$_2$P/SBA-15 As a Hydrodeoxygenation Catalyst with Enhanced Selectivity for the Conversion of Methyl Oleate Into n-Octadecane. *ACS Catal.* **2012**, *2*, 592–598. [CrossRef]

31. Yang, Y.; Ochoa-Hernández, C.; Pizarro, P.; de la Peña O'Shea, V.A.; Coronado, J.M.; Serrano, D.P. Synthesis of Nickel Phosphide Nanorods as Catalyst for the Hydrotreating of Methyl Oleate. *Top. Catal.* **2012**, *55*, 991–998. [CrossRef]

32. Yang, Y.; Ochoa-Hernández, C.; Pizarro, P.; de la Peña O'Shea, V.A.; Coronado, J.M.; Serrano, D.P. Influence of the Ni/P ratio and metal loading on the performance of NixPy/SBA-15 catalysts for the hydrodeoxygenation of methyl oleate. *Fuel* **2015**, *144*, 60–70. [CrossRef]

33. Zarchin, R.; Rabaev, M.; Vidruk-Nehemya, R.; Landau, M.V.; Herskowitz, M. Hydroprocessing of soybean oil on nickel-phosphide supported catalysts. *Fuel* **2015**, *139*, 684–691. [CrossRef]

34. Pan, Z.; Wang, R.; Li, M.; Chu, Y.; Chen, J. Deoxygenation of methyl laurate to hydrocarbons on silica-supported Ni-Mo phosphides: Effect of calcination temperatures of precursor. *J. Energy Chem.* **2015**, *24*, 77–86. [CrossRef]

35. Xue, Y.; Guan, Q.; Li, W. Synthesis of bulk and supported nickel phosphide using microwave radiation for hydrodeoxygenation of methyl palmitate. *RSC Adv.* **2015**, *5*, 53623–53628. [CrossRef]

36. Zheng, Z.; Li, M.-F.; Chu, Y.; Chen, J.-X. Influence of CS_2 on performance of Ni_2P/SiO_2 for deoxygenation of methyl laurate as a model compound to hydrocarbons: Simultaneous investigation on catalyst deactivation. *Fuel Process. Technol.* **2015**, *134*, 259–269. [CrossRef]

37. Lee, S.I.; Kim, D.W.; Jeon, H.J.; Ju, S.J.; Ryu, J.W.; Oh, S.H.; Koh, J.H. Metal Phosphorus Compound for Preparing Biodiesel and Method Preparing Biodiesel Using the Same. U.S. Patent US2014/0150332A1, 5 June 2014.

38. Alvarez-Galvan, M.; Campos-Martin, J.; Fierro, J. Transition Metal Phosphides for the Catalytic Hydrodeoxygenation of Waste Oils into Green Diesel. *Catalysts* **2019**, *9*, 293. [CrossRef]

39. Deliy, I.; Shamanaev, I.; Gerasimov, E.; Pakharukova, V.; Yakovlev, I.; Lapina, O.; Aleksandrov, P.; Bukhtiyarova, G. HDO of Methyl Palmitate over Silica-Supported Ni Phosphides: Insight into Ni/P Effect. *Catalysts* **2017**, *7*, 298. [CrossRef]

40. Shamanaev, I.V.; Deliy, I.V.; Aleksandrov, P.V.; Gerasimov, E.Y.; Pakharukova, V.P.; Kodenev, E.G.; Ayupov, A.B.; Andreev, A.S.; Lapina, O.B.; Bukhtiyarova, G.A. Effect of precursor on the catalytic properties of Ni_2P/SiO_2 in methyl palmitate hydrodeoxygenation. *RSC Adv.* **2016**, *6*, 30372–30383. [CrossRef]

41. He, L.; Wu, C.; Cheng, H.; Yu, Y.; Zhao, F. Highly selective and efficient catalytic conversion of ethyl stearate into liquid hydrocarbons over a Ru/TiO_2 catalyst under mild conditions. *Catal. Sci. Technol.* **2012**, *2*, 1328. [CrossRef]

42. Shamanaev, I.; Aleksandrov, P.; Kodenev, E.; Pakharukova, V.; Gerasimov, E.; Deliy, I.; Bukhtiyarova, G. Synergetic Effect of Ni_2P/SiO_2 and γ-Al_2O_3 Physical Mixture in Hydrodeoxygenation of Methyl Palmitate. *Catalysts* **2017**, *7*, 329. [CrossRef]

43. Deliy, I.; Shamanaev, I.; Aleksandrov, P.; Gerasimov, E.; Pakharukova, V.; Kodenev, E.; Yakovlev, I.; Lapina, O.; Bukhtiyarova, G. Support Effect on the Performance of Ni2P Catalysts in the Hydrodeoxygenation of Methyl Palmitate. *Catalysts* **2018**, *8*, 515. [CrossRef]

44. Laurent, E.; Delmon, B. Influence of water in the deactivation of a sulfided NiMo?-Al2O3 catalyst during hydrodeoxygenation. *J. Catal.* **1994**, *146*, 281–291. [CrossRef]

45. Şenol, O.İ.; Viljava, T.-R.; Krause, A.O.I. Hydrodeoxygenation of methyl esters on sulphided NiMo/γ- Al_2O_3 and CoMo/γ-Al_2O_3 catalysts. *Catal. Today* **2005**, *100*, 331–335. [CrossRef]

46. Shamanaev, I.V.; Deliy, I.V.; Aleksandrov, P.V.; Reshetnikov, S.I.; Bukhtiyarova, G.A. Methyl palmitate hydrodeoxygenation over silica-supported nickel phosphide catalysts in flow reactor: Experimental and kinetic study. *J. Chem. Technol. Biotechnol.* **2019**. [CrossRef]

47. Prins, R.; Bussell, M.E. Metal Phosphides: Preparation, Characterization and Catalytic Reactivity. *Catal. Lett.* **2012**, *142*, 1413–1436. [CrossRef]

48. Liu, X.; Xu, L.; Zhang, B. Essential elucidation for preparation of supported nickel phosphide upon nickel phosphate precursor. *J. Solid State Chem.* **2014**, *212*, 13–22. [CrossRef]

49. Peroni, M.; Mancino, G.; Baráth, E.; Gutiérrez, O.Y.; Lercher, J.A. Bulk and γ-Al_2O_3 supported Ni_2P and MoP for hydrodeoxygenation of palmitic acid. *Appl. Catal. B Environ.* **2016**, *180*, 301–311. [CrossRef]

50. Li, K.; Wang, R.; Chen, J. Hydrodeoxygenation of Anisole over Silica-Supported Ni_2P, MoP, and NiMoP Catalysts. *Energy Fuels* **2011**, *25*, 854–863. [CrossRef]

51. Bui, P.; Cecilia, J.A.; Oyama, S.T.; Takagaki, A.; Infantes-Molina, A.; Zhao, H.; Li, D.; Rodríguez-Castellón, E.; Jiménez López, A. Studies of the synthesis of transition metal phosphides and their activity in the hydrodeoxygenation of a biofuel model compound. *J. Catal.* **2012**, *294*, 184–198. [CrossRef]

52. Montesinos-Castellanos, A.; Zepeda, T.A.; Pawelec, B.; Fierro, J.L.G.; De Los Reyes, J.A. Preparation, characterization, and performance of alumina-supported nanostructured Mo-phosphide systems. *Chem. Mater.* **2007**, *19*, 5627–5636. [CrossRef]

53. Newsam, J.M.; Treacy, M.M.J.; Koetsier, W.T.; Gruyter, C.B.D. Structural Characterization of Zeolite Beta. *Proc. R. Soc. A Math. Phys. Eng. Sci.* **1988**, *420*, 375–405. [CrossRef]

54. Flanigen, E.M.; Lok, B.M.; Patton, R.L.; Wilson, S.T. Aluminophosphate molecular sieves and the periodic table. *Pure Appl. Chem.* **1986**, *58*, 1351–1358. [CrossRef]

55. Peroni, M.; Lee, I.; Huang, X.; Baráth, E.; Gutiérrez, O.Y.; Lercher, J.A. Deoxygenation of Palmitic Acid on Unsupported Transition-Metal Phosphides. *ACS Catal.* **2017**, *7*, 6331–6341. [CrossRef]

56. Gosselink, R.W.; Hollak, S.A.W.; Chang, S.W.; Van Haveren, J.; De Jong, K.P.; Bitter, J.H.; van Es, D.S. Reaction pathways for the deoxygenation of vegetable oils and related model compounds. *ChemSusChem* **2013**, *6*, 1576–1594. [CrossRef] [PubMed]

57. File, P.D. *JCPDS—International Centre for Diffraction Data*; ICDD: Swarthmore, PA, USA, 1997.

 *catalysts*

Article

Catalytic Pyrolysis of Aliphatic Carboxylic Acids into Symmetric Ketones over Ceria-Based Catalysts: Kinetics, Isotope Effect and Mechanism

Tetiana Kulik [1,*], **Borys Palianytsia** [1] **and Mats Larsson** [2]

[1] Chuiko Institute of Surface Chemistry, NAS of Ukraine, 17 General Naumov Str., 03164 Kyiv, Ukraine; borys.palianytsia@gmail.com

[2] Department of Physics, AlbaNova University Center, Stockholm University, SE-106 91 Stockholm, Sweden; mats.larsson@fysik.su.se

* Correspondence: tanyakulyk@i.ua; Tel.: +38-044-422-9676

Received: 29 November 2019; Accepted: 21 January 2020; Published: 3 February 2020

Abstract: Ketonization is a promising way for upgrading bio-derived carboxylic acids from pyrolysis bio-oils, waste oils, and fats to produce high value-added chemicals and biofuels. Therefore, an understanding of its mechanism can help to carry out the catalytic pyrolysis of biomass more efficiently. Here we show that temperature-programmed desorption mass spectrometry (TPD-MS) together with linear free energy relationships (LFERs) can be used to identify catalytic pyrolysis mechanisms. We report the kinetics of the catalytic pyrolysis of deuterated acetic acid and a reaction series of linear and branched fatty acids into symmetric ketones on the surfaces of ceria-based oxides. A structure–reactivity correlation between Taft's steric substituent constants Es* and activation energies of ketonization indicates that this reaction is the sterically controlled reaction. Surface D_{3-n}-acetates transform into deuterated acetone isotopomers with different yield, rate, $E^{\neq}$, and deuterium kinetic isotope effect (DKIE). The obtained values of inverse DKIE together with the structure–reactivity correlation support a concerted mechanism over ceria-based catalysts. These results demonstrate that analysis of Taft's correlations and using simple equation for estimation of DKIE from TPD-MS data are promising approaches for the study of catalytic pyrolysis mechanisms on a semi-quantitative level.

Keywords: carboxylic acids upgrading; ketonization; deuterated acetic acid; acetone D-isotopomers distribution; H/D exchange; inverse deuterium kinetic isotope effect; kinetic parameters; activation energy; catalytic pyrolysis of biomass; bio-oil

---

1. Introduction

Catalytic pyrolysis can effectively convert second-generation feedstocks (lignocellulose, waste oils, fats, algae, agricultural and forest residues, etc.) into bio-oil or pyrolysis oil [1–7]. The potential for the use of biomass resources of second-generation feedstocks in Europe is very high [8]. At the same time, the cost of second-generation raw materials is low [8]. One of the major components of bio-oil [1–7,9–14] is organic acids. Currently, some of them, including acetic, valeric, levulinic (γ-ketovaleric acid), 2,5 furan dicarboxylic acids, etc., are considered as key-building platforms in biomass conversion technologies [1–7,12,14]. The upgrading of bio-derived carboxylic acids has great economic, social and environmental advantages compared with the traditional use of fossil hydrocarbon resources.

The ketonization reaction, or ketonic decarboxylation is one of the attractive ways to convert carboxylic acids into sustainable biofuels and valuable industrial products with high added value [15–23]. Ketonization provides for the formation of a new C-C bond in one step, preserves the initial functional group C = O, and significantly reduces the oxygen content in the molecule due to decarboxylation and dehydration. In addition, the carbon chain almost doubles as a result of the symmetric ketone formation. The obtained ketones can be upgraded by catalytic hydrogenation/dehydration to liquid long-chain hydrocarbons [12,14].

Despite the close attention to the ketonization reaction, last year's publications [15–18], including numerous studies and reviews [16,17,19–23] suggested that the mechanism of this reaction is not completely clear. Over the past few decades, several basic mechanisms of this reaction have been proposed [19–41]: (i) free-radical mechanism [24], (ii) direct concerted mechanism [20,25], (iii) mechanism involving an acyl intermediate [26], (iv) ketene based mechanisms [27–29], (v) and a beta-keto-acid-intermediate based mechanisms [30–35] (Scheme 1). The free-radical mechanism, the acyl-based mechanism, and ketene-based mechanism involve the participation of free radicals, acylium cation and ketene species as key intermediates of ketonization, respectively. The concerted mechanism was proposed by Rand et al. [25] for ketonization of adipic and 2,2,5,5-tetramethyladipic acids. This mechanism involves (a) decarboxylation with carbanion formation, (b) nucleophilic attack of carbanion with C-C bond formation, and (c) formation of ketone and a hydroxide anion (Scheme 1).

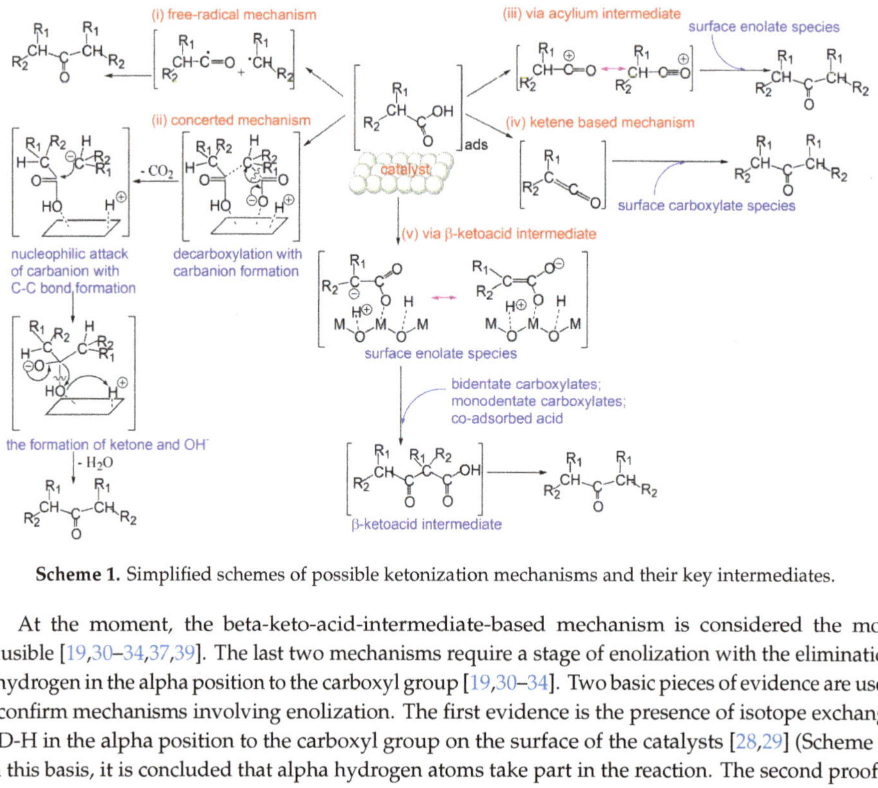

Scheme 1. Simplified schemes of possible ketonization mechanisms and their key intermediates.

At the moment, the beta-keto-acid-intermediate-based mechanism is considered the most plausible [19,30–34,37,39]. The last two mechanisms require a stage of enolization with the elimination of hydrogen in the alpha position to the carboxyl group [19,30–34]. Two basic pieces of evidence are used to confirm mechanisms involving enolization. The first evidence is the presence of isotope exchange of D-H in the alpha position to the carboxyl group on the surface of the catalysts [28,29] (Scheme 2). On this basis, it is concluded that alpha hydrogen atoms take part in the reaction. The second proof is that a symmetric ketone pivalone is not formed from pivalic acid $(CH_3)_3C\text{-COOH}$ [28,30,35]. It was postulated that the reason for the lack of ketonization of pivalic acid is the absence of α-hydrogen, which excludes enolization, so ketonization is impossible [19,30,33]. However, this fact may be due to the steric effect of the substituents at the reaction center in the transition state (TS) for pivalic acid. Accordingly, it was assumed that enolisation is the rate-limiting step of the ketonization reaction.

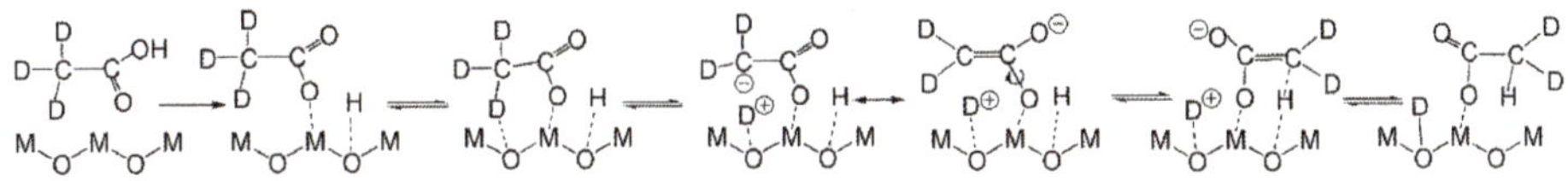

Scheme 2. H/D exchange and enolization on the surface of the catalyst.

However, on the basis of a detailed kinetic analysis [33], quantum chemical calculations [32], and an investigation of the isotopic effect [26,36], it was shown that enolization is not a rate-limiting stage. Rather the loss of the CO_2 or the formation of a new C-C bond are the likely rate-determining steps [26,32,33,36]. Despite the enormous efforts spent on the establishment of the mechanism of this reaction, kinetic studies, isotope effect studies [26,27,36], quantum chemical calculations [30–33,36], studies of cross-ketonization reactions [28,30,33,38], $^{13}C/^{12}C$ exchange studies [34], H/D exchange studies on the catalyst surfaces [28,29,39], experiments with ^{13}C-labeled acetic acids [28,33,34], Fourier Transform Infrared Spectroscopy FT-IR studies [36,37,39–41], etc., the question of establishing the full details of the ketonization reaction mechanism is still not completely closed [39]. However, linear free energy relationships (LFERs) have never been applied to study of the mechanism of ketonization. LFER is a very informative experimental approach because it can give information about the TS and the reaction mechanism on a semiquantitative basis [42,43]. Therefore, in our work, we decided to apply three of the most important experimental methods [42] to establish the mechanism of this reaction over ceria-based catalysts [44], which were not used or insufficiently used in previous works: 1) kinetic study and analysis of kinetics (n, T_{max}, $E^{\neq}$, v_0, $dS^{\neq}$), 2) analysis of substituent effects (Taft plots) or LFERs between activation free energies change induced by substituents and the steric, the electron donating or electron withdrawing characteristics of the substituent and 3) study of the magnitude and the origin of deuterium kinetic isotope effect (DKIE).

The novelty of this study is due to the fact that using the LFERs principle, we obtained structure–reactivity correlation between Taft's steric substituent constants Es* and activation energies of ketonization, and evaluated the contribution of the electronic and steric effects of the substituents for a wide reaction series of carboxylic acids (acetic, propionic, butyric, isobutyric, valeric, pivalic, hexanoic, heptanoic, nonanoic, octanoic, decanoic). In addition, a novel simple approximate equation was proposed for a quick estimate of the DKIE from temperature-programmed desorption mass spectrometry (TPD-MS) data. The values of inverse DKIE of the formation of a series of acetone D-isotopomers have been obtained using this simple equation. The obtained values of the inverse DKIE allowed us to get new physical insights about the sterically controlled mechanism of catalytic pyrolysis of aliphatic carboxylic acids into symmetric ketones over ceria-based catalysts.

2. Results and Discussion

2.1. Kinetic Study

It is well known that on the surfaces of various oxide catalysts, different pathways of the conversion of aliphatic acids may take place [18,44–46] (Scheme 3). Depending on the surface properties of the catalyst, reactions can occur such as ketenization, ketonization, decarboxylation, decarbonylation, cracking, reduction to aldehydes, and others.

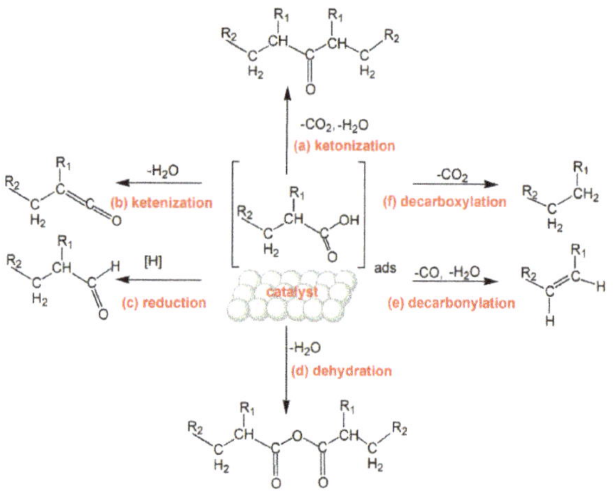

Scheme 3. The possible pathways of the conversion of aliphatic acids on the surfaces of oxide catalysts.

As was previously established [43,47–49], the ketenization reaction (pathway b, Scheme 3) proceeds on the silica at 300–450 °C with a high level of selectivity. Ketenes formation has also been observed over Al_2O_3/SiO_2, TiO_2/SiO_2 [41] and aluminosilicates due to the presence of Brönsted acid sites on their surfaces [50]. Cinnamic acids on the silica surface transform into corresponding ketenes, vinyl- and acetylene-benzenes via three parallel pathways b, e and f: ketenization, decarboxylation, and decarbonylation, [51–53]; whereas over the nanoceria surface, only the decarboxylation (f) takes place [54]. A comparative study [41] of pyrolytic decomposition of valeric acid over SiO_2, Al_2O_3, CeO_2/SiO_2, Al_2O_3/SiO_2, and TiO_2/SiO_2 catalysts showed that propylketene forms on the surfaces of all these oxides except for CeO_2/SiO_2. The formation of both ketene and ketone was observed on the surface of Al_2O_3. The ceria-based catalyst converted valeric acid into dibutylketone most efficiently in the investigated series of oxides due to the presence of basic sites on its surface [41]. As was noted earlier [40], ceria is the most basic oxide in the series of oxides CeO_2, TiO_2, and Al_2O_3.

The acid-base and redox properties of ceria-based materials have been extensively investigated because the unique catalytic activity enables ceria to catalyze a variety of organic reactions [55–58]. The surface of ceria is characterized by very low Lewis and Brönsted acid site density and weak acid strength [39], while at the same time extremely high density [59] of strong Lewis-base sites [60], which interact with carboxylic acids to form different types of carboxylates (monodentate, bidentate: chelate and bridge) [40,41,61]. Subsequent transformation of the surface carboxylates leads to the formation of ketones [37,40,41,61].

The main pyrolysis products of aliphatic carboxylic acids on the CeO_2/SiO_2 surface are the corresponding ketones and CO_2. The typical TPD-MS data of the catalytic ketonization of aliphatic carboxylic acids are presented in Figure 1A,B with the example of propionic acid. Figure 1A,B show the mass spectrum and TPD curves obtained during the catalytic pyrolysis of propionic acid immobilized on the CeO_2/SiO_2 surface. The TPD profiles of the molecular ion of the ketonization product (diethyl ketone or 3-pentanone) with $m/z = 86$ and its fragment ions with $m/z = 72$ $C_2H_5COCH_2^+$), 57 ($C_2H_5CO^+$), and 29 ($C_2H_5^+$) reflect the shape of each other (Figure 1B). These ions are present in the spectra at the temperature of the maximum rate of the 3-pentanone formation, Figure 1A. It is known that ketonization is always accompanied by the release of water and CO_2. Intensive decarboxylation accompanying the formation of the corresponding symmetrical dialkyl ketone is observed during propionic acid pyrolysis (Figure 1A,B) and all studied acids (C2–C10), except pivalic acid. As was previously established [19,35], and also observed it in this work, pivalic acid forms two other major products, isobutene and 2-dimethyl-4-methyl-3-pentanone (Table 1).

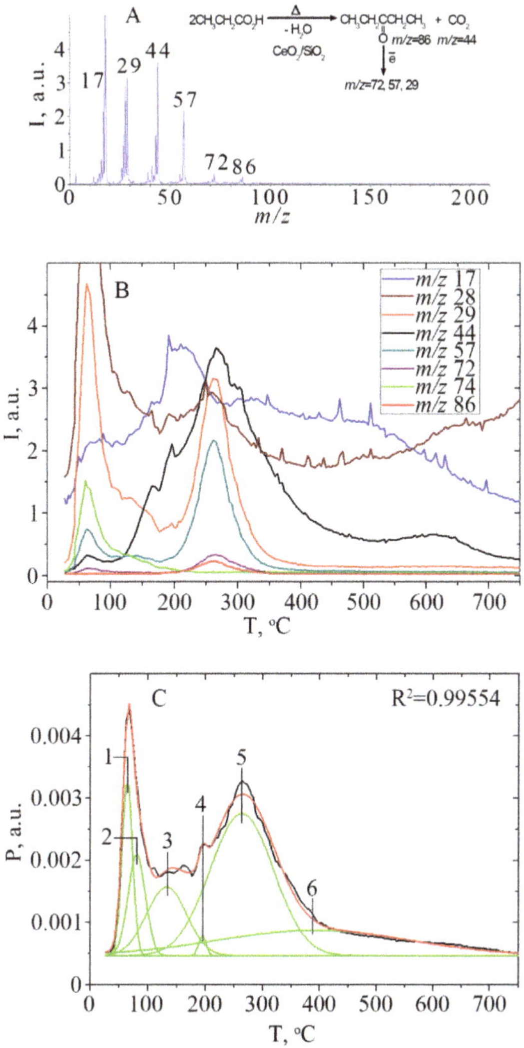

Figure 1. Catalytic pyrolysis of propionic acid supported on the nanocomposites CeO_2/SiO_2 with low amount of CeO_2 (lCeSi) surface (0.6 mmol g^{-1}). (**A**) Mass-spectrum of pyrolysis products at 265 °C, obtained after electron impact ionization. (**B**) temperature-programmed desorption curves (TPD curves) of the molecular ion of propionic acid with $m/z = 74$ and its fragment ions with $m/z = 72, 44, 29, 28, 17$ attributed to the desorption of the physically adsorbed and hydrogen-bonded acid. TPD curves attributed to the ketonization products: molecular ion of CO_2 with $m/z = 44$, and molecular ion of 3-pentanone with $m/z = 86$ and its fragment ions with $m/z = 72, 57, 29$. TPD curve attributed to water desorption: fragment ion of molecular ion of water with $m/z = 17$. (**C**) Vapor pressure measured as a function of temperature (P/T curve); deconvolution of P/T curve ($R^2 = 0.99554$); experimental data is black line, fitting curve—Red line, Gaussians—Green line.

Table 1. Kinetic parameters (temperature of the maximum desorption rate T_{max}, reaction order n, activation energy $E^{\neq}$, pre-exponential factor v_0 and activation entropy $dS^{\neq}$) of ketonization of reaction series of carboxylic acids on the lCeSi surface (0.6 mmol/g). Sum of the Taft Induction Constants ($\sum\sigma^*$) and Sum of the Taft Steric Constants ($\sum E_s$) of the Substituents at the Reaction Center.

Acid	ΣE_s	$\Sigma\sigma^*$	T_{max}, °C	m/z	n	$E^{\neq}$, kJ·mol^{-1}	v_0, s^{-1}, (n = 1); mol^{-1}s^{-1}, (n = 2)	$dS^{\neq}$, cal·K^{-1} mol^{-1}	$D\pm$ %,	R^2 [a]	$E^{\neq}$ [b], kJ·mol^{-1}
Acetic C2	3.72	1.2	247	58	1	119	5.3×10^9	−15	2.5	0.982	115.1
					2	212	3.5×10^{19}	30	11	0.957	-
Propionic C3	2.48	0.8	266	86	1	122	4.4×10^9	−16	4	0.970	119.4
					2	271	6.7×10^{24}	54	43	0.879	-
Butyric C4	2.41	0.7.	262	114	1	113	2.4×10^9	−17	3.4	0.972	118.5
					2	252	9.7×10^{22}	45	21	0.922	-
Isobutyric C4	1.24	0.4	284	114	1	128	7.4×10^9	−15	0.4	0.995	123.6
					2	241	8.7×10^{20}	36	6	0.962	-
Valeric C5	2.12	0.685	268	142	1	123	4.8×10^9	−15	2.5	0.964	119.9
					2	254	5.2×10^{22}	44	10	0.930	-
Pivalic C5	0	0	290	57 [c]	1	130	5.2×10^8	−20	4.8	0.962	-
					2	251	4.2×10^{19}	30	40	0.853	-
Hexanoic C6	2.09	0.67	269	85	1	120	1.3×10^9	−18	4.7	0.958	120.1
					2	205	1.0×10^{17}	18	13	0.933	-
Heptanoic C7	2.08	0.55	272	142	1	122	1.6×10^9	−18	6	0.947	120.8
					2	282	1.7×10^{25}	56	20	0.920	-
Octanoic C8	-	0.49	282	85	1	122	1.4×10^9	−18	9.0	0.940	123.1
					2	250	5.3×10^{21}	40	32	0.900	-
Nonanoic C9	-	0.43	290	85	1	119	1.5×10^9	−18	4.9	0.966	124.9
					2	209	2.0×10^{17}	18	14	0.944	-
Decanoic C10	-	0.45	292	72	1	124	5.2×10^9	−15	3.5	0.960	125.4
					2	295	5.2×10^{26}	62	21	0.903	-

[a] The coefficient of determination. [b] $E^{\neq} = ln(B/lnB)\, RT_{max}$. [c] The fragment ion of 2-dimethyl-4-methyl-3-pentanone.

The possible side products were monitored but were not definitively detected, including pathways b, c, d, e, Scheme 3. The complex shape of the TPD curve for the ion with m/z = 44 may be evidence of the presence of path (f) at ~200 °C. Our assumption confirms the presence of peaks in the TPD curves at ~200 °C for ions with m/z = 28 and 29, which are the most intense in the ethane EI mass-spectrum [62]. The P/T curve was decomposed into individual Gaussians in order to semi-quantify the contribution of ketonization and decarboxylation with the formation of ethane in the process of propionic acid conversion, Figure 1C.

A comparative analysis of the peaks on the TPD curves (Figure 1B) with the peaks on the P/T curve (Figure 1C) shows that the P/T curve is the result of a superposition of several peaks that correspond to the desorption of water, physically adsorbed and hydrogen-bonded propionic acid, and desorption of pyrolysis products. The decomposition of the P/T curve into individual Gaussians allowed us to identify peaks 1 and 3 with desorption of propionic acid in molecular form, peaks 2 and 6 with processes of desorption of water and peaks 4 and 5 with desorption of products of catalytic transformations, because they are centered at the same temperature as corresponding TPD peaks, Table 2, Figure 1B,C.

The calculated integral intensities of peaks 1, 3, 4, and 5 make it possible to semi-quantify the efficiency of the catalytic conversion of propionic acid, Table 2. The degree of conversion of propionic acid during the pyrolysis is about 0.67, while the selectivity for the formation of 3-pentanone is above 0.90 and ethane is about 0.02. The obtained data are in good agreement with previously published data [63,64] regarding the catalytic activity of catalytic systems based on CeO_2/SiO_2 and catalytic systems based on ZrO_2 in the ketonization of propionic acid [37]. The selectivity of 3-pentanone is higher than 97.5% on the ZrO_2–based catalysts [37] and >93% on the CeO_2/SiO_2–based catalysts [63]. Deng et al. [64] have also confirmed a high catalytic activity of CeO_2/SiO_2 for acetic acid ketonization because it could convert acetic acid to acetone with high selectivity 96.63%. The activation energy $E^{\neq}$ for propionic acid ketonization with 3-pentanone formation is 122 $kJ \cdot mol^{-1}$ (calculated by Arrhenius plot method), and 119 $kJ \cdot mol^{-1}$ (calculated by the modified equation), Table 1. These data also are in good agreement with the previous reports of 117 $kJ \cdot mol^{-1}$ on m-ZrO_2 [33] and 124 $kJ \cdot mol^{-1}$ on m-ZrO_2 and 100 $kJ \cdot mol^{-1}$ on t-ZrO_2 [37].

Conversion and selectivity are highly dependent on reaction conditions, especially temperature. The effect of the reaction temperature on the degree of conversion and selectivity to ketone was studied [16,37,64]. It was shown that an increase in temperature to 425 °C allows the conversion of acetic acid to be achieved at about 93%–100% [16,64]. Importantly, the highest conversion of propionic acid on the zirconium surface was observed at a temperature of the highest ketonization rate [37]. Therefore, we can assume that the obtained temperature values of T_{max} (Table 1), which are attributed to the temperature of the maximum rate of ketonization reaction of $C_nH_{2n+1}COOH$ (n = 1–9), can be very useful for choosing the optimal conditions for the catalytic conversion of carboxylic acids into symmetric ketones. T_{max} for the investigated series of carboxylic acids is in the range 247–292 °C. We suggest that optimal temperature conditions for the efficient catalytic ketonization of carboxylic acids over CeO_2/SiO_2 catalyst could also be in this range.

It is known that cerium-based oxides are effective catalysts of ketonization [21,38,56,63,64]. Therefore catalytic systems for industrial applications have been developed on their basis. Gliński et al. [63] investigated over 20 metal oxides supported on SiO_2, Al_2O_3, and TiO_2. It was demonstrated, that CeO_2 was proved to be the most efficient as an active phase for ketonization reaction [63,64]. Gliński and coworkers [63] have suggested that high activity of CeO_2 supported on SiO_2, TiO_2, and Al_2O_3 are a result of the synergy due to the interaction between the CeO_2 and oxide support. We also assume synergies between the support and the active phase of the catalyst that contains strong basic centers. This determines the high catalytic activity of CeO_2 materials.

Table 2. The calculated integral intensities for the peaks obtained by deconvolution of the P/T curve ($R^2 = 0.99554$, Figure 1C).

Peak [a]	Process/Reaction	Product, $m\backslash z$	T_{max} (°C) TPD-Curve	T_{max} (°C) P/T Curve	Peak Area (a.u.), Σ (Peak Areas)	%
		Desorption of C_2H_5COOH in molecular form				
1	Desorption physically adsorbed acid	C_2H_5COOH m/z 74	62 63	65	0.06329	13
3	Desorption of H-bonded acid	C_2H_5COOH m/z 29	131	133	0.09624	20
					Σ (1+3) = 0.15953	33
		Catalytic conversion of C_2H_5COOH				
4	Decarboxylation with ethane formation	C_2H_6 m/z 29 CO_2 m/z 44	196 196	195	0.00511	~1
5	Ketonic decarboxylation	$(C_2H_5)_2CO$ m/z 86 CO_2 m/z44	265 267	264	0.31606	66
					Σ (4+5) = 0.32117	67
					Σ (1+3+4+5) = 0.48070	100
		H_2O desorption				
2	Desorption physically adsorbed water	H_2O m/z 17	87	81	0.06071	26
6	Dehydration, dehydroxylation	H_2O m/z 17	391	391	0.17421	74
					Σ (2+6) = 0.23492	100

[a] Peak on the P/T curve (Figure 1C).

2.2. Analysis of Kinetics Data

Detailed and targeted kinetic studies for reaction series of aliphatic carboxylic acids C2–C10 (acetic, propionic, butyric, isobutyric, valeric, pivalic, hexanoic, heptanoic, nonanoic, octanoic, decanoic) on the surface of nanosized CeO_2/SiO_2 have been investigated by temperature-programmed desorption mass spectrometry (TPD MS), Table 1. The obtained data and our previous study [41] showed that CeO_2/SiO_2 is suitable catalysts for ketonization.

The kinetic parameters (n, T_{max}, $E^{\neq}$, v_0, $dS^{\neq}$) of the reaction of ketone formation were calculated, as described previously [43] (Table 1). The obtained values of the kinetic parameters show that the reaction proceeds according to the first order, since the determination coefficients (R^2) are much higher, and the calculated deviation values are much smaller for the first order than for the second order. In addition, the first order of the reaction was also confirmed by us earlier [41] by studying the influence of the degree of surface coverage on the reaction rate. Namely, the absence of T_{max} displacement towards low temperatures with increasing acid concentration on the surface indicates the first order [65].

The pre-exponential factor for this reaction series has close values within the limits of the same order. This indicates that the transition state for this reaction series has a similar structure. Thus, the entropy of activation for this series is constant ($dS^{\neq}$ = const), so this reaction series is an isentropic series. This fact has allowed us to apply a Redhead peak maximum method to calculate the activation energy [66]. This method avoids the effect of various distortions in the form of a peak on the calculated values of the activation energy that can be produced by using complete analysis of TPD curves and the Arrhenius plot method [65–68]. In this work, we used the modified Equation (1) as suggested by Kislyuk and Rozanov on the basis of Redhead equation [67]. This relation directly includes the temperature of the peak maximum T_{max}, [67]:

$$E^{\neq} = R\,T_{max}\,ln(B/lnB) \tag{1}$$

where

$$B = (nv_0 T_{max} C_{max}^{\,n-1})b \tag{2}$$

In this formula, n is the reaction order, v_0 is the pre-exponential factor, $C^{n-1}_{\,max}$ is the concentration of the adsorbate at T_{max} and b is the value of the sample heating rate.

A study of a number of acids shows an increase in the T_{max} of reaction with an increase in the number of the methylene units and the branching in the molecule (Table 1). The corresponding increase in the activation energy was also obtained by using modified Equation (1) and the arithmetic average of the pre-exponential factor $v_{0average} = 1/10\ \Sigma v_{0\ acid} = 3.53 \pm 1.89 \times 10^9$ s^{-1} for the reaction series of 10 acids.

Thus, as a result of the kinetic analysis carried out, it is established that the reaction is of first order, the activation entropy has a negative value, i.e., the number of degrees of freedom in the transition state decreases and the transition state has a highly ordered structure. Thus, the next task is to establish how changes in the structure of the acid molecule affect the activation energy. This task can be solved by investigating the effect of the substituent using the approaches of LFERs.

2.3. Analysis of Substituent Effects (Taft Plots) or Linear Free Energy Relationships (LFERs)

We were unable to find literature data on the use of LFERs to study the mechanism of ketonization, despite the fact that more in-depth data about the structure of the transition state can be obtained from studies of substituent effects [42]. So the next step was to search for correlations between the kinetic parameters and the thermodynamic ones: the activation energy and the Taft's constants of substituents, respectively. An important step in the search for such correlations is the $E^{\neq}$-σ^* and $E^{\neq}$-Es analysis, which allows us to determine the reaction center in the molecule. Such correlations can be obtained only if the reaction center is chosen correctly. Taft's correlation was obtained (Figure 2A) only if the α-carbon atom was chosen as the reaction center, and not, for example, the β-carbon atom or the carbon atom of the carboxyl group. So, a linear correlation between the activation energies and the sum of the Taft steric constants at the α-carbon atom ($E^{\neq}$-ΣEs) with a high squared correlation coefficient $R^2 = 0.9488$ indicates that the ketonization reaction is sterically sensitive, Figure 2A. The correlation $E^{\neq}$-$\Sigma\sigma^*$ sum of induction Taft's constants (Figure 2B) has a low value of the squared correlation coefficient $R^2 = 0.8227$. These data indicate that the activation energy of ketonization is predominantly dependent on the steric effects of the substituents. That is, the transition state is less sensitive to the inductive effect of the substituents, although it is possible that electron-donor substituents slow down this reaction. The Taft correlation $E^{\neq}$-ΣEs indicates that the LFER principle is applicable to this reaction series. The presence of large substituents in the reaction center may prevent the interaction of atoms and thereby slow down the reaction or make it impossible. The steric effect does not allow to achieve the necessary conformation in the transition state for the realization of ketonization, the so-called "tert-butyl effect". Although pivalic acid does not form a pivalon, it is an important member of this reaction series. Pivalic acid is the standard for this reaction series since the sum of both the induction and steric Taft's constants for the α-carbon atom of the pivalic acid is 0 ((ΣEs = $3 \times$ Es(CH$_3$)=$3 \times 0 = 0$; $\Sigma\sigma^* = 3x\sigma^*$(CH$_3$) = $3 \times 0 = 0$). Thus, the greatest steric effect should be observed, namely, for pivalic acid.

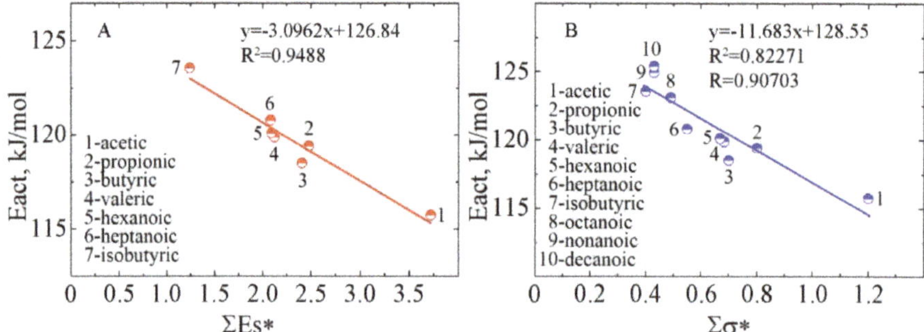

Figure 2. (**A**) Linear correlation between the activation energies of ketonization of reaction series of carboxylic acids (C2–C10) on lCeSi and the sum of the Taft steric constants (ΣEs*) at the α-carbon atom of carboxylic acids. (**B**) The correlation between the activation energies and the sum of the Taft induction constants ($\Sigma\sigma^*$) at the α-carbon atom of the carboxylic acids.

2.4. Study of DKIE During Catalytic Ketonization of CD3COOH (D3)

The isotopic effect is one of the most powerful tools for establishing the mechanism of reactions along with LFER and kinetic studies [42]. The presence of a primary H/D isotopic effect indicates that the bond containing the H atom being broken or formed in a kinetically relevant step [42]. However, there are contradictory data in the literature. Thus, for example, Hendren and coworkers [27] obtained the values of the primary isotopic effects k(H) = k(D) between 1.4 and 6.7 for the ketonization reaction. The absence of a primary isotopic effect was demonstrated in the study of the ketonization of deuterated acetic acid [26]. A very low value of the primary isotopic effect k(H)/k(D) = 1.1 was obtained in the work of Wang and Iglesia [36]. Thus, recent experimental data of the DKIE [26,36], as well as quantum chemical calculations [32,33] completely exclude enolization as a rate-limiting stage. However, this is not enough to completely exclude it as a necessary stage. In this regard, TPD-MS was used to an elucidation of DKIE by studying of the kinetics of the thermal transformations of deuterated acetic acid (CD$_3$COOH) on the surface of the nanocomposites CeO$_2$/SiO$_2$ with higher amount of CeO$_2$ (hCeSi).

2.4.1. The Distribution of Acetone Isotopomers

The results of the TPD-MS studies confirm the presence of isotopic exchange on the surface of the catalysts, (Figure 3A,B). The peaks on the TPD curves corresponding to the isotopomers of acetone have a different intensity, and different T_{max} localization, (Figure 3A,B). Earlier, in 1978, Munuera et al. [29], in their extremely important work, had already estimated the relative yield of isotopomers of acetone as result of the ketonization of a mixture of acids CD_3COOH (76%) and CH_3COOH (24%) adsorbed on the surface of titanium dioxide. However, due to the then instrumental imperfection, these estimates were not accurate enough, since they were made only on the basis of the peak height in the mass spectra at a fixed temperature ~250 °C. It is necessary to take into account the integrated intensity of the peaks for correct estimation.

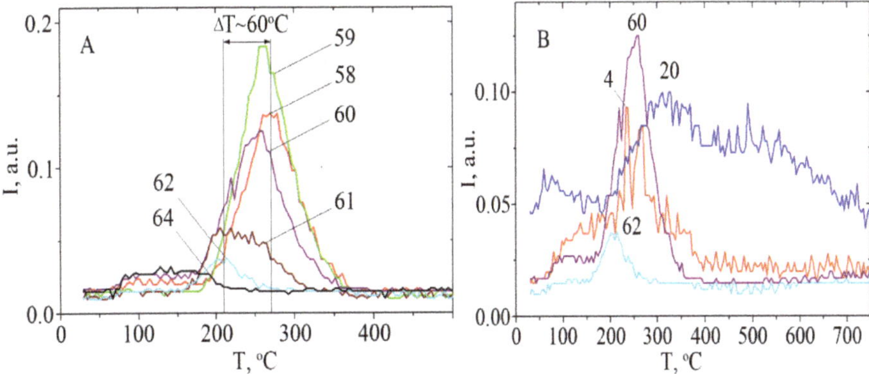

Figure 3. (**A**) TPD curves for the ions of acetone isotopomers with m/z = 64, 62, 61, 60, 59, 58; (**B**) TPD curves for the ions of acetone isotopomers with m/z = 62, 60; TPD curves for the ions of D_2O^+ with m/z = 20 for the ions of D_2^+ with m/z = 4 were obtained during the pyrolysis of CD_3COOH on the hCeSi surface.

In this regard, the mathematical procedure has been conducted to obtain the correct values of T_{max} and integral intensities of the isotopomers peaks (see Figure 4, Figures S1–S5 and Table 3). Thus, the integral intensities of TPD peaks and distribution (%) for ions of acetone isotopomers with m/z = 62 (4%), 61 (11%), 60 (24%), 59 (35%), 58 (26%) were obtained during the pyrolysis of isotopic-labeled acetic acid CD_3COOH on the hCeSi surface (Figure 4 and Table 3). It is noteworthy that the sum of the integral intensities of TPD peaks of ions of isotopomers with m/z = 62, 61, 60, 59, 58 is very close to integral intensities of TPD peaks of molecular ions of $(CH_3)_2CO$ and $(CH_3)_2{}^{13}CO$ that were obtained during the pyrolysis of nonlabeled acetic acid and ^{13}C-labeled acetic acid (Figure 5, Figures S6 and S7, and Table 3).

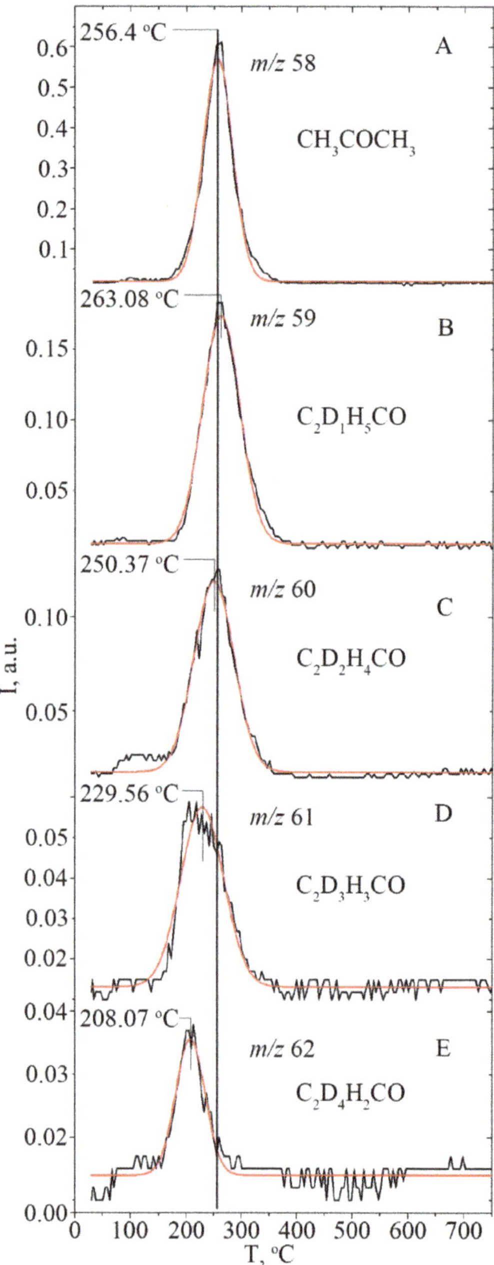

Figure 4. Deconvolution of TPD curve for the ion of acetone with m/z = 58 that was obtained during the pyrolysis of nonlabeled acetic acid on the hCeSi surface (**A**). Deconvolution of TPD-curves for the ions of acetone isotopomers with *m/z* = 62, 61, 60 and 59 that were obtained during the pyrolysis of deuterated acetic acid on the hCeSi surface (**B–E**). Experimental data is the black line; the fitting curve is the red line.

Table 3. Temperature of the maximum desorption rate T_{max}, integral intensity ($\int I(m/z)$) and distribution (%) of acetone isotopomers with m/z = 64, 62, 61, 60, 59, 58 obtained during the catalytic pyrolysis of isotopic-labeled acetic acid CD_3COOH, $CH_3{}^{13}COOH$ and nonlabeled acetic acid CH_3COOH on the hCeSi surface (0.6 mmol·g^{-1}).

m/z Isotopomer of Acetone	T_{max}	$\int I(m/z)$, a.u.	Distr., %	R^2 [a]
		CD_3COOH		
58, $C_2D_0H_6CO$	269.9	10.21	26.20	0.9906
59, $C_2D_1H_5CO$	263.1	13.58	34.85	0.9929
60, $C_2D_2H_4CO$	250.4	9.17	23.53	0.9817
61, $C_2D_3H_3CO$	229.5	4.32	11.09	0.9774
62, $C_2D_4H_2CO$	208.1	1.44	3.70	0.9641
64, $C_2D_6H_0CO$	-	<0.25	<0.64	-
$\Sigma \int I(m/z)$		~38.97	~100	-
		$CH_3{}^{13}COOH$		
$(CH_3)_2{}^{13}CO$	255.7	~37.30	~100	0.9955
		CH_3COOH		
$(CH_3)_2CO$	256.4	~36.79	~100	0.9912

[a] The coefficient of determination.

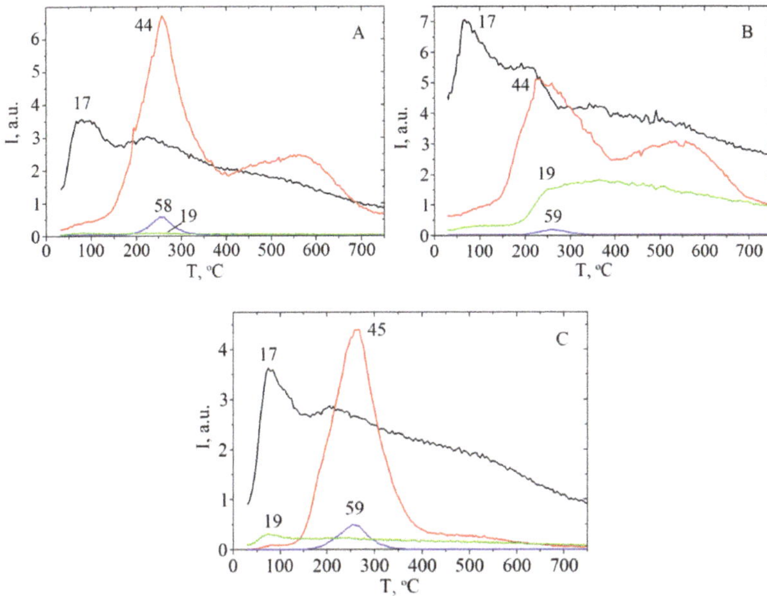

Figure 5. TPD curves for the molecular ion of acetone with m/z = 58, molecular ion of ^{13}C-labeled acetone with m/z = 59, molecular ion of CO_2 with m/z = 44, molecular ion of $^{13}CO_2$ with m/z = 45, ion of H_3O^+ with m/z = 19, fragment ion of deuterated water HDO with m/z = 19 and fragment ion of water with m/z = 17 were obtained during the pyrolysis of CH_3COOH (**A**); CD_3COOH (**B**); $CH_3{}^{13}COOH$ (**C**) on the hCeSi surface.

The maximum yield is observed for monodeuterated acetone with m/z = 59, (Figures 3A, 4B and 5A). The increase in the number of deuterium atoms decreases the yield of the isotopomer. Unfortunately, we were unable to estimate the amount of fully deuterated (m/z = 64, $C_2D_6H_0CO$) and mono-deuterated acetone (m/z = 63, $C_2D_5H_1CO$) because their molecular weights are equal to the molecular weights of the molecular ions CD_3CO_2D and CD_3CO_2H, respectively. As a consequence, overlapping of the peaks of these isotopomers of acetone ($C_2D_6H_0CO$ and $C_2D_5H_1CO$) with desorption peaks of physically adsorbed CD_3CO_2D and CD_3CO_2H acid molecules could occur.

It is most likely that the H/D isotope exchange on the surface and the ketonization reaction are two independent processes that are not related to each other. It seems that isotopic exchange easily proceeds on the oxide surface at temperatures much lower than a temperature at the start of ketonization. The adsorption layer of isotopomers of surface acetates $-O-C(=O)-CD_nH_{3-n}$ with a different percentage distribution is formed in the isotopic exchange result, which is due to the efficiency of H/D exchange on the surface of a particular oxide. The efficiency of H/D exchange depends on the nature of the oxide surface. Based on the experimental data obtained, it can be said that on the surface of CeO_2/SiO_2, the H/D exchange starts at a quite low temperature ≤ 150 °C (Figure 3A,B). This process can likely proceed very easy through the tunneling mechanism, or through an enolization (see Scheme 2). H/D exchange was observed by Kozhevnikov et al. as a result of the exchange between surface acetates and the Brønsted acid sites [39].

Upon further heating, these exchanged deuterium atoms are desorbed in the form of deuterated water molecules DHO ($m/z = 19$) and D_2O ($m/z = 20$) as a result of dehydration of surface, (Figures 3B and 5B). The differences in the processes of water desorption during the pyrolysis of isotope-labeled and nonlabeled acetic acids are clearly seen in Figure 5A–C.

2.4.2. Study of the Magnitude and the Origin of DKIE

We assume that initially an adsorption layer from the surface complexes of acetic acid isotopomers $D_nH_{3-n}COOH$ is formed on the surface as a result of the H–D exchange. Their percentage distribution is due to the amount and strength of the surface active centers on which the H/D exchange occurs. Furthermore, these surface complexes of acetic acid isotopomers $D_nH_{3-n}COOH$ are transformed upon heating. As a result of ketonization, seven isotopomers and three isomers of isotope-labeled acetone can be formed: $m/z = 64, 63, 62$ (2 isomers), 61 (2 isomers), 60 (2 isomers), 59 and 58 (Scheme 4). The same value of the rate of formation of various isotopomers and accordingly, the same value of the primary DKIE, should be observed if the "beta-ketoacid" mechanism works.

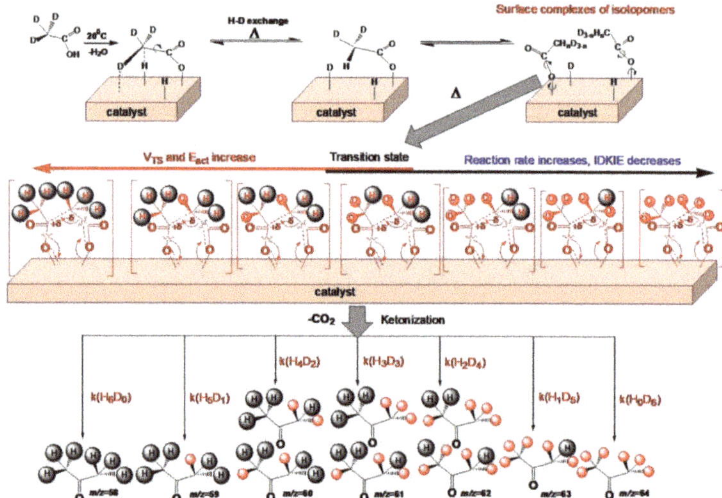

Scheme 4. The possible mechanism of deuterated acetic acid ketonization into the acetone isotopomers involves the simultaneous carbon–carbon bond and carbon dioxide formations in one single step.

Interestingly, the TPD peaks for the isotopomers of acetone were observed at different temperatures of the maximum desorption rate (T_{max}) on the CeO_2/SiO_2 surface (see Figures 3 and 4, Table 3). Most importantly, TPD peaks are shifted to higher temperatures with an increase in the numbers of H-atoms in a molecule of an isotopomer. As was shown earlier in the work performed by Leardini with co-authors, this may indicate the presence of an inverse isotope effect [69]. Determination of the exact value of the isotopic effect is a difficult experimental and calculation problem [42,69]. However, establishing even the type of isotope effect—primary, secondary or inverse—can play a key role in determining the reaction mechanism.

Therefore, we decided to use the modified Taft equation for the DKIE calculation. This equation was earlier proposed in the work of one of the coauthors [43] to study the mechanisms of catalytic reactions using the TPD-MS data:

$$\lg k/k_o = \lg(B/2.3\lg B)\cdot\Delta T_{max}/T_{o\,max} \tag{3}$$

After some manipulation, we obtain the simple equations for the elucidation of KIE values from TPD-MS data:

$$KIE = k/k_o = 10^{\lg(B/2.3\lg B)\cdot\Delta T max/To\,max} \tag{4}$$

$$DKIE = k(H)/k(D) = 10^{\lg(B/2.3\lg B)\cdot(T max(D)-T max(H))/T max(D)} \tag{5}$$

The calculating kinetic parameters of ketonization reaction of CD_3COOH, CH_3COOH and $CH_3{}^{13}COOH$ on the hCeSi surface showed that the pre-exponential factors have close values within the limits of the same order ~10^9 as in the cause of reaction series of carboxylic acids on the lCeSi surface (Table 4). This result allows using Equation (5) for calculation of DKIE of formation of acetone isotopomers from deuterated acetic acid in comparison with formation acetone from nonlabeled acetic acid.

Table 4. Kinetic parameters (temperature of the maximum desorption rate T_{max}, reaction order n, activation energy $E^{\neq}$, pre-exponential factor v_0 and activation entropy $dS^{\neq}$) of ketonization of CD_3COOH, $CH_3{}^{13}COOH$ and CH_3COOH on the hCeSi surface; the deuterium kinetic isotope effect (k(H)/k(D) obtained using the suggested Equation (5).

m/z, Isotopomer of Acetone	T_{max}, °C	k(H)/k(D)	n	$E^{\neq}$, kJ mol^{-1}	v_0, s^{-1}	$dS^{\neq}$, cal K^{-1} mol^{-1}	D±, %	R^2 [a]	$E^{\neq}$ [b], kJ mol^{-1}
				CD$_3$COOH					
58, $C_2D_0H_6CO$	269.9	-	1	125	5.1×10^9	−15	11	0.908	120.3
			2	225	8.3×10^{19}	31	24	0.854	-
59, $C_2D_1H_5CO$	263.1	1.3951	1	116	1.1×10^9	−17	3	0.959	118.7
			2	168	3.1×10^{19}	6	8	0.901	-
60, $C_2D_2H_4CO$	250.4	0.7368	1	118	3.8×10^9	−16	4	0.973	116.2
			2	240	2.3×10^{22}	45	30	0.922	-
61, $C_2D_3H_3CO$	229.5	0.2402	1	-	-	-	-	-	111.2
62, $C_2D_4H_2CO$	208.1	0.0689	1	-	-	-	-	-	106.4
				CH$_3{}^{13}$COOH					
59, $CH_3{}^{13}COCH_3$	255.7	-	1	117	1.8×10^9	−17	7	0.960	117.0
			2	204	2.5×10^{18}	25	11	0.906	-
				CH$_3$COOH					
58, CH_3COCH_3	256.4	-	1	122	5.4×10^9	−15	6	0.963	117.2
			2	220	1.1×10^{20}	32	13	0.900	-

[a] The coefficient of determination. [b] $E^{\neq} = ln(B/lnB)RT_{max}$.

We calculated the isotope effect for isotopomers from one to four deuterium atoms in a molecule (Table 4). The reaction rate constant of formation of the deuterated d4-ketone in ~14.5-fold is larger than the reaction rate constant of the formation of nonlabeled ketone from nonlabeled CH_3COOH. In addition, based on the fact that the T_{max} of acetone formation from $CH_3{}^{13}COOH$ is rather close to T_{max} of acetone formation from CH_3COOH, it can be concluded that KIE = k(^{12}C)/k(^{13}C) is about ~1 (Table 4 and Figure 5).

The obtained values of inverse DKIE for $C_2D_2H_4CO$ ($m/z = 60$), $C_2D_3H_3CO$ ($m/z = 61$), and $C_2D_4H_2CO$ ($m/z = 62$) isotopomers formation confirm the concerted mechanism (Table 4). It is known that inverse DKIE is observed for a mechanism involving concerted bonds reorganization in the cyclic TS—for example, the Cope rearrangement [42,70,71]. The reason for this is that the compacting of the cyclic TS is the most important condition and a requirement for this type of reaction. Moreover, the different values of inverse DKIE confirm participating of all deuterium/protium atoms in the cyclic TS. As such, it has been suggested that the concerted mechanism involves the simultaneous single-step formation of a carbon–carbon bond and carbon dioxide release (Scheme 4).

The smaller size of the deuterium atom facilitates the achievement of the planar cyclic TS, in which the synchronous reorganization of bonds can occur easily only due to electron transfer, Scheme 4. Most likely, the reaction is realized in a single surface complex in which two acid residues are monodentate-bonded to two cerium atoms or one cerium atom. Ding and coauthors, by using the in situ Diffuse reflectance infrared Fourier transform spectroscopy (DRIFTS), showed that monodentate carboxylates are much more active then bidentate species during ketonization over ZrO_2-based catalysts [37]. It is known that the deuterium atom is ~15% smaller than the protium atom. Replacement of each next atom of protium by deuterium leads to a decrease in the volume of the TS and, correspondingly, to an increase in the reaction rate and to a decrease in the activation energy. We calculated the activation energies of acetone isotopomers formation by using Equation (1) (Table 4). In point of fact, we can see that an increase in the number of deuterium atoms in the isotopomer molecules lead to a decrease in the activation energy of ketonization. This explains the presence of different isotope effects during the formation of acetone isotopomers.

3. Materials and Methods

3.1. Materials

The research was carried out for a deuterated acetic acid D_3CCOOD (Sigma-Aldrich, St.Louis, USA, >99.5 atom %D), ^{13}C-labeled acetic acid $H_3C^{13}COOH$ (Sigma–Aldrich, St.Louis, USA, 99 atom %^{13}C), and a reaction series of aliphatic carboxylic acids $C_nH_{2n}COOH$ (n = 1–9): acetic (Merck, Schuchardt, Germany, 99.7%), propionic (Merck, 99.5%), butyric (Merck, Schuchardt, Germany, 99.5%), isobutyric (Merck, Schuchardt, Germany, 99.5%), valeric (Merck, Schuchardt, Germany, 99.5%), hexanoic (Merck, Schuchardt, Germany, 99.5%), heptanoic (Merck, Schuchardt, Germany, 99.5%), octanoic (Merck, Schuchardt, Germany, 99.5%), nonanoic (Merck, Schuchardt, Germany, 99.5%), decanoic (Merck, Schuchardt, Germany, 99.5%), and pivalic (Fluka, 99.5%). Nonporous nanocomposites CeO_2/SiO_2 with low amount of CeO_2—lCeSi and with higher amount of CeO_2—hCeSi were prepared by using powdery fumed silica and cerium (III) acetylacetonate (Sigma-Aldrich). The modification of silica with cerium acetylacetonate and the characterization of obtained CeO_2/SiO_2 oxides using X-ray diffraction XRD and low-temperature Ar adsorption/desorption were described previously [72] (Table 5).

Table 5. Characteristics of CeO_2/SiO_2 catalysts: specific surface area, amount and crystal size of CeO_2.

Catalysts [1]	S_a, m^2g^{-1}	Amount of CeO_2,%	d_{av},nm
lCeSi	230	12.3	5
hCeSi	212	18.3	5

[1] Catalysts preparation and characterization were described previously by Borysenko et al. [72].

Samples of aliphatic carboxylic acids on the surfaces of the nanosized oxides with loadings of 0.6 mmol g^{-1} were obtained by impregnation. In a typical sample preparation session, 1 g of oxide was gradually mixed with 25 mL of a solution of acid. The samples were then aged at room temperature for 24 h, dried and stored in a desiccator until its use in the TPD-MS experiment. Ethanol solutions were used to prepare samples of reaction series of aliphatic carboxylic acids C2–C10 on the lCeSi surface. Aqueous solutions were used to prepare samples of deuterated acetic acid, nonlabeled acetic acid and ^{13}C-labeled acetic acid on the hCeSi surface. Due to the fact that the deuterium atom of the hydroxyl group of the acid is subjected to very fast H/D exchange in aqueous solutions; in fact, the impregnation was performed with d3-acetic acid (D_3CCOOH). This is what we wanted to achieve, in order to eliminate the distortion of the results due to the rapid exchange of acidic hydrogen atoms of the carboxyl group.

3.2. Temperature-Programmed Desorption Mass Spectrometry and Kinetic Parameters

The experiments were performed in an electron ionization MKh-7304A monopole mass spectrometer (Sumy, Ukraine) adapted for temperature-programmed desorption [33]. Sample weighing ~10 mg was first placed in a molybdenum-quartz ampule (5.4 mm in diameter, 20 cm in length, and 12 mL in volume) and evacuating to ~5 × 10^{-5} Pa at ~20 °C. For all measurements, the molybdenum-quartz ampule was filled to only approximately <<1/16 of the full volume to limit interparticle-diffusion effects [43]. All TPD measurements were carried out by heating the samples to ~750 °C at a constant rate of 0.17 °C s^{-1} using a programmed linear heating schedule ($dT/dt = b$, in which b is the heating rate). The mass spectra and the P/T curves (pressure-temperature curves; P, the pressure of volatile pyrolysis products; T, the temperature of samples) were recorded and analyzed using a computer-based data acquisition and processing setup. The pressure of volatile pyrolysis products was measured with a pressure vacuum gauge VMB-14. The mass spectra were recorded for a mass range of 1–210 amu. The kinetic parameters (temperature of the maximum desorption rate T_{max}, reaction order n, activation energy $E^{\neq}$, pre-exponential factor v_0 and activation entropy $dS^{\neq}$) were calculated from the well-resolved TPD peaks for which the shape and position on the temperature scale were repeated with good resolution in several measurements. The procedure of obtaining kinetic parameters from TPD-MS data and full details of this equipment has been extensively described elsewhere [43].

4. Conclusions

In this study, the applicability of TPD-MS together with LFERs for the elucidation of the mechanism of catalytic pyrolysis of carboxylic acids into ketones over CeO_2/SiO_2 was investigated. The kinetic parameters of catalytic pyrolysis of $C_nH_{2n}COOH$ (n = 1–9), CD_3COOH, and $CH_3{}^{13}COOH$ into ketones were calculated and analyzed. On their bases, the structure–reactivity correlation between Taft's steric substituent constants Es* and activation energies of catalytic ketonization for the reaction series of fatty acids on the surface of nanosized CeO_2/SiO_2 was obtained. This correlation shows that ketonization is a sterically sensitive reaction. The study of the pyrolysis kinetics of deuterated acetic acid on the CeO_2/SiO_2 surface showed that H/D exchange starts at low temperature, <150 °C, and led to the formation layer of surface acetates isotopomers —O-C(=O)-CD_nH_{3-n}. It seems that H–D exchange on the surface and the ketonization are two independent processes and isotopic exchange proceeds easily at temperatures lower than the temperature of the start of ketonization reaction. Surface acetates during heating transform into deuterated acetone isotopomers with different yield, T_{max}, rate, $E^{\neq}$, and DKIE. Activation energy and DKIE decrease with increasing of D-atoms number in the acetone isotopomer molecule. The obtained values of inverse DKIE do not confirm enolization as a rate-limiting step of ketonization over CeO_2/SiO_2. Moreover, the obtained values of inverse DKIE together with the structure–reactivity correlation support the concerted mechanism involving the synchronous reorganization of bonds in the cyclic TS. A proposed simple equation (5) can be a useful tool for the elucidation of KIE values from TPD-MS data. Its applicability has been demonstrated on this experimental data.

Supplementary Materials: The following are available online at http://www.mdpi.com/2073-4344/10/2/179/s1, Figures S1–S5: Fitting of the TPD peaks for the molecular ion of acetone isotopomer with m/z = 58–62 obtained during the pyrolysis of isotopic-labeled acetic acid CD_3COOH on the hCeSi surface (0.6 mmol·g^{-1}). Figure S6: Fitting of the TPD peak for the molecular ion of acetone obtained during the pyrolysis of nonlabeled acetic acid CH_3COOH on the hCeSi surface (0.6 mmol·g^{-1}). Figure S7: Fitting of the TPD peak for the molecular ion of acetone obtained during the pyrolysis of isotopic-labeled acetic acid $CH_3{}^{13}COOH$ on the hCeSi surface (0.6 mmol·g^{-1}).

Author Contributions: T.K. conducted samples preparation, results analysis, paper writing and project administration; B.P. conducted kinetics experiments; M.L. contributed to results discussion, editing of the paper and project administration. All authors have read and agreed to the published version of the manuscript.

Funding: This publication is based on work supported by the Swedish Research Council (VR, 348-2014-4250), by STCU (Grant P707), by the Volkswagen Foundation and by NAS of Ukraine (Program "New functional substances and materials of chemical production"). T.K. acknowledges the Swedish Institute for research scholarship for senior researchers within Visby Programme [03804/2016].

Acknowledgments: We are grateful to Borysenko M.V. and Kulyk K.S. for providing the CeO_2/SiO_2 samples.

Conflicts of Interest: The authors declare no conflict of interest.

References

1. Bridgwater, A.V. Review of fast pyrolysis of biomass and product upgrading. *Biomass Bioenergy* **2012**, *38*, 68–94. [CrossRef]
2. Kataki, R.; Chutia, R.S.; Mishra, M.; Bordoloi, N.; Saikia, R.; Bhaskar, T. Feedstock suitability for thermochemical processes. In *Recent Advances in Thermo-Chemical Conversion of Biomass*, 1st ed.; Pandey, A., Bhaskar, T., Stöcker, M., Sukumaran, R., Eds.; Elsevier: Amsterdam, The Netherlands, 2015; pp. 31–74.
3. Zhang, S.; Yang, X.; Zhang, H.; Chu, C.; Zheng, K.; Ju, M.; Liu, L. Liquefaction of biomass and upgrading of bio-oil: A review. *Molecules* **2019**, *24*, 2250. [CrossRef] [PubMed]
4. Lin, Y.C.; Huber, G.W. The critical role of heterogeneous catalysis in lignocellulosic biomass conversion. *Energy Environ. Sci.* **2009**, *2*, 68–80. [CrossRef]
5. Albrecht, K.O.; Olarte, M.V.; Wang, H. Upgrading Fast Pyrolysis Liquids. In *Thermochemical Processing of Biomass: Conversion into Fuels, Chemicals and Power*, 2nd ed.; Brown, R.C., Ed.; Wiley Online Library: Hoboken, NJ, USA, 2019. [CrossRef]
6. Sheldon, R.A. Green chemistry, catalysis and valorization of waste biomass. *J. Mol. Catal. A Chem.* **2016**, *422*, 3–12. [CrossRef]
7. Climent, M.J.; Corma, A.; Iborra, S. Conversion of biomass platform molecules into fuel additives and liquid hydrocarbon fuels. *Green Chem.* **2014**, *16*, 516–547. [CrossRef]
8. De Wit, M.; Faaij, A. European biomass resource potential and costs. *Biomass Bioenergy* **2010**, *34*, 188–202. [CrossRef]
9. Nanda, S.; Mohanty, P.; Kozinski, J.A.; Dalai, A.K. Hydrothermal and Thermochemical Synthesis of Bio-Oil from Lignocellulosic Biomass: Composition, Engineering and Catalytic Upgrading. In *Industrial Biotechnology*, 1st ed.; Thangadurai, D., Sangeetha, J., Eds.; Apple Academic Press: New York, NY, USA, 2017; pp. 345–390.
10. Nanda, S.; Mohanty, P.; Kozinski, J.A.; Dalai, A.K. Physico-Chemical Properties of Bio-Oils from Pyrolysis of Lignocellulosic Biomass with High and Slow Heating Rate. *Energy Environ. Res.* **2014**, *21*, 21–32. [CrossRef]
11. Novakovskiy, D.J.; Jones, J.M. Uncatalysed and potassium-catalysed pyrolysis of the cell-wall constituents of biomass and their model compounds. *J. Anal. Appl. Pyrol.* **2008**, *83*, 12–25.
12. Serrano-Ruiz, J.C.; Braden, D.J.; West, R.M.; Dumesic, J.A. Conversion of cellulose to hydrocarbon fuels by progressive removal of oxygen. *Appl. Catal. B* **2010**, *100*, 184–189. [CrossRef]
13. Zhang, Y.; Cui, H.; Yi, W.; Song, F.; Zhao, P.; Wang, L.; Cui, J. Highly effective decarboxylation of the carboxylic acids in fast pyrolysis oil of rice husk towards ketones using $CaCO_3$ as a recyclable agent. *Biomass Bioenergy* **2017**, *102*, 13–22. [CrossRef]
14. Lange, J.P.; Price, R.; Ayoub, P.M.; Louis, J.; Petrus, L.; Clarke, L.; Gosselink, H. Valeric Biofuels: A Platform of Cellulosic Transportation Fuels. *Angew. Chem. Int. Ed.* **2010**, *49*, 4479–4483. [CrossRef] [PubMed]
15. Jiang, B.; Xi, Z.; Lu, F.; Huang, Z.; Yang, Y.; Sun, J.; Liao, Z.; Wang, J.; Yang, Y. Ce/MgAl mixed oxides derived from hydrotalcite LDH precursors as highly efficient catalysts for ketonization of carboxylic acid. *Catal. Sci. Technol.* **2019**, *9*, 6335–6344. [CrossRef]
16. Ling, H.; Wang, Z.; Wang, L.; Stampfl, C.; Wang, D.; Chen, J.; Huang, J. Composition-structure-function correlation of Ca/Zn/AlOx catalysts for the ketonization of acetic acid. *Catal. Today* **2019**. [CrossRef]
17. Ignatchenko, A.V. Multiscale approach for the optimization of ketones production from carboxylic acids by the decarboxylative ketonization reaction. *Catal. Today* **2019**, *338*, 3–17. [CrossRef]

18. Psarras, A.C.; Michailof, C.M.; Iliopoulou, E.F.; Kalogiannis, K.G.; Lappas, A.A.; Heracleous, E.; Triantafyllidis, K.S. Acetic acid conversion reactions on basic and acidic catalysts under biomass fast pyrolysis conditions. *J. Mol. Catal.* **2019**, *465*, 33–42. [CrossRef]

19. Pham, T.N.; Sooknoi, T.; Crossley, S.P.; Resasco, D.E. Ketonization of carboxylic acids: Mechanisms, catalysts, and implications for biomass conversion. *ACS Catal.* **2013**, *3*, 2456–2473. [CrossRef]

20. Renz, M. Ketonization of carboxylic acids by decarboxylation: Mechanism and scope. *Eur. J. Org. Chem.* **2005**, *6*, 979–988. [CrossRef]

21. Kumar, R.; Enjamuri, N.; Shah, S.; Al-Fatesh, A.S.; Bravo-Suárez, J.J.; Chowdhury, B. Ketonization of oxygenated hydrocarbons on metal oxide based catalysts. *Catal. Today* **2018**, *302*, 16–49. [CrossRef]

22. Dooley, K.M. Catalysis of acid/aldehyde/alcohol condensations to ketones. *Catalysis* **2004**, *17*, 293–319.

23. Simakova, I.L.; Murzin, D.Y. Transformation of bio-derived acids into fuel-like alkanes via ketonic decarboxylation and hydrodeoxygenation: Design of multifunctional catalyst, kinetic and mechanistic aspects. *J. Energy Chem.* **2017**, *25*, 208–224. [CrossRef]

24. Hites, R.A.; Biemann, K. Mechanism of ketonic decarboxylation. Pyrolysis of calcium decanoate. *J. Am. Chem. Soc.* **1972**, *94*, 5772–5777. [CrossRef]

25. Rand, L.; Wagner, W.; Warner, P.O.; Kovac, L.R. Reactions catalyzed by potassium fluoride. II. The conversion of adipic acid to cyclopentanone. *J. Org. Chem.* **1962**, *27*, 1034–1035. [CrossRef]

26. Gumidyala, A.; Sooknoi, T.; Crossley, S. Selective ketonization of acetic acid over HZSM-5: The importance of acyl species and the influence of water. *J. Catal.* **2016**, *340*, 76–84. [CrossRef]

27. Hendren, T.S.; Dooley, K.M. Kinetics of catalyzed acid/acid and acid/aldehyde condensation reactions to non-symmetric ketones. *Catal. Today* **2003**, *85*, 333–351. [CrossRef]

28. Pestman, R.; Koster, R.M.; van Duijne, A.; Pieterse, J.A.Z.; Ponec, V. Reactions of Carboxylic Acids on Oxides. 2. Bimolecular Reaction of Aliphatic Acids to Ketones. *J. Catal.* **1997**, *168*, 265–272. [CrossRef]

29. González, F.; Munuera, G.; Prieto, J.A. Mechanism of Ketonization of Acetic Acid on Anatase TiO2 Surfaces. *J. Chem. Soc. Faraday Trans.* **1978**, *74*, 1517–1529. [CrossRef]

30. Pulido, A.; Oliver-Tomas, B.; Renz, M.; Boronat, M.; Corma, A. Ketonic decarboxylation reaction mechanism: A combined experimental and DFT study. *ChemSusChem* **2013**, *6*, 141–151. [CrossRef]

31. Pham, T.N.; Shi, D.; Resasco, D.E. Kinetics and Mechanism of Ketonization of Acetic Acid on Ru/TiO2 Catalyst. *Top. Catal.* **2014**, *57*, 706–714. [CrossRef]

32. Ignatchenko, A.V.; McSally, J.P.; Bishop, M.D.; Zweigle, J. Ab initio study of the mechanism of carboxylic acids cross-ketonization on monoclinic zirconia via condensation to beta-keto acids followed by decarboxylation. *J. Mol. Catal.* **2017**, *441*, 35–62. [CrossRef]

33. Ignatchenko, A.V.; Kozliak, E.I. Distinguishing enolic and carbonyl components in the mechanism of carboxylic acid ketonization on monoclinic zirconia. *ACS Catal.* **2012**, *2*, 1555–1562. [CrossRef]

34. Ignatchenko, A.V.; Cohen, A.J. Reversibility of the catalytic ketonization of carboxylic acids and of beta-keto acids decarboxylation. *Catal. Commun.* **2018**, *111*, 104–107. [CrossRef]

35. Oliver-Tomas, B.; Renz, M.; Corma, A. Ketone formation from carboxylic acids by Ketonic decarboxylation: The exceptional case of the tertiary carboxylic acids. *Chem. Eur. J.* **2017**, *23*, 12900–12908. [CrossRef] [PubMed]

36. Wang, S.; Iglesia, E. Experimental and theoretical assessment of the mechanism and site requirements for ketonization of carboxylic acids on oxides. *J. Catal.* **2017**, *345*, 183–206. [CrossRef]

37. Ding, S.; Zhao, J.; Yu, Q. Effect of Zirconia Polymorph on Vapor-Phase Ketonization of Propionic Acid. *Catalysts* **2019**, *9*, 768. [CrossRef]

38. Nagashima, O.; Sato, S.; Takahashi, R.; Sodesawa, T. Ketonization of carboxylic acids over CeO2-based composite oxides. *J. Mol. Catal. A Chem.* **2005**, *227*, 231–239. [CrossRef]

39. Almutairi, S.T.; Kozhevnikova, E.F.; Kozhevnikov, I.V. Ketonisation of acetic acid on metal oxides: Catalyst activity, stability and mechanistic insights. *Appl. Catal. A* **2018**, *565*, 135–145. [CrossRef]

40. Hasan, M.A.; Zaki, M.I.; Pasupulety, L. Oxide-catalyzed conversion of acetic acid into acetone: An FTIR spectroscopic investigation. *Appl. Catal. A* **2003**, *243*, 81–92. [CrossRef]

41. Kulyk, K.; Palianytsia, B.; Alexander, J.D.; Azizova, L.; Borysenko, M.; Kartel, M.; Larsson, M.; Kulik, T. Kinetics of valeric acid ketonization and ketenization in catalytic pyrolysis on nanosized SiO_2, γ-Al_2O_3, CeO_2/SiO_2, Al_2O_3/SiO_2 and TiO_2/SiO_2. *ChemPhysChem* **2017**, *18*, 1943–1955. [CrossRef]

42. Anslyn, E.V.; Dougherty, D.A. *Modern Physical Organic Chemistry*; University Science Books: Sausalito, CA, USA, 2006. [CrossRef]

43. Kulik, T.V. Use of TPD-MS and linear free energy relationships for assessing the reactivity of aliphatic carboxylic acids on a silica surface. *J. Phys. Chem. C* **2012**, *116*, 570–580. [CrossRef]

44. Dupre, G.D.; Schlosberg, R.H.; Pancirov, R.J.; Ashe, T.R.; Baset, Z. Pyrolysis studies of organic oxygenates VI. thermal Chemistry of long chain aliphatic acids. *Liq. Fuels Technol.* **1985**, *3*, 323–344. [CrossRef]

45. Kim, K.S.; Barteau, M.A. Pathways for carboxylic acid decomposition on titania. *Langmuir* **1988**, *4*, 945–953. [CrossRef]

46. García-Sánchez, M.; Sales-Cruz, M.; Lopez-Arenas, T.; Viveros-García, T.; Pérez-Cisneros, E.S. An Intensified Reactive Separation Process for Bio-Jet Diesel Production. *Processes* **2019**, *7*, 655. [CrossRef]

47. Martinez, R.; Huff, M.C.; Barteau, M.A. Synthesis of ketenes from carboxylic acids on functionalized silica monoliths at short contact time. *Appl. Catal. A* **2000**, *200*, 79–88. [CrossRef]

48. Libby, M.C.; Watson, P.C.; Barteau, M.A. Synthesis of Ketenes with oxide catalysts. *Ind. Eng. Chem. Res.* **1994**, *33*, 2904–2912. [CrossRef]

49. Azizova, L.R.; Kulik, T.V.; Palyanytsya, B.B.; Lipkovska, N.A. Thermal and hydrolytic stability of grafted ester groups of carboxylic acids on the silica surface. *J. Therm. Anal. Calorim.* **2015**, *122*, 517–523. [CrossRef]

50. Brei, V.V.; Brichka, A.V. A Study of the Brönsted Site Acidity of Crystalline and Amorphous Aluminosilicates: 2. Thermal Decomposition of Grafted Acetyl Groups. *Adsorp. Sci. Technol.* **1996**, *14*, 359–362. [CrossRef]

51. Kulik, T.V.; Barvinchenko, V.N.; Palyanitsa, B.B.; Smirnova, O.V.; Pogorelyi, V.K.; Chuiko, A.A. A desorption mass spectrometry study of the interaction of cinnamic acid with a silica surface. *Russ. J. Phys. Chem.* **2007**, *8*, 83–90. [CrossRef]

52. Kulik, T.V.; Lipkovska, N.A.; Barvinchenko, V.N.; Palyanytsya, B.B.; Kazakova, O.A.; Dovbiy, O.A.; Pogorelyi, V.K. Interactions between bioactive ferulic acid and fumed silica by UV–vis spectroscopy, FT–IR, TPD MS investigation and quantum chemical methods. *J. Colloid Interface Sci.* **2009**, *339*, 60–68. [CrossRef]

53. Kulik, T.V.; Lipkovska, N.O.; Barvinchenko, V.M.; Palyanytsya, B.B.; Kazakova, O.A.; Dudik, O.O.; Menyhárd, A.; László, K. Thermal transformation of bioactive caffeic acid on fumed silica seen by UV–Vis spectroscopy, thermogravimetric analysis, temperature programmed desorption mass spectrometry and quantum chemical methods. *J. Colloid Interface Sci.* **2016**, *470*, 132–141. [CrossRef]

54. Nastasiienko, N.; Palianytsia, B.; Kartel, M.; Larsson, M.; Kulik, T. Thermal transformation of caffeic acid on the nanosized cerium dioxide studied by temperature programmed desorption mass-spectrometry, thermogravimetric analysis and FTIR-spectroscopy. *Colloids Interfaces* **2019**, *3*, 34. [CrossRef]

55. Vlasenko, N.V.; Kyriienko, P.I.; Yanushevska, O.I.; Valihura, K.V.; Soloviev, S.O.; Strizhak, P.E. The Effect of Ceria Content on the Acid–Base and Catalytic Characteristics of ZrO_2–CeO_2 Oxide Compositions in the Process of Ethanol to n-Butanol Condensation. *Catal. Lett.* **2019**, 1–9. [CrossRef]

56. Vivier, L.; Duprez, D. Ceria-Based Solid Catalysts for Organic Chemistry. *ChemSusChem* **2010**, *3*, 654–678. [CrossRef] [PubMed]

57. Wu, Z.; Mann, A.K.; Li, M.; Overbury, S.H. Spectroscopic investigation of surface-dependent acid–base property of ceria nanoshapes. *J. Phys. Chem.* **2015**, *C 119*, 7340–7350. [CrossRef]

58. Wu, Z.; Li, M.; Mullins, D.R.; Overbury, S.H. Probing the surface sites of CeO2 nanocrystals with well-defined surface planes via methanol adsorption and desorption. *ACS Catal.* **2012**, *2*, 2224–2234. [CrossRef]

59. Martin, D.; Duprez, D. Evaluation of the acid-base surface properties of several oxides and supported metal catalysts by means of model reactions. *J. Mol. Catal. A Chem.* **1997**, *118*, 113–128. [CrossRef]

60. Binet, C.; Daturi, M.; Lavalley, J.C. IR study of polycrystalline ceria properties in oxidised and reduced states. *Catal. Today* **1999**, *50*, 207–225. [CrossRef]

61. Calaza, F.C.; Chen, T.L.; Mullins, D.R.; Xu, Y.; Overbury, S.H. Reactivity and reaction intermediates for acetic acid adsorbed on CeO2 (111). *Catal. Today* **2015**, *253*, 65–76. [CrossRef]

62. Stein, E. *NIST Chemistry WebBook. IR and Mass Spectra, Database Number 69*; Mallard, W.G., Linstrom, P.J., Eds.; National Institute of Standards and Technology: Gaithersburg, MD, USA, 2000.

63. Gliński, M.; Kijeński, J.; Jakubowski, A. Ketones from monocarboxylic acids: Catalytic ketonization over oxide systems. *Appl. Catal. A Gen.* **1995**, *128*, 209–217. [CrossRef]

64. Deng, L.; Fu, Y.; Guo, Q.X. Upgraded acidic components of bio-oil through catalytic ketonic condensation. *Energy Fuels* **2008**, *23*, 564–568. [CrossRef]

65. Woodruff, D.P.; Delchar, T.A. *Modern Techniques of Surface Science*; Cambridge University Press: London, UK, 1986. [CrossRef]
66. Redhead, P.A. Thermal Desorption of Gases. *Vacuum* **1962**, *12*, 203–211. [CrossRef]
67. Kislyuk, M.U.; Rozanov, V.V. Temperature-programmed desorption and temperature-programmed reaction - methods of studying of kinetics and mechanisms of catalytic processes. *Kinet. Catal.* **1995**, *36*, 89–98.
68. Carter, G. Thermal resolution of desorption energy spectra. *Vacuum* **1962**, *12*, 245–254. [CrossRef]
69. Leardini, F.; Fernández, J.F.; Bodega, J.; Sanchez, C. Isotope effects in the kinetics of simultaneous H and D thermal desorption from Pd. *J. Phys. Chem. Solid* **2008**, *69*, 116–127. [CrossRef]
70. Houk, K.N.; Gustafson, S.M.; Black, K.A. Theoretical secondary kinetic isotope effects and the interpretation of transition state geometries. 1. The Cope rearrangement. *J. Am. Chem. Soc.* **1992**, *114*, 8565–8572. [CrossRef]
71. Hoque, M.E.U.; Guha, A.K.; Kim, C.K.; Lee, B.S.; Lee, H.W. Concurrent primary and secondary deuterium kinetic isotope effects in anilinolysis of O-aryl methyl phosphonochloridothioates. *Org. Biomol. Chem.* **2009**, *7*, 2919–2925. [CrossRef]
72. Borysenko, M.V.; Kulyk, K.S.; Ignatovych, M.V.; Poddenezhny, E.N.; Boiko, A.A.; Dobrodey, A.O. Sol–gel synthesis of silica glasses, doped with nanoparticles of cerium oxide. In *Nanomaterials and Supramolecular Structures*; Shpak, A.P., Gorbyk, P.P., Eds.; Springer: Dordrecht, The Netherlands, 2009; pp. 227–233. [CrossRef]

MDPI

St. Alban-Anlage 66

4052 Basel

Switzerland

Tel. +41 61 683 77 34

Fax +41 61 302 89 18

www.mdpi.com

Catalysts Editorial Office

E-mail: catalysts@mdpi.com

www.mdpi.com/journal/catalysts